Learning to Listen, Listening to Learn

Learning to Listen, Listening to Learn

Music Perception and the Psychology of Enculturation

Marcus T. Pearce

OXFORD
UNIVERSITY PRESS

OXFORD
UNIVERSITY PRESS

Great Clarendon Street, Oxford, OX2 6DP,
United Kingdom

Oxford University Press is a department of the University of Oxford.
It furthers the University's objective of excellence in research, scholarship,
and education by publishing worldwide. Oxford is a registered trade mark of
Oxford University Press in the UK and in certain other countries

Published in the United States of America by Oxford University Press
198 Madison Avenue, New York, NY 10016, United States of America

British Library Cataloguing in Publication Data

Data available

Library of Congress Control Number: 2024951659

ISBN 9780198848004

DOI: 10.1093/oso/9780198848004.001.0001

Printed and bound by
CPI Group (UK) Ltd, Croydon, CR0 4YY

The manufacturer's authorised representative in the EU for product safety is Oxford University Press España S.A. of El Parque Empresarial San Fernando de Henares, Avenida de Castilla, 2 – 28830 Madrid (www.oup.es/en or product.safety@oup.com). OUP España S.A. also acts as importer into Spain of products made by the manufacturer.

Contents

Preface

The importance of musical experience is so evident for so many people in so many contexts across so many different cultures that it requires little introduction. And yet the nature of musical experience varies widely from person to person, context to context, and culture to culture. This book presents an extended programme of research aiming to understand how music perception and cognition arise through a process of enculturation relying on general-purpose psychological mechanisms of statistical learning and probabilistic prediction. Going beyond nuts and bolts (important though these are) to investigate perception and cognition in real-world domains such as music can enable a deep and sophisticated understanding of psychological processing as it operates in the untidy complexity of everyday life. Music, like language, is an ubiquitous component of culture and this book attempts to answer the question of how individuals achieve maturity in music perception (which, after all, most of us do) through their experience of developing within a musical culture, just as they achieve maturity in speech perception, through developing within a linguistic culture.

The book has several distinct points of origin, including, most recently a broad theoretical perspective developed in preparation for a grant proposal. Earlier inspiration came from Carol Krumhansl's *Cognitive foundations of musical pitch*, discovered as a wide-eyed undergraduate browsing the Experimental Psychology Library at Oxford University (very probably when I should have been reading something else), which introduced me to the possibility of understanding music from a psychological perspective, and doing so using mathematical models. Subsequently, lectures given by Kim Plunkett and the book *Rethinking innateness* that he wrote with Jeffrey Elman, Elizabeth Bates, Mark Johnson, Annette Karmiloff-Smith, and Domenico Parisi opened my eyes to the use of computational modelling to understand psychological processing. The modelling involved machine learning (simple recurrent neural networks in those days) and proved capable of simulating developmental trajectories as a process of learning the structure of the sensory environment. This empiricist perspective made a deep impression on me and underlies the research programme presented in this book. The work of Darrell Conklin, whose paper *Multiple viewpoint systems for music prediction* was pointed out to me by Alan Smaill at Edinburgh University, pioneered the use of probability and information theory as a basis for sophisticated modelling of learning and prediction of music. Darrell's ground-breaking innovations, building on those of his mentors John Cleary and Ian Witten, underlie the IDyOM model which is used to simulate music perception throughout the book. Finally, I have lost count of the number of times I have had the experience of thinking I had a good idea, only to discover that Leonard Meyer had got there several decades earlier, standing as a healthy reminder in an age of rapid technological innovation not to forget the bright minds of the past, whose writings often transcend time.

Regarding the audience I have in mind for this book, those working on music perception and cognition will have a core interest in the subject matter but I hope that the ideas and results presented will have wider interest amongst psychologists and neuroscientists investigating learning and perception, including cognitive and neural processing of speech, as the other prominent auditory component of culture (where prediction and statistical learning also play an important role). The details of the computational modelling presented in Chapter 2 will not be relevant to some readers and are not necessary beyond the overview provided at the beginning of the chapter. That said, I believe the modelling itself, as well as the perceptual evaluation of it, will be of interest to researchers in music informatics and perhaps other areas of computer science, including statistical language modelling. Although the topic is perception of music, rather than music per se, I hope that the book will be of interest to musicologists; after all, music is composed, performed, interpreted, and appreciated by human minds in a cultural context, so understanding music perception should also tell us something about music itself. However, a high level of musical training should not be required to read this book. Although musical terms will be used to refer to some musical structures, especially in Chapters 2–4, the goal is to understand perception of these structures, rather than the structures themselves, and sufficient context should be provided in the book for an interested non-musician. The musical examples given, using score notation, are genuinely for purposes of illustration. Finally, because the book is fundamentally concerned with understanding enculturation, the concepts and evidence developed should be relevant to those interested in the psychology of enculturation beyond music, either from the general perspective of cross-cultural or developmental psychology, or in specific domains such as language, dance, the visual arts, and beyond.

The existence of this book depends on the efforts of many people to whom I owe a debt of gratitude. This is especially true for those colleagues who generously gave their time to read large chunks of material, think deeply about it, and provide critical but always constructive feedback. This includes David Huron who read an entire draft; Ana Clemente, Roger Dean, and Carol Krumhansl who covered most of the book; and those who gave feedback on individual chapters either in writing or during lively discussions that took place in periodic lab meetings devoted to specific chapters as I completed more or less presentable first drafts: David Baker, Roberta Bianco, Vincent Cheung, Rémi de Fleurian, Adam Garrow, Edward Hall, Madeline Hamilton, Niels Chr. Hansen, Ilana Harris, Peter Harrison, Tom Kaplan, Amira Korkor, David Quiroga-Martinez, Bastiaan van der Weij, Chris Winnard, Adrien Ycart, and Farida Yusuf. I am also indebted to Justin London and David Sears for their insightful comments on Chapter 4 and related material, and to David Temperley for detailed feedback on several chapters.

Beyond this, I am grateful to all those colleagues and collaborators who not only contributed to the research presented in this book but also developed my thinking through shared discussion of ideas. For conciseness, I have adopted the approach of highlighting the lead author(s) for each study discussed in the book, usually the junior scientists who did most of the actual work, rather than including long lists of contributing authors. However, this is certainly not to downplay the often significant contributions of those authors, whose names are included in the references. I would like to highlight longstanding collaborations with Maria Chait, Fred Dick, and Andrea Halpern, which have had a significant impact on my thinking over the past decade or so. Working with Peter Vuust, Elvira Brattico, Morten Kringelbach, and their team at the Centre for Music in the Brain at Aarhus University—before, during and after the period of time that I was physically present at the Centre—has provided ongoing opportunities to deepen my understanding of the musical brain. I am also grateful for the rich intellectual culture provided by my colleagues at the Centre for Digital Music and Cognitive Science Research Group at Queen Mary University of London.

External collaboration with David Sears, William Caplin, and Stephen McAdams had a major impact on the computational modelling of harmony perception covered in Chapter 4 which was then taken forward by Peter Harrison during his doctoral research. Working with Ben Gold and Robert Zatorre as well as Vincent Cheung, Lars Meyer, and Stefan Koelsch developed my understanding of neural processing of reward, presented in Chapter 5, while a fruitful collaboration with Daniel Müllensiefen underpins some of the research covered in Chapter 6. I would like to remember two colleagues who very sadly passed away during the writing of this book: Steven M. Demorest and Fabia Franco. Fabia, together with Nina Politimou and Lauren Stewart, introduced me to developmental experimentation in a study covered in Chapter 7 while the work on cultural distance also presented in Chapter 7 emerged from collaboration with Steve together with Steven Morrison. Over the years, I have also benefited immensely from intellectually deep and rewarding discussions with Joydeep Bhattacharya, Ian Cross, Tuomas Eerola, Henkjan Honing, Nori Jacoby, David Poeppel, Martin Rohrmeier, and Richard Widdess.

I am thankful to Martin Baum at OUP for his continuous editorial support and, in particular, his patience during the hiatus in writing imposed by the Covid-19 pandemic. Also at OUP, Charlotte Holloway and Jade Dixon provided helpful additional support in the early and late stages of writing respectively. Finally, I am eternally grateful to my family for giving me time to write: a book is rarely completed during office hours and frequently tests the recursive depth of Hofstadter's law.

Cambridge, UK M.T.P.
August 2024

1 Introduction

Picture yourself listening to a piece of music you know well when the power on your device suddenly fails and you find yourself imagining, perhaps even humming or whistling, how the music would have continued. You may well find you can conjure up quite a vivid image of how the music continues, reflecting a representation of the piece stored in your long-term memory, retrieved and cast into a dynamically unfolding auditory image that consumes your conscious experience for a period of time.[1]

Now consider the same scenario differing only in that the piece of music you are listening to is not one you have heard before. Whereas, in the original scenario, your imagined continuation could proceed in only one way, reflecting your memory of that particular piece of music, now there are many ways in which you can imagine the music proceeding. The continuation you imagine likely depends on several things. First, the preceding context of the music you heard before it stopped—if this context (or one similar to it) had appeared before in the piece then you might use that repetition to reconstruct a likely continuation. For example, if the music stopped in the middle of a repeated melodic theme, refrain, or chorus, you could use this knowledge to predict the continuation of the passage. If you had inferred a regular temporal pulse or beat in the piece, your imagined continuation would be likely to preserve this temporal regularity.

As well as the preceding context, your imagined continuation will also depend on your lifetime auditory experience, including the music (and possibly also languages) to which you have had significant exposure. Just as you will have implicitly acquired an internal model of at least one natural language, enabling speech comprehension, you almost certainly also possess an internal model of at least one musical style, enabling you to perceive and appreciate music in that style. As with natural language, this model does not tell you exactly what will appear next in your imagined continuation but it does place constraints on which

[1]A neuroimaging experiment conducted by Halpern and Zatorre (1999) had listeners imagine continuations to familiar tunes, finding activation in planum temporale, associated with complex sound analysis (Griffiths & Warren, 2002), and inferior frontal gyrus, associated with auditory working memory (Kumar et al., 2016).

sounds are likely to appear and the order in which you are likely to imagine them. Without formal education, these internal models of music—as for natural language—are implicit and opaque to introspection. The long-term process of *enculturation* by which such internal models are acquired through exposure is both implicit and obligatory.

Although we have conceptualized this as an explicit task, the thesis defended in this book is that the brain automatically and implicitly generates predictions for what will happen next in a wide range of contexts, including auditory perception. I will argue that these predictions are derived from internal models acquired through statistical learning during enculturation. The psychological processes of statistical learning and model-based prediction will be conceived mechanistically, allowing for computational implementation and comparison with human perception to test the theory. Just as you are able to imagine different attributes of the music (lyrics, melody, harmony, rhythm, timbre, and so on) in the thought experiment above, the processes of statistical learning and prediction can apply to and account for perception of a broad range of different attributes of music. Less obviously, statistical learning and prediction provide a computational foundation for modelling a wide range of psychological processes at play during musical listening including perceptual grouping, attention, memory, similarity perception, complexity perception, as well as affective and aesthetic experience.

1.1 Perception and culture

Culture consists of the accumulated phenomena—knowledge, skills, artefacts, values, dispositions, meanings, and so on—that are shared amongst a population of individuals and passed on through social learning. Learning requires communication of such phenomena within the culture and this, in turn, requires perception of those phenomena.

Perception can be viewed as a process of creating an internal reconstruction of the external world, including not just its surface features but also its higher-level structure. This involves extraction of features from incoming sensory data (e.g., firing of frequency sensitive neurons in the auditory nerve in response to hearing a particular sound) in a way that is analogous to a bottom-up signal processing pipeline. In the auditory domain, this entails identifying and fusing together aspects of the auditory scene likely to have been generated by the same objects and segregating them from those aspects generated by other objects (Bregman, 1990). However, especially for higher-level structure, it can also require inference of features that are hidden or ambiguous underlying causes of the sensory input (Barlow, 2001; Friston, 2005; Gregory, 1980; Knill & Pouget, 2004). For example, it is much easier to understand someone speaking in a crowded room full of people talking if we know who the speaker is and what they are talking about. Likewise, the ability to recognize a familiar melody embedded in a noisy auditory environment is greatly enhanced if we know what piece of music we are listening out for (Dowling, 1973).

However, the ability to generate top-down expectations for future events confers distinct advantages that go beyond perceptual inference (Huron, 2006; Schultz et al., 1997). If an organism is able to correctly anticipate future events, decisions can be made and actions taken in advance to increase the chances of advantageous outcomes and reduce the risk of disadvantageous outcomes. As Dennett (1991, p. 177) puts it, "The key to control is the ability to track or even anticipate the important features of the environment, so all brains are, in essence, anticipation machines." Even when the organism does not have the agency to influence future events, being able to predict them allows it to prepare an appropriate physiological and psychological receptive and regulatory state (e.g., in terms of attention or arousal) tailored to processing the expected event. It can also prepare appropriate behavioural responses, so as to be able to react more quickly and adaptively to future events. Finally, prediction provides an extraordinarily powerful cost function for learning: representations and processes that generate accurate predictions can be selected over those that generate inaccurate predictions. This allows optimization of predictive mechanisms without an external tutor that tells the learner the correct answer—often referred to as labelled data in the context of supervised machine learning—by virtue of the fact that the passage of time inexorably brings predictions into contact with reality.

Perception seems likely to involve a combination of bottom-up processes based on mechanisms that may be universal and top-down anticipatory processes based on learned internal models of a sensory domain.[2] One might reasonably hypothesize that aspects of the external sensory world that remain fixed over long periods of time (e.g., those reflecting constants of the physical world) would be good candidates for universal mechanisms while more variable aspects would be good candidates for learned mechanisms (Huron, 2006). An extreme example of the latter scenario is given by cultural domains in which knowledge and behaviour are learned and transmitted between individuals. As a result, perception of artefacts and behaviour created within a cultural domain is much more dependent on learning through enculturation than perception of the physical world. Examples of cultural

[2]Of course, the picture is not as simple as this because bottom-up processes feed the acquisition of top-down processes while the latter also modulate the operation of the former.

domains range from fundamental psychological phenomena like spoken language through technology, science, and mathematics to the visual, performing, and literary arts.

Within cultures, domain knowledge is passed between individuals from generation to generation. *Iterated learning* is a particular kind of cultural transmission in which a behaviour arises in one individual through observation of that behaviour in another individual who acquired the behaviour in the same way (Kirby et al., 2014). Changes in behaviour can arise through two mechanisms: cultural selection, whereby some behaviours or artefacts are preferred and so increasingly dominate the cultural pool; and biased transformation, whereby inaccurate transmission arises due to psychological constraints in perception, memory, and performance (Mesoudi, 2021). Behavioural experiments and computational simulations of the cultural evolution of language have shown that iterated learning can lead to artificial languages becoming more learnable across generations, acquiring compositional and combinatorial structure that reflect the learning biases of the agents involved (whether human or artificial) and emulate characteristics of natural languages. For example, Verhoef et al. (2014) showed experimentally that 10 generations of learning from another participant followed by recall of sounds made on a slide whistle led to sounds that became more easily learnable (i.e., more predictable or compressible, with lower entropy) and more accurately recalled across generations, reflecting increasing compositionality. However, compositional structure only arises when the pressure for compressibility during learning is balanced with pressure for expressivity during communication (Kirby et al., 2015).

Similar experiments with musical stimuli have shown that rhythms become simpler and more learnable through iterated learning (Ravignani et al., 2017). These changes reflect both universal biases towards low integer ratio rhythms and idiosyncrasies specific to particular cultures that are systematically related to the music of those cultures; for example, Malian musicians and dancers show stronger bias than people from other cultures towards rhythms with interval ratios of 3:3:2 and 7:2:3, corresponding to prominent rhythmic patterns in their musical repertoire (Jacoby et al., 2024). Physical constraints of performance related to the configuration of the instrument can also bias the rhythms that emerge through iterated learning (Miton et al., 2020). Iterated learning experiments with pitched melodic stimuli show similar biasing of transmission chains by psychological processing, motor constraints, and cultural influence (Anglada-Tort et al., 2023; Lumaca & Baggio, 2017).

Iterated learning suggests a process of cumulative cultural evolution in which cultural innovations are accumulated over successive generations (Derex & Mesoudi, 2020). However, rather than reflecting a passive process of accumulation, an emphasis on creativity in human cultural domains means that cultural artefacts increase exponentially across generations (Enquist et al., 2008). This results in a process in which generations of creators and their audiences (who may be one and the same) experience an ever-growing set of inputs and learn from it during enculturation. Creators both recreate the existing repertoire and also add to the repertoire, attempting to push beyond (or transform) existing limits but can only push as far as their audiences will tolerate. Their output goes together with the existing repertoire (with some attrition) into the enculturation process of the next generation. Together these processes of creativity and attrition produce a changing cultural repertoire over time which, in turn, leads to changes in the perception of the cultural domain. While the present work focuses on the psychological mechanisms of perception, an emphasis on enculturation situates it within this broader understanding of cultural evolution.

In spite of the importance of enculturation in cultural domains such as music, it remains possible that perception of artefacts from these domains could reflect, in part, innately specified inherited mechanisms. Because music, like language, is a cultural phenomenon (Repp, 1991) and not subject to biological evolution, it makes sense to focus evolutionary enquiry on the set of psychological traits making up the spontaneously developing capacity for creating and perceiving music, which has been referred to as *musicality* (Honing et al., 2015). In other words, it is important to distinguish the set of psychological mechanisms—learning included—involved in perception of cultural artefacts like music from the artefacts themselves, which are the object of cultural evolution. It is interesting to ask whether there are aspects of musicality that reflect biologically evolved traits and, if so, the context in which they evolved (e.g., specifically to support music or in another context such as communication or scene analysis) as well as their comparative relationship with potentially homologous traits in other species (which are generally considered not to exhibit musicality). However, there are multiple challenges involved in demonstrating generational change in genetic variation due to natural selection of heritable traits in any psychological domain. Because psychological traits do not fossilize, it is impossible to assess the selection pressures or demonstrate generational change in genetic variation on the timescales over which these traits were evolving (Lewontin, 1998). The challenges become even more formidable in complex cultural domains such as music and language, where there is a very real possibility of gene-culture coevolution (Patel, 2022). As a result, research on the evolution of human musicality must triangulate between indirect evidence from multiple sources (Fitch, 2006; Honing & Ploeger, 2012; Justus & Hutsler, 2005; Tinbergen, 1963). In this context, computational modelling can identify traits that *can*, in principle, be acquired through cultural experience while psychological experimentation can assess whether such models provide accurate simulations of these traits and their dependence on experience. The research presented in this book provides an example of such combined modelling and experimentation.

1.2 Statistical learning and probabilistic prediction

This book presents an attempt to understand the psychological mechanisms involved in perception of one particular cultural domain, music. The approach is theoretically grounded in two hypotheses.

Statistical learning hypothesis (SLH): enculturation is a process of implicit statistical learning in which individuals progressively acquire internal models of the statistical regularities present in the exposure domain, over multiple timescales ranging from short (e.g., an individual artefact) to long (e.g., an entire lifetime of exposure).

Probabilistic prediction hypothesis (PPH): while experiencing a novel artefact, an enculturated individual applies models learned via the SLH to generate probabilistic predictions that enable them to organize and process their mental representations of the artefact and generate culturally appropriate responses.

Statistical learning is a process of apprehending the regularities or dependencies between elements within a sensory domain. By dependency, I mean that the identity of one element of a sensory stimulus is not statistically independent from the identity of one or more other elements. A domain that has no dependencies (e.g., white noise) affords no possibility of statistical learning.[3] The nature of the dependencies will vary between domains. For static images, dependencies can be formed between spatially distinct elements but we will be particularly concerned with dependencies between elements separated in time. In particular, for sequential cultural domains such as speech and music, the identity of an element at one point in time can depend on the identity of the elements at other timepoints in the sequence (both earlier and later). Because such auditory sequences are experienced (and learned) in time, it makes sense to focus on dependencies between an event and the preceding elements in the sequence. In the cognitive science literature, statistical learning is sometimes associated specifically with *transitional probabilities*—the first-order conditional probability of one element based on its immediate predecessor in the sequence—introduced in a prominent

[3]Though this doesn't stop the brain trying (Huettel et al., 2002).

tradition of research (following Saffran et al., 1996). However, the theoretical conception of statistical learning adopted here is much broader, encompassing any dependencies between an event and the preceding context. Statistical learning is generally considered to be an implicit, obligatory psychological process, taking place without conscious awareness and is outside volitional control. As such it is sometimes referred to as implicit statistical learning (Christiansen, 2019).

As well as adults, statistical learning has been observed in human infants (Saffran & Kirkham, 2018) and non-human species, where it seems likely to be responsible for vocal learning in songbirds (Fehér et al., 2016; Santolin & Saffran, 2018) and generalized pattern learning in certain non-human primate species (e.g., the cotton-top tamarin; Neiworth et al., 2017). In humans, statistical learning has been demonstrated for sequences of spoken syllables (Saffran et al., 1996), tones (Krishnan et al., 2022; Saffran et al., 1999), timbres (Tillmann & McAdams, 2004), visual stimuli (Fiser & Aslin, 2002; Hunt & Aslin, 2001; Turk-Browne et al., 2005), and tactile stimulation (Conway et al., 2005). However, rather than being a unitary cross-modal psychological construct, experimental work suggests that statistical learning has components that are specific to particular modalities and potentially even particular kinds of dependency within a modality (Siegelman et al., 2017). This does not necessarily preclude the transfer of acquired statistical knowledge across domains. For example, Patel (2008, pp. 224–225) argued that statistical learning of prosodic patterns in speech could influence perception and creation of music (see also § 7.3.1).

Probabilistic prediction is the ability to use the knowledge about statistical dependencies acquired through statistical learning to make a prediction for a future event relative to a given context. A non-probabilistic prediction is a deterministic expectation for a single outcome without considering any alternative possibilities. In contrast, a probabilistic prediction entertains the possibility of more than one alternative outcome and associates each with a relative likelihood of occurrence. As I will use the term here, it is not necessary for the psychological process of prediction to be formally probabilistic—forming a distribution over all possible outcomes that sums to unity—but it is often useful to assume as much for the purposes of running and analysing computational simulations. The probabilities associated with the different outcomes depend on the domain from which knowledge is acquired through statistical learning. A probabilistic prediction is, therefore, a conditional probability distribution for an event yet to occur given the preceding context. In principle, the prediction may be conditionalized upon the context in any way imaginable, reflecting the dependencies that are acquired during statistical learning.[4]

Probabilistic prediction plays a key role in many psychological domains including visual perception (Summerfield & Egner, 2016) and language (Kuperberg & Jaeger, 2016; Pickering & Gambi, 2018). For example, detection and localization of visually presented objects is enhanced by knowing the object category (e.g., cat or guitar) in advance (Stein & Peelen, 2015). Furthermore, the brain generates representations of predicted visual stimuli that are highly similar to representations of the actual stimulus prior to the arrival of the corresponding afferent sensory information (Blom et al., 2020). This could help the visual system to accommodate neural processing delays when interacting in dynamic environments such as team sports or ensemble musical performance. Going beyond vision to sensorimotor processing, it is thought that the inability to tickle oneself arises from predicting the perceptual consequences of one's own motor commands resulting in dampening of those signals (Blakemore et al., 2000). The predictive nature of speech perception is convincingly demonstrated by so-called *visual world* experiments. For example, if participants are shown a visual scene depicting a boy, some toys, and a cake while listening to sentences, they look more towards the cake when hearing "The boy will eat the ..." than "The boy will move the ..." (Altmann & Kamide, 1999). Critically the eye movements are recorded before the final noun is heard, demonstrating prospective prediction. In addition to semantic processing, visual world experiments have demonstrated predictive processing of phonology (Allopenna et al., 1998) and syntax (Arai & Keller, 2013). Finally, predictive processing of natural language appears to be probabilistic rather than all-or-none. In reading tasks, the predictive surprisal of words—which varies continuously—shows significant correlations with processing difficulty reflected in reading times (Smith & Levy, 2013) and associated neural activation (Frank et al., 2015). Furthermore, these findings generalize to speech comprehension (Willems et al., 2016).

In summary, the theoretical hypothesis proposed is that through statistical learning, individuals internalize the structure of the cultural domains to which they are exposed, resulting in internal cognitive models of that structure, stored in various memory systems (see Chapters 3 and 6). This process of learning is implicit and dynamic; it proceeds both at long timescales across the lifespan, resulting in models that reflect the entire cultural

[4]A distinction is sometimes made between learning of distributional statistics (i.e., unconditional frequencies of occurrence) and transitional statistics (i.e., frequencies conditional on the occurrence of previous events) (Thiessen, 2017). For present purposes, no such distinction is made with distributional statistics treated as a degenerate form of conditional statistics.

domain, but also at shorter timescales whilst experiencing individual instances from the domain. These internal models allow individuals to predict forthcoming events, given the preceding context, when experiencing new exemplars from the cultural domain. These predictions are hypothesized to feed other aspects of perception including inference of hidden structure, segmentation of a perceptual scene, comparison of auditory objects, perception of complexity, as well as emotional and aesthetic experience. Although they will be applied to music perception, the two hypotheses underlying this research programme invoke general psychological mechanisms of statistical learning and probabilistic prediction that apply, in principle at least, more widely to perception of other sequential communicative cultural domains including natural language and sequential motor domains such as dance.

1.3 Music perception: A case study in the psychology of enculturation

Music is an important component of human culture, which shares many features in common with language. It consists of human-initiated patterns of sound created and perceived in association with a wide range of contexts and purposes. It is communicative, generally being created with the intention that others will listen to it and derive pleasure (and, often, meaning) from the experience. It is widely thought to be universal across human cultures, and is passed from generation to generation, showing cultural evolution over time and exhibiting significant differences between cultural traditions. Individuals develop within a musical tradition, acquiring through experience (and, sometimes, explicit training) internal models of the structure of the music to which they are exposed. These internal models allow listeners to appreciate the compositional structure of the music of their culture and respond appropriately. Suitably enculturated listeners can, for example, recognize the emotional expression of a piece of music, know how to dance to it, understand its referential significance (e.g., military, religious, or social connotations), and take pleasure in its construction. By contrast, they may fail to respond appropriately in these ways to the music of other cultures.

The goal of the present enquiry is to understand the psychological processes involved in music perception and how they are acquired through enculturation. This goal is approached through an extended programme of research testing the statistical learning and probabilistic prediction hypotheses. While these hypotheses are domain general, music perception makes an interesting case study for their examination. In some ways, music is simpler than language, with less emphasis on referential meaning in addition to syntax, making it a more tractable object of study. Furthermore, it is more common with music than language for perceptual fluency to exist in the absence of productive fluency, making it more natural to study perception in isolation. Finally, enculturated music perception is often achieved purely through exposure to music with less emphasis on formal education than is often the case for language, allowing for fewer potential confounds in the effects of implicit statistical learning on perception. In other words, from a pragmatic perspective, it is far easier to find individuals who are enculturated musical listeners without (much) formal music training than it is to find individuals who comprehend speech fluently without (much) education in reading and writing.

1.4 A historical perspective

The present research programme builds on a longstanding history of scholarship aiming to develop a general understanding of music perception. One particular branch of this tradition drew from theoretical perspectives developed by the musicologists Eduard Hanslick (1854), Leonard Meyer (1956, 1957), and Eugene Narmour (1977, 1990, 1992) before transitioning towards cognitive science and psychology with the work of Carol Krumhansl (1990) and David Huron (2006). Rather than presenting an exhaustive review, my goal here is to capture the primary themes of this tradition, which provide the contextual foundations of the present work, establishing the theoretical origins of the statistical learning and probabilistic prediction hypotheses, as they apply to music perception.

1.4.1 Musicological roots

The idea that expectation plays an important role in music perception and appreciation has a venerable history. The 18th-century British music teacher, John Holden, wrote that "Our enjoyment of music depends, in a great measure, on a faculty of retaining the ideas of former sounds, and anticipating those which are to follow" (Holden, 1766, p. 4, cited in Raz, 2018). If a piece of familiar music is interrupted before it ends, the experience would be "quite unsatisfactory, and a plain disappointment to the hearer. This natural expectation, which never ceases till some proper succession of notes occur, leading to, and terminating upon one certain sound, is not peculiar to a few tones only, but is common to all" (Holden, 1770, part 1, § 29, pp. 23–24, cited in Raz, 2018).

In 19th-century Europe, the flourishing of program music gave rise to a debate between *referentialists*, including the composers Richard Wagner and Franz Liszt, and *formalists*, such as the Austrian music critic Eduard Hanslick and the composer Johannes Brahms. The debate centred on the question of whether music possesses referential content such that musical structures designate non-musical entities such as physical objects, scenes, individuals, or feelings. Within the context of this debate, Hanslick (1854) argued that the aesthetic effects of music are specifically musical, introducing the psychological mechanism of unconscious expectation:

> The most significant factor in the mental process which accompanies the comprehending of a musical work and makes it enjoyable [...] is the mental satisfaction which the listener finds in continuously following and anticipating the composer's designs, here to be confirmed in his expectations, there to be agreeably led astray. It goes without saying that this mental streaming [...] occurs unconsciously and with the speed of lightning. Only such music as brings about and rewards this mental pursuing [...] will provide fully artistic satisfaction.[5]
>
> (Hanslick, 1854, p. 64)

Taking ideas from both sides of this debate, Leonard Meyer (1956) developed Hanslick's formalist account of expectation as a vehicle for emotional expression in music into a fully fledged account of musical appreciation (see Chapter 5). Inspired by John Dewey (1894, 1895), the central thesis of this account is that "Emotion or affect is aroused when a tendency to respond is arrested or inhibited" (Meyer, 1956, p. 14). According to Meyer, expectations for forthcoming musical structures create patterns of tension and resolution, which, in turn, elicit affective states such that violations of expectation and states of uncertainty are negatively valenced and vice versa. In principle, this allows for a range of different emotional states by varying the way in which expectations are elicited and subsequently confirmed or denied. For example, differences in the strength of expectation can produce different states of anticipation while an event that is delayed might produce a different kind of tension from an event that confounds expectations outright.

Meyer (1956) couched his theory of expectation in probabilistic terms. Inspired by developments in information theory (Shannon, 1948), he proposed that "styles in music are basically complex systems of probability relationships in which the meaning of any term or

[5]This passage bears a striking resemblance to the following portrayal of the improvisational form *gat* in North Indian classical music: "The essence of this style of improvisation is the combination of changing rhythmic groupings and unexpected twists and turns of the melody, all at a fast speed that keeps the listener from guessing what will happen next" (Widdess, 2015, p. 145).

series of terms depends upon its relationships with all other terms possible within the style system" (Meyer, 1956, p. 54) and "the norms and deviants of a style upon which expectation and consequently meaning are based are to be found in the habit responses of listeners who have learned to understand these relationships" (p. 61). Taking these ideas further, Meyer (1957) suggested that a musical style may be conceived as a Markov process and related predictive uncertainty, failure, and success to the information-theoretic concepts of entropy, information, and redundancy respectively:

> Musical meaning arises when an antecedent situation, requiring an estimate as to the probable modes of pattern continuation, produces uncertainty as to the [...] nature of the expected consequent [...] the weaker the probability of a particular consequent in any message, the greater the uncertainty (and information) involved in the antecedent-consequent relationship.
>
> (Meyer, 1957, p. 416)

Meyer also distinguished between desirable and undesirable uncertainty, the former arising "within and as a result of the structured probabilities of a style system in which a finite number of antecedents and consequents become mutually relevant through the habits, beliefs, and attitudes of a group of listeners" and the latter "when the probabilities are not known either because the listeners' habit responses are not relevant to the style (cultural noise), or because external interference obscures the structure of the situation being considered (acoustical noise)" (pp. 420–421).

The idea that expectation plays a foundational role in music perception was extended and developed by the music theorist Eugene Narmour (1977, 1990, 1992), who was mentored by Meyer during his doctoral research at the University of Chicago. Narmour developed a detailed account of melody perception known as the *Implication Realization* (IR) theory. According to this theory, melodic expectations arise from two independent perceptual systems—the bottom-up and top-down systems of melodic implication—the first reflecting automatic, unconscious, and universal psychological mechanisms and the second reflecting learned, culture dependent mechanisms.[6]

In the bottom-up system, melodic intervals vary in the degree of *closure* they convey, depending on the properties of the interval. While strong closure signifies the termination of ongoing melodic structure, an unclosed interval is termed an *implicative interval* and generates an expectation for the following interval which is termed the *realized interval*.

[6] Experimental research testing the predictions of the IR theory with listeners is covered in Chapter 3.

The expectations generated by implicative intervals are described by Narmour in terms of principles of continuation, influenced by the Gestalt principles of proximity, similarity, good continuation, and symmetry. These principles reflect the size of the implicative interval and the size and direction of the realized interval relative to the implicative interval. Narmour (1992) presented a detailed analysis of how these two-interval melodic structures combine together to form longer and more complex *chains*, depending on the closure implied by the constituent structures. Furthermore, tones that are emphasized by strong closure (e.g., those beginning or ending a melodic structure or chain) at one level can be *transformed* to a higher level. The same bottom-up principles of implication operate on sequences of (possibly non-contiguous) tones at higher transformational levels and, theoretically, a tone may be transformed to any number of higher levels. Although Narmour restricted his attention to melodic intervals, the same bottom-up principles of implication were intended to apply to other musical attributes such as duration, metrical position, timbre, and dissonance.

The IR theory also stressed the importance of top-down influences on melodic expectation which are learned and therefore specific to musical cultures and dependent on musical knowledge and experience. These influences include both *extra-opus* knowledge about style specific norms, such as tonal and metrical systems, and *intra-opus* knowledge about aspects of a particular composition such as distinctive motivic and rhythmic patterns. Since the top-down system in Narmour's IR theory is largely independent from the bottom-up system, it may generate implications which conflict with those generated by the bottom-up system.

1.4.2 Psychological developments

The theoretical proposals of Meyer and Narmour provided inspiration to a generation of psychologists. As a prominent example, Carol Krumhansl (1990) set out to understand the knowledge of musical structure possessed by listeners and how it affects the way music is perceived and remembered; the focus was on pitch perception and, in particular, psychological representations of the tonal properties of tones, chords, and keys. Krumhansl's research plays a key role in motivating the use of tonal pitch representations in computational simulations of music perception throughout the present work. Although her research emphasized geometric representations of musical pitch, the results also highlighted aspects of music perception that were not easily explainable in geometric terms. Rather they seemed to reflect knowledge that was acquired through learning of statistical structure in music.

In particular, Krumhansl examined whether tonal expectations could plausibly be derived from statistical learning of tone frequencies in music. Expectations were assessed using the *probe-tone paradigm* in which participants rate how well a final tone (the probe tone) fits in the context of a preceding musical passage—in this case, the probe tones were the 12 chromatic tones in an octave which were presented in major or minor tonality-defining contexts (Krumhansl & Kessler, 1982). The results showed that the ratings correlate strongly with distributions of scale degrees in a large sample of European vocal music from the classical and romantic periods (Knopoff & Hutchinson, 1983; Youngblood, 1958)[7] as well as a smaller sample of English-language nursery rhymes (Pinkerton, 1956). Since the perceptually immediate contexts used in the experiment do not show such distributions, Krumhansl argued that listeners must have internalized statistical information about the distribution of tones through their prior experience of music.

These results were limited to perception of individual tones. However, the approach was extended to perception of relationships between pairs of tones. The experimental data came from an experiment in which two pitches were presented following a tonality-defining context and listeners rated how well the second tone followed the first. The average ratings from a group of musicians correlated significantly with the statistical co-occurrence of those pairs in 20 songs by Schubert, Mendelssohn, and Schumann (Youngblood, 1958) and nursery rhymes (Pinkerton, 1956) but not a measure of sensory dissonance (Hutchinson & Knopoff, 1978). Krumhansl concluded from these results that:

> Listeners appear to be very sensitive to the frequency with which the various elements and their successive combinations are employed in music. It seems probable, then, that abstract tonal and harmonic relations are learned through internalizing distributional properties characteristic of the style. This sensitivity to distributions of elements was central to L. B. Meyer's account (1956) of how cultural norms determine the listener's experience of the music. He claims the feeling of tension and resolution, activity and rest, instability and stability, and the pointing of certain events to others, are a consequence of the probability relations in the style system that listeners have learned through experience.
>
> (Krumhansl, 1990, p. 286)

As discussed further in Chapter 7, Krumhansl's research also demonstrated that statistical learning is at play when listening to the musical contexts heard in probe-tone experiments, even when the listener is unfamiliar with the musical style.

---

[7] Including pieces by Johann Adolph Hasse, Wolfgang A. Mozart, Franz Schubert, Robert Schumann, Felix Mendelssohn, and Richard Strauss.

Palmer and Krumhansl (1990) extended this approach to event timing in an experiment that elicited goodness-of-fit judgements for probe events whose timing was varied systematically, presented in the context of a regular sequence of context events (with a lower pitch) whose periodicity implied one of four metres (2/4, 4/4, 3/4, and 6/8). The mean ratings correlated significantly with the distributions of onset timing for piano compositions in the same four metres by Bach, Mozart, Brahms, and Shostokovich. Once again, these distributions were not present in the context sequences, so they must have derived from internalized metrical models reflecting the statistics of event occurrence in previously experienced music. Studies of Malian drumming (London et al., 2017) and Turkish makam music (Holzapfel, 2015), characterized by repeated contrametrical rhythmic patterns, demonstrate that onset frequency does not always show a strong correspondence with metrical accent. Rather, "it is precisely because there is a stable differentiation of the distribution of note onset frequencies relative to their metric framework that statistical learning processes can gain traction" (London et al., 2017, pp. 497–479).

The latest large-scale contribution to this line of thinking was made by David Huron (2006), who presented a general psychological theory of expectation and examined in detail how it can account for various aspects of music perception and experience. According to the theory, ability to anticipate the future has strong implications for survival and the psychological mechanisms underlying expectation arose through biological evolution and are automatic, ubiquitous, and mostly unconscious. Although auditory expectations do exhibit innate components that are immutable to experience (e.g., the azimuth component of auditory localization), the theory emphasizes an evolved capacity for statistical learning as a mechanism behind expectation. However, Huron notes that perceptual expectations sometimes reflect simpler approximations to statistical patterns present in musical corpora—for example, while melodic phrases tend to show arch-shaped contours, listeners show expectations for pitch descent at phrase endings but not ascent at phrase beginnings. While Krumhansl (1990) emphasized zeroth-order probability distributions, for Huron, perceptual expectations reflect statistical learning of higher-order conditional probabilities, though the analyses presented are mostly limited to first-order probabilities.[8]

Huron (2006) differentiated expectations arising from different kinds of memory. Semantic memories formed through long-term experience of a musical style produce schematic expectations while short-term memory provides the basis for dynamic expectations reflecting patterns repeated locally within a listening episode. Episodic memory for particular familiar

[8]To be fair, Krumhansl notes this limitation, which reflects statistical corpus analyses available at the time.

pieces of music gives rise to veridical expectations based on knowledge of how the piece continues. These kinds of expectation are distinguished from conscious expectation based on explicit knowledge, perhaps resulting from formal training or study, of how musical patterns are likely to continue. To avoid overgeneralization of schematic expectations, the theory proposes the existence of *cognitive firewalls* between memory stores of style-specific knowledge such that listeners with experience of more than one style can switch between stylistic schemas analogous to the way in which multilingual speakers switch between natural languages depending on the context. Huron also addresses the representations of musical stimuli that are used to generate expectations, proposing that multiple candidates are considered and evaluated concurrently, with those having the best predictive accuracy retained. New representations are created and evaluated throughout development with a general preference for simplicity over complexity. Some of these representations will be discarded and others retained on the basis of their predictive accuracy in the face of experience, reflecting the structure of the auditory environment.

A large part of the theory relates to the role of expectation in emotional and aesthetic responses to music. Huron proposes five components of an anticipatory episode. Two of these are pre-outcome responses occurring before the predicted event: the *imagination response* involves conscious anticipation of what might occur in the future, how it might feel, and what actions might be taken; the *tension response* tailors the psychological and physiological state of the organism to match the uncertainty and implications of the different outcomes. There are three post-outcome responses: the *prediction response* is positive for correct predictions and negative for incorrect predictions; the *reaction response*, a fast, defensive response to the outcome, reflecting a worst-case scenario; the *appraisal response*, a slower, reflective response involving conscious assessment of the outcome.

The prediction response can create affective states on its own; because expectation is an evolved capacity, predictive successes are rewarded and elicit positive emotions, while predictive failures are penalized and elicit negative affective states. These emotions are misattributed to the event that was predicted and the overall phenomenon is referred to as the *prediction effect*. Examples in music perception include perception of tonally and metrically predictable and unpredictable events. However, surprising events can be experienced as pleasurable due to *contrastive valence* between the appraisal response and the prediction or reaction responses. For example, if the appraisal response for a surprising event is positive, or even neutral, the contrast relative to the negative effects of the prediction response will produce a positive experience overall.

1.4.3 Comparison with rule-based approaches

A separate tradition of research conceives music perception as the application of symbolic rules that specify preferred structural interpretations of musical input (Lerdahl & Jackendoff, 1983; Schenker, 1935; Temperley, 2001, 2007, 2009). Inspired by Chomskian generative linguistics, the composer and music theorist Fred Lerdahl and the linguist Ray Jackendoff developed the first large-scale cognitive theory of music perception in their book *A Generative Theory of Tonal Music*, invariably referred to as GTTM, aiming to capture the implicit knowledge that enables a listener enculturated in Western tonal music:

> to organize and make coherent the surface patterns of pitch, attack, duration, intensity, timbre, and so forth. Such a listener is able to identify a previously unknown piece as an example of the idiom, to recognize elements of a piece as typical or anomalous, to identify a performer's error as possibly producing an 'ungrammatical' configuration, to recognize various kinds of structural repetitions and variations, and, generally, to comprehend a piece within the idiom.
>
> (Lerdahl & Jackendoff, 1983, p. 3)

Central to the theory is the *reduction hypothesis*, associated with the music theorist Heinrich Schenker (1935), according to which all the pitch events in a piece of music are organized perceptually into a strict hierarchy of relative importance.

Lerdahl and Jackendoff expressed their theory in terms of both *well-formedness rules* that distinguish legal from illegal structural interpretations (like a formal grammar of natural language) and *preference rules* that specify preferences for the set of legal interpretations, arguing that whereas the former are more important than the latter for language, the converse is true of music. The theory defines four types of hierarchical structure: first, *grouping structure*, corresponding to the identification of segment boundaries (using Gestalt-like principles of event proximity and similarity); second, *metrical structure* corresponding to the pattern of periodically recurring strong and weak beats; third, *time-span reduction* representing the relative structural importance of pitch events within contextually established units derived from the metrical and grouping structure; and finally, *prolongational reduction* reflecting patterns of tension and relaxation amongst pitch events at various levels of structure. According to the theory, grouping and metrical structure are largely derived in a bottom-up fashion, directly from the musical surface; these structures are used by top-down processes that generate a time-span reduction which is, in turn, used in generating a prolongational reduction. Many aspects of GTTM are assumed to be universal and innate.

Wide-ranging though it is, GTTM was not specified in sufficient detail to enable a computational implementation. In his book, *Analysis and cognition of basic musical structures*, David Temperley (2001)—who had studied with Lerdahl—presented a programme of research to quantify and implement preference rule systems for metre and grouping, and extend them to other musical domains, including counterpoint, pitch spelling, harmonic structure, and key structure. The computational models take as input a list of notes each with a pitch (integers reflecting a division of the octave into 12 semitone steps, where middle C = 60) as well as start and end times (in milliseconds). The overall approach was to use dynamic programming to compute preferred analyses for the input (e.g., time signature, grouping structure, tonal pitch, key signature) using the preference rules specified for each domain. The result is an appealingly consistent and interpretable framework, applied across multiple aspects of music perception. This work also addressed GTTM's self-imposed pragmatic limitation to the final state of an individual having listened to an entire piece by extending the preference rule approach to real-time listening using the concept of revision—updating an initial interpretation on the basis of subsequent events (see also Jackendoff, 1991). This research dropped GTTM's commitment to hierarchical reduction and remained agnostic on the question of whether the preference rules are innate or learned.

In his later work, Temperley developed a probabilistic interpretation of preference rules by embedding them in the framework of Bayesian inference. Temperley (2007) presented Bayesian models of melodic metre and key finding as well as a polyphonic key-finding model while Temperley (2009) constructed an integrated model of stream segregation, metrical and harmonic analysis for polyphonic music. The overall strategy was to define a generative model for the likelihood of the surface attribute of interest (e.g., rhythm or pitch) in the musical input given a structural description (e.g., metre or key). The structural description that maximizes the product of the likelihood and a prior distribution over the possible structural descriptions is chosen. These models share much common ground with the preference-rule approach of Temperley (2001), replacing preference rules with probability distributions defining the generative model and often using dynamic programming as an efficient way of inferring the optimal structural interpretation. To give a flavour of the approach, the pitch model of Temperley (2007) specifies distributions (whose parameters are estimated from a corpus) governing the pitch range of a melody (range profile), the pitch proximity of successive notes (proximity profile), and the conditional probability of chromatic scale degrees given a major or minor key (the key profile). The generative process involves selecting a key and central pitch (from distributions estimated from a corpus) and then sampling successive pitches from an RPK profile (obtained by taking the product of the range, proximity, and key profiles). Therefore, the probability of a given pitch in a melody is the normalized product of its probability in the range profile (given the central pitch), its probability in the proximity profile (given the previous pitch), and its probability in the key profile (given the selected key). This generative process can be used to estimate the likelihood of a pitch sequence given a key, providing a model of pitch expectation in melody. For key finding, the key that maximizes the likelihood of a given pitch sequence is selected.

Despite arising from rather different foundations, it is interesting that the tradition of research connecting Schenker, Lerdahl and Jackendoff, and Temperley has evolved towards a focus on probabilistic modelling, bearing more than a passing resemblance to research in the tradition of Meyer, Krumhansl, and Huron. A key difference is an emphasis in the former on structural analysis and in the latter on expectations derived from statistical learning. An outcome of the research presented in this book (and, from a different direction, the work of Temperley, 2007, 2009) is that no such dichotomy need be drawn: structural analysis can benefit from powerful models of expectation based on statistical learning and, conversely, prediction performance can be improved by models of structural inference.

1.4.4 Information-theoretic models of music

Research on computational modelling of music evolved alongside these developments in research on music perception with occasional points of contact such as the adoption of ideas from information theory by Meyer (1956, 1957) and the use of statistical corpus analyses by Krumhansl (1990). The publication of Claude Shannon's mathematical theory of communication in 1948 inspired a wave of interest in applying information-theoretic models to music (amongst other fields, psychology included). In one of the first such studies, Pinkerton (1956) wrote that "the composer of a melody must make the entropy of his music low enough to give it an apparent pattern and at the same time high enough so that it has sufficient complexity to be interesting" (p. 78). Pinkerton computed a zeroth-order distribution of diatonic scale degrees in a corpus of 39 nursery rhymes yielding a redundancy estimate of 9%. Youngblood (1958) used zeroth- and first-order models to compare Gregorian chant with songs from the Romantic period (composed by Schubert, Mendelssohn, and Schumann), finding that the former exhibited higher redundancy than the latter. Cohen (1962) applied the same methods to two songs by Elvis Presley. This approach to music analysis was extended by Lejaren Hiller, who had pioneered the use of computers to generate music (Hiller & Isaacson, 1959), in studies of music by Mozart, Beethoven, Berg and Hindemith (Hiller & Bean, 1966), and Webern (Hiller & Fuller, 1967), the latter using second-order distributions of both pitch and timing. Meanwhile, David Kraehenbuehl and Edgar Coons developed an approach that reflected more naturally the dynamic listening experience, in which predictions for musical events depend only on learning from the preceding context rather than a statistical analysis of the entire piece of music (Coons & Kraehenbuehl, 1958; Kraehenbuehl & Coons, 1959).[9]

The *cognitive revolution* of the late 1950s and early 1960s (Gardner, 1987; Miller, 2003) saw information-theoretic methods fall out of favour and with isolated exceptions (e.g., Knopoff & Hutchinson, 1981, 1983), three decades passed before they returned to centre stage in computational modelling of musical structure. Instrumental in this regard were the sophisticated statistical models developed by Darrell Conklin which addressed the limitations of work conducted in the 1950s and 1960s through three technical innovations (Conklin, 1990; Conklin & Cleary, 1988; Conklin & Witten, 1995). First, the use of smoothing to combine predictions from models of different order (Chen & Goodman, 1999; Cleary & Witten, 1984) which allowed prediction to benefit from both the structural specificity of higher-order models and the statistical robustness of lower-order models. Second, the combination of predictions from a long-term model, trained on a large corpus, and a short-term model, trained incrementally afresh for each individual composition. Third, the development of the multiple viewpoint framework, allowing the combination of predictions from models of multiple representations of different musical attributes, including pitch and rhythm. Chapter 2 covers these techniques in detail.

1.5 Scope, structure, and methodology

In the language of Zenon Pylyshyn (1989), the present research takes the *high road* (the algorithmic level of Marr, 1982) to understanding the cognitive representation and processing involved in music perception. In other words, cognitive representations and processes are approached via the development and testing of formal computational models of statist-

[9]Joel E. Cohen went on to become a distinguished population biologist (e.g., Cohen, 2003) whereas Edgar Coons switched fields from music into neuroscience, completing his PhD at Yale with Neil Miller, who (inadvertently) inspired Al Bregman to run experiments that ended up giving birth to the field of auditory scene analysis—in spite of seeming like a distraction at the time—with the advice: "If a phenomenon is strong enough to get in the way of something that you are trying to study, then perhaps it is worth studying in itself" (Bregman, 1990, p. xi).

ical learning and probabilistic prediction, rather than via detailed understanding of the underlying neural substrate (Pylyshyn's *low road*).[10]

A model is presented in Chapter 2 that implements the statistical learning and probabilistic prediction hypotheses. At Marr's (1982) computational level, the model can tell us something about the task to be solved (in this case, generating expectations derived from learning), the inputs and outputs, and the overall logic of how it can be solved efficiently and effectively. This is a useful starting point but the model cannot in and of itself tell us very much about human perception. In an insightful article on the role of computational modelling in music cognition research, Desain et al. (1998, p. 153) proposed an approach "in which a working computational model is seen much more as the starting point of analysis and research, rather than as an end product." The present research follows this direction, starting with a model of statistical learning and probabilistic prediction and then developing an extended programme of empirical research with human listeners that tests and refines the predictions of the model.

This involves systematically comparing the way in which the model responds to tasks involving musical stimuli with responses of human listeners to those same tasks and stimuli. In many cases, this is supplemented by comparisons between different models, to benchmark the evaluation, and different model variants, to assess which parameter settings best simulate perception. These cognitive scientific studies, which benefit from the use of ecologically valid stimuli, are combined with complementary experimental studies in which stimuli are systematically manipulated to test particular hypotheses. Finally, since any given set of empirical data is necessarily limited in scope—covering a restricted range of participants, stimuli, or tasks—the theory is evaluated through a sustained programme of research comparing its responses with human responses across a wide range of experimental contexts.

This programme of research is presented in Chapters 3–8. Chapter 3 uses the model introduced in Chapter 2 to simulate the results of empirical experiments on perceptual expectation for the pitches of a melody, while Chapter 4 extends this to expectations for the timing of melodic events and harmonic movement, as well as metrical and tonal inference. The next two chapters go beyond expectation to simulate emotional and aesthetic experience of music (Chapter 5) as well as other psychological processes involved in perception of music: memory, grouping, and similarity perception (Chapter 6). Chapters 3 to 6 focus on simulat-

[10] My use of this terminology is not intended to derogate the low road in any way and evidence from neuroimaging experiments can and will be used to test models developed along the high road.

ing the perception of non-musicians enculturated in Western music.[11] In the subsequent chapters, the scope is extended to the question of how music perception and appreciation are affected by two key components of musical enculturation—musical exposure (Chapter 7) and musicianship (Chapter 8). Throughout Chapters 3 to 8, the statistical learning and probabilistic prediction hypotheses are tested through comparison of computational simulations with human perception in empirical experiments. Finally, Chapter 9 concludes by examining the limitations of the approach, anticipating the future development of the research programme and discussing implications for neural processing, cultural evolution, creativity, and cultural domains beyond music, including language.

The present work inherits from Meyer (1956), Narmour (1977, 1990, 1992), Krumhansl (1990), and Huron (2006) an emphasis on expectation as a fundamental psychological mechanism in music perception, including its role in the aesthetic experience of music (Huron, 2006; Meyer, 1956). Narmour (1990, 1992) additionally emphasized the distinction between bottom-up and top-down processes and distinguished extra-opus from intra-opus top-down knowledge, which map onto Huron's distinction between schematic and dynamic knowledge. In these terms, the present approach is strongly top-down and addresses both schematic and dynamic influences on expectation. Krumhansl (1990) introduced the idea of perceptual expectations for musical events deriving from statistical learning and, while her monograph focused on pitch perception, the present work is inspired by the more expansive scope of Huron (2006), Lerdahl and Jackendoff (1983), and Temperley (2001, 2007, 2009). The probabilistic approach differs from that of Temperley (2007, 2009) primarily in emphasizing auditory expectation as a process of probabilistic prediction based on higher-order statistical learning of schematic and dynamic regularities. Whereas all of these authors have addressed to varying degrees the perceptual experience of musicians, the present work focuses explicitly on understanding the perception of non-musicians—enculturated but not trained—with the exception of Chapter 8, which examines effects of musicianship.

In terms of methodological approach, Temperley (2001) distinguished between introspective, experimental, and computational methods within cognitive scientific approaches to music. Meyer (1956), Narmour (1977, 1990, 1992), and Lerdahl and Jackendoff (1983) followed an introspective approach, similar to that often used in linguistics. They aimed to

[11] In the present context, the term *Western* is used to denote popular, folk, and art music influenced by European musical traditions, in which diatonic tonality often plays an important role. However, in the present day, the term is both geographically inaccurate and associated with unhelpful geopolitical connotations while also suggesting a misleading dichotomization between Western and non-Western music. On the other hand, it is widely used and understood while alternatives such as *Eurogenetic* (Benetos & Holzapfel, 2013) felt awkward in many contexts. Therefore, with some reluctance, I have kept with it throughout the book.

develop psychological theories reflecting their own expert music-theoretic understanding of music, motivated by analysis of musical examples. Temperley (2001) characterized his own approach as computational and introspective, reflected in the sophisticated nature of the modelling and analytical approach to evaluation. Krumhansl (1990) took an approach that was primarily experimental complemented by mathematical modelling and statistical analysis. Huron (2006) combined introspective, computational, and experimental perspectives. The present research is firmly computational and experimental in terms of its approach. It aims to understand music perception rather than music itself.[12]

While aiming for generality, it is important to be clear about the limits of the research presented in this book, so as to carefully circumscribe its scope. Rather than taking raw audio as input, the modelling assumes an input of symbolic representations of music consisting of discrete events with discrete pitches and onset times, assuming that auditory processing delivers such a representation (or an approximation of it). This means that the research is limited to what one might call note-based music (Landy, 1994, p. 47), in which the structure is primarily carried by discrete auditory events with definite temporal and spectral properties (e.g., pitch, onset time, and duration); possibilities for extending the approach to sound-based music (e.g., electroacoustic music or Tibetan throat singing) are considered in the concluding chapter. In terms of musical attributes, the focus is on perception of pitch and rhythm in melody with harmonic movement considered in Chapters 4–5 and stream segregation given preliminary treatment in Chapter 6. Extending the approach to perception of other aspects of polyphonic music remains a topic for future research. Although some of the experimental stimuli contain expressive variations in performance timing, dynamics, articulation, and timbre, perception of these phenomena are not generally captured in the modelling or analysis, with very few exceptions. Chapter 5 examines emotional and aesthetic experience of music from the perspective of the statistical learning and probabilistic prediction hypotheses. The focus on expectation results from its close relation to musical structure, making it a good candidate for investigating effects of enculturation. However, expectation is just one of many mechanisms involved in the emotional and aesthetic experience of music, which can also depend on other mechanisms such as individual experience with particular pieces of music and referential imagery for non-musical entities (Juslin, 2013). Other important contributors to emotional and aesthetic experience of music are the current attentional, emotional, and motivational state of the listener, their stylistic preference, their personality and musical

[12]This is not to say that music perception has nothing to contribute to music theory or vice versa—the work of Huron, Krumhansl, Lerdahl, Meyer, Narmour, and Temperley, discussed above, stands testament to the mutual benefits that can arise.

identity, as well as the influence of the listening context (e.g., social or environmental factors). The omission of the factors noted above does not reflect a judgement that they are not important components of music perception and experience, only that one has to simplify in the interests of tractability.

The development of cultural domains across many generations has resulted in phenomena that are no less extraordinary than the results of biological evolution. However, understanding the psychological mechanisms underlying this development poses a significant challenge due to the timescales involved and the fact that it often leaves no observable trace, meaning that only the present-day residue of the ongoing process can be studied *in vivo*. By combining computational modelling and empirical experimentation, the research programme presented in this book intends to shed light on the perceptual component of cultural evolution, by testing the psychological mechanisms proposed by the statistical learning and probabilistic prediction hypotheses. Placed in this wider context, the outcomes of the research have implications that go beyond music to other cultural domains and beyond perception to the dynamic cyclical process in which cultural artefacts are perceived, appreciated, created, and passed on to future generations.

2 A Computational Model

It is striking that music appears very early in the historical development of artificial intelligence and cognitive science. Computer programs were developed to compose and analyse music from the mid-1950s (Brooks Jr. et al., 1957; Hiller & Isaacson, 1959; Sowa, 1956) around the same time as the seminal Dartmouth workshop of 1956 (attended by Marvin Minsky, John McCarthy, Claude Shannon, Alan Newell, and Herb Simon amongst others) which is often hailed as the birthplace of the field of artificial intelligence. In 1979, the year the Cognitive Science Society was founded, Christopher Longuet-Higgins (who had coined the term Cognitive Science in 1973) wrote the following in a paper published in the journal *Nature*:

> You are browsing, let us imagine, in a music shop, and come across a box of faded pianola rolls. One of them bears an illegible title, and you unroll the first foot or two to see if you can recognise the work from the pattern of holes in the paper. Are there four beats in the bar, or only three? [...]. Eventually, you decide that the only way of finding out is to buy the roll, take it home, and play it on the pianola. Within seconds, your ears have told you what your eyes were quite unable to make out [...]. How does the brain solve the problem you were unable to solve in the music shop? Can we construct a precise and plausible theory of the cognitive processes that must be involved?
>
> (Longuet-Higgins, 1979, pp. 307–308)

By "precise and plausible theory", Longuet-Higgins means that the theory should be expressed in sufficient detail that a computer program can be written implementing the hypothesized cognitive processes which can then be used to simulate the perception of a listener.

This approach has proved to be a powerful method for understanding cognition for two primary reasons (Johnson-Laird, 1983; Longuet-Higgins, 1981; Newell & Simon, 1976). First, implementing an executable computer program requires that the hypothesized mechanisms (and the assumptions underlying them) are fully specified. Second, an implemented theory can be evaluated by systematically comparing the behaviour of the computer program with that of humans presented with the same stimuli and tasks. In comparison with the traditional approach in experimental psychology, this allows for greater ecological validity since it is more straightforward to use genuine musical pieces as stimuli, replacing controlled manipulation of stimulus properties with precise characterization of the stimulus (and its psychological processing) in the computational model.

In this chapter, I will present a computational model that implements the statistical learning and probabilistic prediction hypotheses outlined in Chapter 1. The model is based on methods developed in data compression and computational linguistics and the focus for evaluation in this chapter is on optimal prediction performance. This means seeking parameters that enable the model to predict unseen music most accurately—in other words, the parameters that assign the highest probability to notes that actually occur in music (meaning that notes that do not occur consequently have lower probabilities). Human listeners do not necessarily achieve optimal prediction performance due both to specific limitations in the cognitive and neural machinery underlying statistical learning and probabilistic prediction and to limitations on general psychological processes such as memory and attention. However, optimizing prediction performance serves three functions. First, it allows clear and direct comparison between different models and model parameterizations. Second, it provides a yardstick, allowing human performance on a task to be compared with that of an optimal ideal observer, given equivalent perceptual input and experience (rather than oracular prior knowledge of the experimental task). Third, it gives assurance that any failure of the model to simulate human performance on a task is not due to the model performing the task to an inferior level within the scope defined by its parameters. Furthermore, to the extent that human performance is inferior to the ideal observer, constraints can be applied to the model in a systematic way to bring its performance in line with human performance. The present chapter is concerned only with the first of these functions; in subsequent chapters, the behaviour of the model will be systematically compared with listeners' perception of music.

As a prelude, § 2.1 provides an informal presentation of the model that should be sufficient for understanding all the material presented in the remainder of the book. Many readers will not need to go beyond this section and can turn instead to Chapter 3. For those interested in the computational details, a formal presentation follows in §§ 2.2 and 2.3 including coverage of the representation of musical structure, statistical learning of the syntactic structure of a musical style, and the estimation of probabilities using a learned model. Section 2.4 presents a comparative performance evaluation of different parameterizations of the model, with the goal of determining optimal parameters for prediction performance. Section 2.5 compares the performance of the model with other competing models, including neural networks, for predicting melodic pitch, temporal structure, and harmonic movement. The chapter closes in § 2.6 with a discussion of whether more powerful models are required both to capture adequately the complexity of musical syntax itself and the cognitive representations of syntactic structure possessed by listeners.

2.1 An informal sketch

The computational model presented here was originally developed by Pearce (2005) based on techniques first introduced by Cleary and Witten (1984), Bunton (1996), and Conklin (1990) and has subsequently become known as *IDyOM* (for Information Dynamics of Music).[1] IDyOM uses statistical learning and probabilistic prediction to acquire and process internal representations of the syntactic structure of a musical style and repeated structure within a particular piece of music. Given exposure to music, IDyOM learns statistical regularities present in the music in terms of sequential patterns that determine the probability of particular events appearing in particular contexts (e.g., the pitch or timing of a note at a particular point in a melody).

IDyOM takes as input discrete, symbolic representations of music: sequences of discrete events (e.g., notes or chords) each having a finite set of basic attributes (representing, for example, timing or pitch content), each of which assumes values from a finite *alphabet* of discrete elements. For present purposes, this means an event representation in which musical notes have a pitch, an onset time, and a duration: pitch is represented using integers corresponding to the keys on a piano keyboard (in semitone steps where middle C is 60) while onset time and duration are represented numerically such that a crotchet (or quarter note) corresponds to 24 basic time units. Chords are considered to be sets of notes overlapping in time.[2] Figure 2.2 shows an example of the representation scheme applied to a chorale melody from the collection harmonized by J. S. Bach. Given such a representation of

[1] IDyOM is publicly available as free and open source software, along with documentation, at https://www.marcus-pearce.com/idyom.

[2] From the perspective of prediction, duration is not very useful because note endings often coincide with the beginning of the next note, leaving no time after perceiving the duration of a note to form a prediction for the onset of the next note. However, duration is important for computing overlaps between notes in a polyphonic texture, which is used for modelling harmonic movement in Chapter 4.

music, IDyOM is designed to capture several empirically motivated intuitions about human predictive processing of music: first, that expectations depend on knowledge drawn both from long-term listening to music and from short-term listening to the current piece of music; second, that listeners' expectations are sensitive to the musical context; third, that listeners generate expectations both for the pitch and for the timing of musical events; and finally, that listeners entertain multiple musical features in generating expectations for continuation. I will consider these design features in turn.

IDyOM combines two sources of input for statistical learning (see Figure 2.1): first, given a large body of music, it constructs a *long-term model*, intended to reflect the generalized, stylistic structure of the music; second, given exposure to an individual piece of music, IDyOM learns an initially empty *short-term model* dynamically and incrementally while processing the piece. In psychological terms, the long-term model is intended to reflect schematic stylistic knowledge about music acquired through a lifetime of listening to a musical style (Huron, 2006; Krumhansl, 1990). The short-term model is intended to reflect dynamic learning of the internal structure of repeated material within an individual piece of music (Castellano et al., 1984; Kessler et al., 1984; Oram & Cuddy, 1995). IDyOM can be configured to use the short-term model or long-term model only, which are referred to as the STM and LTM configurations respectively. Once trained, the LTM is completely static, it will never learn from subsequently presented music. However, a variant of this configuration allows the LTM to continue learning while predicting stimuli presented after its initial training. This is known as the LTM+. The BOTH configuration combines the LTM and STM while the BOTH+ configuration combines the LTM+ and STM. Table 2.1 summarizes these different ways of configuring IDyOM's long- and short-term models.

Both the long-term and the short-term model store information about musical structure in the form of n-grams—subsequences of length n—plus frequency counts for the number of times each n-gram has occurred. An n-gram can be thought of as consisting of a *context* of length $n-1$ followed by a single element *prediction*. The long- and short-term models store all n-grams for all possible values of n appearing in the music they experience. The stored frequency counts can be used to estimate the probability of any given symbol continuing any given context that has appeared in the music presented as input to the model. This accords with evidence that listeners' musical expectations are dependent on the preceding context such that different expectations are generated when the context changes (Krumhansl, 2000). In modelling terms, the length of the context used to make a prediction is called the *order* and is equal to $n-1$. For example, a model that predicts continuations based on

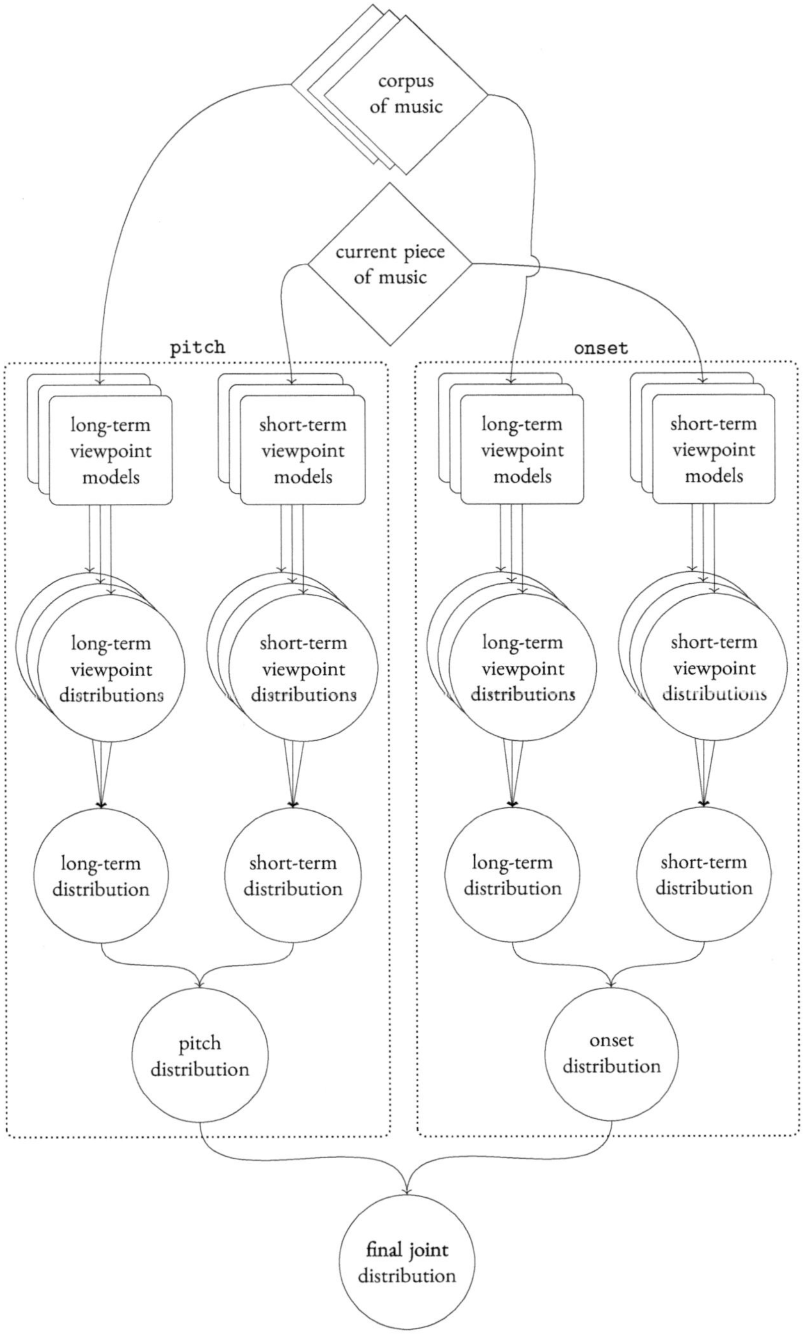

Figure 2.1: The architecture of IDyOM. Diamonds show the music that is input to IDyOM distinguishing between the large corpus of music which is used to train the long-term model and the current piece being processed used to train the short-term model. Squares indicate trained models and circles represent probability distributions generated by those models. The architecture is illustrated for two target viewpoints: `pitch` processed on the left and `onset` processed on the right.

Configuration	Description
STM	The short-term model: incrementally learns each auditory sequence from an initially empty state
LTM	The long-term model: learns from a collection of auditory sequences and, after this initial training, generates predictions for subsequently presented sequences without further learning
LTM+	The long-term model configured to continue learning across all sequences experienced after its initial training
BOTH	The STM combined with the LTM
BOTH+	The STM combined with the LTM+

Table 2.1: The five different ways of configuring IDyOM's long- and short-term models. The auditory sequences referred to could be pieces of music but could also be other kinds of auditory stimulus.

the preceding two events is a second-order model (sometimes called a trigram model since $n = 3$).

Since it is possible that not every element of the alphabet has appeared in a context of a given order, probabilities for the missing elements are estimated using lower-order distributions. This combination of probabilities from models of different order is known as *smoothing*. Smoothing also has the advantage of combining higher-order predictions which are structurally very specific to the context but may be statistically unreliable (because longer contexts appear less frequently, with fewer distinct continuations), with lower-order predictions (based on shorter contexts) that are more structurally generic but also more statistically robust (since they appear more frequently with a wider range of continuations). Rather than having a fixed maximum order for prediction, IDyOM is a *variable-order model* that adaptively varies the order used for each context encountered during prediction. The estimated probability distribution for a given context is computed as a weighted average of the predictions made by models of all orders lower than the adaptively selected maximum order. This distribution represents the estimated probability of different continuations to the context in question.

Listeners generate expectations simultaneously for multiple basic attributes of musical events, including their pitch and onset time (Jones & Boltz, 1989; Prince et al., 2009b). Correspondingly, IDyOM can predict different attributes of the events making up an auditory sequence using distinct models for each attribute. The attributes used by IDyOM are referred to as *viewpoints* and they appear in `typewriter` font. For example, IDyOM can be configured to predict pitch alone, onset time alone, or both pitch and onset time. In the latter case, IDyOM can process pitch and temporal attributes independently (using separate models whose probabilistic output may be subsequently combined) or interactively using a single model of an attribute that links the two domains (e.g., by representing notes as a pair of scale degree and inter-onset interval ratio, see `scale-degree⊗ioi-ratio` in the lower panel of Figure 2.2). When configured to predict both pitch and onset time, IDyOM computes the final probability of the note as the joint probability of its pitch and onset time (see Figure 2.1).

Furthermore, listeners process music using multiple psychological representations of pitch (e.g., pitch height, pitch class, pitch interval, pitch contour, scale degree; Krumhansl, 1990; Shepard, 1982) and time (e.g., absolute duration-based and relative beat-based representations; Bouwer et al., 2020; Teki et al., 2011). Accordingly, IDyOM is able to create models for multiple viewpoints present in, or derived from, the musical surface and combine the predictions made by these models. A distinction is made between *basic viewpoints* present in the musical surface (here `onset`, `duration`, and `pitch`) and *derived viewpoints* representing attributes derived from the basic viewpoints. For example, from `pitch` IDyOM derives the melodic pitch interval in semitones between a note and its immediate predecessor (`interval`), the direction or contour of the interval (`contour`), and the chromatic scale degree (`scale-degree`) as the interval in semitones from the tonal centre (represented by `keysig` and `mode`), assuming that one has been encoded or inferred. Similarly, from the basic viewpoint `onset` (onset time), IDyOM derives the inter-onset interval (IOI) between a note and its immediate predecessor (`ioi`) and the ratio of successive IOIs (`ioi-ratio`). Assuming that a metrical interpretation (represented by the viewpoints `pulses` and `barlength` which relate to a time signature) has been encoded or inferred, event timing can also be represented with respect to the start of the current bar (`posinbar`). Each viewpoint is associated with an alphabet of possible values that it can assume.

Two or more basic or derived viewpoints can be combined into a *linked viewpoint* by computing the Cartesian product of their alphabets, indicated by ⊗, meaning that each element of the alphabet is a tuple of values. For example a link between `interval` and `scale-degree` is denoted `interval⊗scale-degree` and has an alphabet consisting of pairs of pitch intervals and scale degrees. Table 2.2 shows the basic, derived, and linked viewpoints that will be used in this book while Figure 2.2 shows a chorale melody harmonized by J. S. Bach represented in terms of each of these viewpoints.

IDyOM can be configured to predict a single basic viewpoint (e.g., `pitch` only or `onset` only) or more than one (e.g., both `pitch` and `onset`). During prediction, the se-

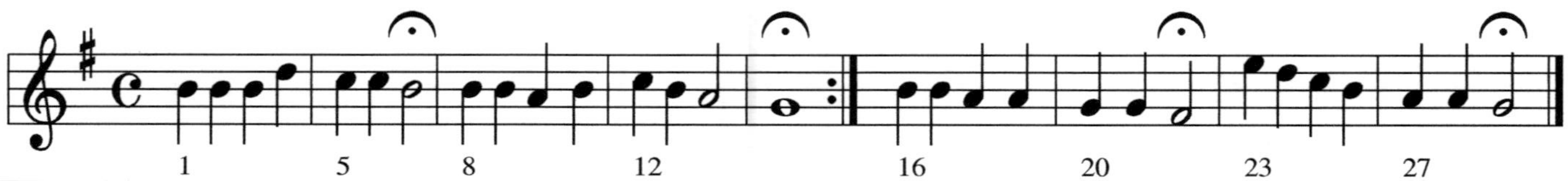

Note position	1	2	3	4	5	6	7	8	9	10	11	12	13	14	15	16	17	18	19	20	21	22	23	24	25	26	27	28	29
`pitch`	71	71	71	74	72	72	71	71	71	69	71	72	71	69	67	71	71	69	69	67	67	66	76	74	72	71	69	69	67
`onset`	0	24	48	72	96	120	144	192	216	240	264	288	312	336	384	480	504	528	552	576	600	624	672	696	720	744	768	792	816
`duration`	24	24	24	24	24	24	48	24	24	24	24	24	24	48	96	24	24	24	24	24	24	48	24	24	24	24	24	24	48
`pulses`	4	4	4	4	4	4	4	4	4	4	4	4	4	4	4	4	4	4	4	4	4	4	4	4	4	4	4	4	4
`barlength`	96	96	96	96	96	96	96	96	96	96	96	96	96	96	96	96	96	96	96	96	96	96	96	96	96	96	96	96	96
`keysig`	1	1	1	1	1	1	1	1	1	1	1	1	1	1	1	1	1	1	1	1	1	1	1	1	1	1	1	1	1
`mode`	0	0	0	0	0	0	0	0	0	0	0	0	0	0	0	0	0	0	0	0	0	0	0	0	0	0	0	0	0
`interval`	⊥	0	0	3	-2	0	-1	0	0	-2	2	1	-1	-2	-2	4	0	-2	0	-2	0	-1	10	-2	-2	-1	-2	0	-2
`contour`	⊥	0	0	1	-1	0	-1	0	0	-1	1	1	-1	-1	-1	1	0	-1	0	-1	0	-1	1	-1	-1	-1	-1	0	-1
`scale-degree`	4	4	4	7	5	5	4	4	4	2	4	5	4	2	0	4	4	2	2	0	0	11	9	7	5	4	2	2	0
`ioi`	⊥	24	24	24	24	24	24	48	24	24	24	24	24	24	48	96	24	24	24	24	24	24	48	24	24	24	24	24	24
`ioi-ratio`	⊥	⊥	1	1	1	1	1	2	1/2	1	1	1	1	1	2	2	1/4	1	1	1	1	1	2	1/2	1	1	1	1	1
`ioi-contour`	⊥	⊥	0	0	0	0	0	1	-1	0	0	0	0	0	1	1	-1	0	0	0	0	0	1	-1	0	0	0	0	0
`posinbar`	0	24	48	72	0	24	48	0	24	48	72	0	24	48	0	0	24	48	72	0	24	48	0	24	48	72	0	24	48
`interval⊗scale-degree`	⊥	(0 4)	(0 4)	(3 7)	(-2 5)	(0 5)	(-1 4)	(0 4)	(0 4)	(-2 2)	(2 4)	(1 5)	(-1 4)	(-2 2)	(-2 0)	(4 4)	(0 4)	(-2 2)	(0 2)	(-2 0)	(0 0)	(-1 11)	(10 9)	(-2 7)	(-2 5)	(-1 4)	(-2 2)	(0 2)	(-2 0)
`ioi-ratio⊗posinbar`	⊥	⊥	(1 48)	(1 72)	(1 0)	(1 24)	(1 48)	(2 0)	(1/2 24)	(1 48)	(1 72)	(1 0)	(1 24)	(1 48)	(2 0)	(2 0)	(1/4 24)	(1 48)	(1 72)	(1 0)	(1 24)	(1 48)	(2 0)	(1/2 24)	(1 48)	(1 72)	(1 0)	(1 24)	(1 48)
`scale-degree⊗ioi-ratio`	⊥	⊥	(4 1)	(7 1)	(5 1)	(5 1)	(4 1)	(4 2)	(4 1/2)	(2 1)	(4 1)	(5 1)	(4 1)	(2 1)	(0 2)	(4 2)	(4 1/4)	(2 1)	(2 1)	(0 1)	(0 1)	(11 1)	(9 2)	(7 1/2)	(5 1)	(4 1)	(2 1)	(2 1)	(0 1)

Figure 2.2: A melody from the chorales harmonized by J. S. Bach (BWV 379) illustrating the viewpoints in Table 2.2. The first vertical panel shows the basic event space: musical events are represented in terms of their chromatic pitch (as a midi note number; 60 = middle C), onset time and duration (where 24 corresponds to a crotchet or quarter note duration). Basic viewpoints also represent the prevailing metre (`barlength`, `pulses`) and tonality (`keysig`, `mode`) used in turn to derive the viewpoints `scale-degree` and `posinbar` respectively. Here, the metre is common time (4/4): `pulses` indicates that each bar has four pulses while `barlength` indicates that a bar has a duration of a semibreve (whole note, 96 basic time units). The tonality is G major: `keysig` indicates one sharp in the key signature while `mode` = 0, corresponding to major. The second panel shows viewpoints derived from pitch: the pitch interval in semitones between a note and its immediate predecessor (`interval`), the direction of the interval (`contour`), and chromatic scale degree (`scale-degree`) as the distance in semitones from the tonic pitch (G or 67). The third panel shows viewpoints derived from `onset`: the inter-onset interval (`ioi`), the ratio between successive inter-onset intervals (`ioi-ratio`), the direction of change in IOI (`ioi-contour`), and the temporal interval from the start of the current bar (`posinbar`). Note that `ioi` is undefined (⊥) for the first note while `ioi-ratio` and `ioi-contour` are undefined for the first two notes. The final panel shows examples of linked viewpoints affording learning of: joint melodic and tonal structure (`interval⊗scale-degree`; see Chapter 3); joint rhythmic and metrical structure (`ioi-ratio⊗posinbar`; in place of `ioi-ratio⊗posinbar⊗barlength` from Table 2.2 for reasons of space); and joint tonal and rhythmic structure (`scale-degree⊗ioi-ratio`; see Chapter 4).

	Viewpoint τ	Description	Alphabet $[\tau]$
Basic	`pitch`	Chromatic pitch	$\mathbb{Z}$
	`onset`	Onset time	$\mathbb{Z}^*$
	`duration`	Duration	$\mathbb{Z}^*$
	`barlength`	Duration of a bar	$\mathbb{Z}^+$
	`pulses`	No. of pulses in bar	$\mathbb{Z}^+$
	`keysig`	No. of sharps/flats	$\{-7, \ldots, 0, \ldots, 7\}$
	`mode`	Tonal mode	$\{0, \ldots, 11\}$
Derived from `pitch`	`interval`	Pitch interval	$\mathbb{Z}$
	`contour`	Pitch contour	$\{-1, 0, 1\}$
	`scale-degree`	Scale degree	$\{0, \ldots, 11\}$
Derived from `onset`	`ioi`	Inter-onset interval	$\mathbb{Z}^+$
	`ioi-ratio`	IOI ratio	$\mathbb{Q}^+$
	`ioi-contour`	IOI contour	$\{-1, 0, 1\}$
	`posinbar`	Interval from start of current bar	$\mathbb{Z}^*$
Linked	`interval⊗scale-degree` `ioi-ratio⊗posinbar⊗barlength`	See caption	See caption

Table 2.2: The viewpoints used in the IDyOM simulations presented in Chapters 2 to 8. For a viewpoint τ, $[\tau]$ represents the alphabet associated with that viewpoint. $\mathbb{Z}$ indicates the set of integers, $\mathbb{Z}^+$positive integers, $\mathbb{Z}^*$non-negative integers, $\mathbb{Q}$ rational numbers, and $\mathbb{Q}^+$positive rational numbers. Basic viewpoints are irreducible properties of musical events (or the prevailing environment in which they appear, such as the metre or tonality). `keysig` represents the number of sharps (positive numbers) or flats (negative numbers) in a key signature while `mode` represents the starting note of the scale such that 0 indicates a major key and 9 a minor key with the same key signature (while allowing for other diatonic modes). Derived viewpoints are computed from a basic viewpoint and can therefore be used by IDyOM to generate predictions for the corresponding basic viewpoint. In general, any two or more viewpoints can be linked together but here we restrict ourselves to a single linked viewpoint for each of the basic viewpoints `pitch` and `onset`. These allow IDyOM to learn representations of sequential melodic and tonal structure for pitch prediction and rhythmic and metrical structure for onset prediction. The alphabet of a linked viewpoint is the Cartesian product of the alphabets of its constituent viewpoints. The alphabet of `interval⊗scale-degree` is [`interval`] × [`scale-degree`] while the alphabet of `ioi-ratio⊗posinbar⊗barlength` is [`ioi-ratio`] × [`posinbar`] × [`barlength`].

lected basic viewpoints are referred to as the *target viewpoints*. Each target viewpoint can be predicted with one or more *source viewpoints*. For example, IDyOM can be configured to predict `pitch` with a single model where the source viewpoint is the same as the target viewpoint (i.e., `pitch`) or with a single derived or linked viewpoint (e.g., `interval` or `interval⊗scale-degree`) or with a combination of more than one basic, derived, or linked viewpoint. For example, it can be configured to predict `pitch` (the target viewpoint) using a combination of two models for `interval` and `scale-degree` (the source viewpoints). Alternatively, it can be configured to predict note onsets with a combination of two models for `ioi` and `ioi-ratio`. These models can be used separately to predict `pitch` or `onset` independently or they can be combined to produce a joint prediction for the `pitch` and `onset` of the next note (see Chapter 4). In this case, each model generates predictive distributions for a single attribute of the next note (i.e., pitch or onset time), which are combined separately for the long-term and short-term models before being combined into the final pitch distribution, as shown in Figure 2.1.

In many of the IDyOM simulations covered in subsequent chapters, a single linked viewpoint is used consistently. For pitch prediction, this is `interval⊗scale-degree` combining representations of melodic and tonal properties of pitch, which can be motivated from psychological and musical perspectives (see Chapter 3). For onset prediction, the linked viewpoint used is `ioi-ratio⊗posinbar⊗barlength` combining rhythmic and metrical influences on prediction, which again can be motivated both psychologically and musically (see Chapter 4). This has the methodological advantage of enabling direct comparisons between different experiments, since the same viewpoints are used consistently for simulation. The question of which viewpoints reflect psychological representations of music is an interesting one that is considered further in §§ 3.5, 4.2.3, and 9.3.3.

While processing sequences of musical events, IDyOM generates expectations for the next event (e.g., the note that continues a melody) in the form of a probability distribution (P) which assigns a probability to each possible next event, conditioned upon the preceding musical context and the prior musical experience of the model. The information-theoretic quantity *entropy* ($H = -\sum_{p \in P} p \log_2 p$) reflects the uncertainty of the prediction before the next event is heard—if every continuation is equiprobable, entropy will be maximal and the prediction highly uncertain while if one continuation has very high probability, entropy will be low and the prediction very certain. When the next event actually arrives, it may have a high probability making it expected or a low probability making it unexpected. Rather than dealing with raw probabilities, which can become vanishingly small, *information content* ($h = -\log_2 p$) provides a more numerically stable measure of how unexpected the model finds an event in a particular context. Information content reflects the amount of unpredictable or non-redundant information contained in an event. This information need not be stored in memory because it can be reconstructed with the model, thereby producing a compressed representation of the event (see also Chapter 6).

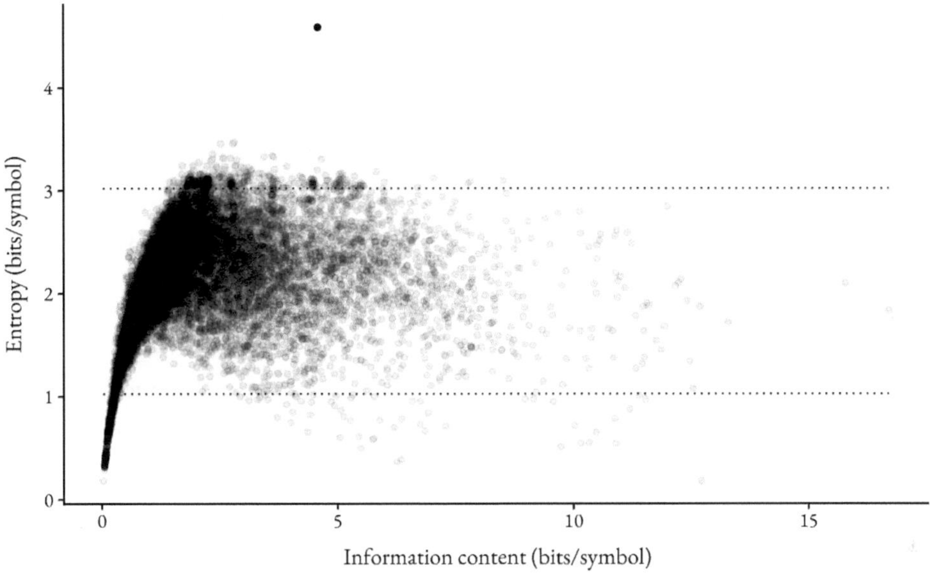

Figure 2.3: The relationship between entropy and information content using predictions from an IDyOM model trained using 10-fold cross-validation to predict the pitch of every note in 370 chorale melodies harmonized by J. S. Bach, comprising 18,472 notes. IDyOM was configured to predict the target viewpoint `pitch` using the source viewpoint `interval⊗scale-degree` and the BOTH+ configuration. For entropy, 5% and 95% quantiles illustrate how information content is constrained when entropy is very low or very high respectively.

Although entropy and information content are conceptually distinct, the two quantities mutually constrain one another. Figure 2.3 illustrates these constraints for an IDyOM model predicting the pitch of every note in a collection of 370 chorale melodies harmonized by J. S. Bach. As can be seen, when entropy is very low, information content is constrained to be either very high or very low. This is because a low entropy distribution corresponds to a very certain prediction in which a single continuation has a very high probability and all other continuations have very low probabilities. If the high-probability continuation does actually follow then information content will be very low (indicating that the continuation was expected), otherwise it will be very high (indicating that the continuation was unexpected). Conversely, when entropy is very high, all possible continuations tend towards equal, moderate probabilities, constraining information content to intermediate values. Figure 2.3 illustrates the absence of intermediate information content when entropy is very low as well as the absence of low and high information content when entropy is very high. In spite of this relationship, entropy and information content are conceptually distinct: entropy measures the uncertainty of a model's prediction in a given context before the outcome is observed, thereby reflecting the model's use of the context to generate a prediction; information content measures how surprising the outcome is once it has been observed relative to the model's prediction, thereby reflecting both the model's predictions *and* the actual events observed by the model (see also § 2.4).

2.2 Modelling statistical learning and probabilistic prediction

2.2.1 The domain of interest

As described in § 2.1, it is assumed that music can be represented as a sequence of discrete events each with a finite number of discrete attributes, each of which may assume one of a finite number of values. For example, melody can be represented as a sequence of notes (the events) each of which has an onset time and a pitch (the properties) drawn from a finite alphabet of possible onset times and pitches. Harmony can be represented as a sequence of chords, each consisting of a finite set of pitched notes that sound simultaneously at a particular point in time. Also of interest are more abstract representations of musical events such as: sequential pitch intervals, contour, and scale degree in melody; metrical strength or relative temporal representations of rhythms that are equivalent under augmentation or diminution for rhythmic sequences; or chord roots, qualities, and functions in harmony. These will be addressed in § 2.3 below. For the purpose of introducing the model of probabilistic prediction, it is assumed that sequences consist of abstract events assuming values drawn from a finite alphabet.

Given a sequence x_1^k of length k composed of symbols drawn from an alphabet $\mathcal{X}$, a model can be defined that makes probabilistic predictions by estimating the conditional probability of each event in x_1^k given the preceding events:

$$p(x \mid \overline{x}) = p(x_i \mid x_1^{i-1}), \forall i \in \{1 \ldots k\}. \quad (2.1)$$

For such a model, the unexpectedness or surprisal of an event x given the preceding context $\overline{x}$ is characterized by its *information content*:

$$h(x \mid \overline{x}) = -\log_2 p(x \mid \overline{x}). \quad (2.2)$$

The uncertainty of the model's prediction for the event is characterized by the *entropy* of the predictive distribution, which reflects the expected value of the information content:

$$H(\overline{x}) = -\sum_{x \in \mathcal{X}} p(x \mid \overline{x})\, h(x \mid \overline{x}). \quad (2.3)$$

Maximum entropy occurs when every symbol in the alphabet has an equal probability, $\forall x \in \mathcal{X}, p(x \mid \overline{x}) = \frac{1}{|\mathcal{X}|}$:

$$H_{max}(\overline{x}) = \log_2 |\mathcal{X}|. \tag{2.4}$$

The notation is flexible enough to accommodate sequences of different musical attributes. If $\mathcal{X}$ is a set of pitches, x_1^k is a melody; if $\mathcal{X}$ is a set of inter-onset intervals, x_1^k is a rhythm; if $\mathcal{X}$ is a set of chords, x_1^k is a harmonic sequence; if $\mathcal{X}$ is a set of letters, x_1^k is printed text. For ease of exposition, I will use examples of predicting pitch in melody and letters in printed English to illustrate the probabilistic models. The use of printed English reflects the fact that many of the techniques to be discussed were developed in the context of language modelling, illustrating the domain-generality of the approach. Predictive modelling of event timing and harmonic movement are covered in Chapter 4.

2.2.2 Fixed-order Markov models

An optimal model implementing Equation 2.1 should emulate as closely as possible the actual source process responsible for generating musical sequences. However, the model is not an oracle with prior knowledge of the source, rather it must be estimated by observing sequences generated by the source. This is analogous to the task faced by listeners learning to appreciate music within a given culture, who must learn about the structural and statistical regularities of the style through exposure to musical examples. It is unlikely that most listeners possess a completely perfect model of the source. Nonetheless, the hypothesis underlying this work is that listeners learn a model that approximates to some extent the regularities present in the music to which they are exposed, thereby allowing more or less accurate predictions to be made. In the following, I will introduce the probabilistic model used within IDyOM by starting with a very simple model and adding features one at a time, motivating each one in terms of its potential for improving prediction performance.

Perhaps the simplest approach to implementing Equation 2.1 is to assume that the probability depends only on a finite portion of the preceding context:

$$p(x \mid \overline{x}) = p(x_i \mid x_{(i-n)+1}^{i-1}) \approx p(x \mid x_1^{i-1}). \tag{2.5}$$

where n is the *order* of the model and the length of the preceding context considered is $n - 1$. This is a fixed-order Markov model or n-gram model. The simplest way of estimating this probability is the *maximum likelihood* method:

$$p(x \mid \overline{x}) = \frac{\text{count}(x, \overline{x})}{\sum_{x \in \mathcal{X}} \text{count}(\overline{x})} \tag{2.6}$$

(a)

0	1	2	3	4	5	6	7	8	9	10	11	Total
1,523	0	1,101	22	1,664	714	2	2,435	3	722	7	200	8,393
0.18	0	0.13	0.0	0.2	0.09	0.0	0.29	0.0	0.09	0.0	0.02	1.0

(b)

	0	1	2	3	4	5	6	7	8	9	10	11	Total
0	497	0	196	2	180	24	0	325	1	60	3	83	1,371
1	0	0	0	0	0	0	0	0	0	0	0	0	0
2	324	0	367	0	212	54	0	88	0	15	0	38	1,098
3	2	0	12	6	0	2	0	0	0	0	0	0	22
4	310	0	346	0	400	151	2	360	0	54	0	6	1,629
5	4	0	118	12	301	163	0	99	0	9	0	8	714
6	0	0	0	0	0	0	0	2	0	0	0	0	2
7	227	0	39	0	546	305	0	972	0	314	0	10	2,413
8	0	0	0	2	0	0	0	0	0	1	0	0	3
9	11	0	4	0	5	13	0	428	1	232	2	25	721
10	2	0	0	0	0	0	0	0	0	3	2	0	7
11	86	0	18	0	2	2	0	32	0	30	0	30	200
Total	1,463	0	1,100	22	1,646	714	2	2,306	2	718	7	200	8,180

Table 2.3: Two fixed-order Markov models trained on a corpus of melodies from 213 German nursery rhymes (of which all but one are in a major key). For ease of illustration, the nursery rhymes are represented as sequences of scale degrees relative to the tonic in the notated key signature (0–11 where 0 = tonic, 2 = supertonic, 4 = mediant, 5 = subdominant, 7 = dominant, 9 = submediant, 11 = the leading tone). This yields an alphabet of 12 elements rather than one of 27 elements for chromatic pitch. (a) A zeroth-order model showing frequency counts for each scale degree in the top row and estimated probabilities in the second row. (b) A first-order model with rows giving the single-tone context and columns showing the frequency with which each scale degree follows each context. Dividing the counts in each row by the total for each row gives the maximum likelihood estimated distribution for the scale degree that will follow the one shown in the first column. Note that the frequency counts for the first-order model sum to less than the zeroth-order context because no prediction can be made for the first note in each nursery rhyme, since there is no preceding context.

where count($\overline{x}$) denotes the number of times the context sequence $\overline{x}$ has appeared while count(x, $\overline{x}$) represents the number of times symbol x has followed the context sequence $\overline{x}$ in the previous experience of the model.

As an example, imagine trying to simulate the musical expectations of a child. We might take a collection of melodies from nursery rhymes, extract the frequency with which each pitch appears and compute maximum likelihood estimates for the probability of each pitch. Indeed, this is exactly what Richard Pinkerton did in his *Scientific American* article from 1956, one of the earliest examples of information theory applied to music. Table 2.3(a) shows the results of such an analysis applied to the melodies from 215 German nursery rhymes.[3] The obvious problem with this model is that the expectations it generates are the same at every point in every melody it ever experiences; it is completely insensitive to the preceding context. We can improve this zeroth-order model by making its predictions conditional upon the preceding note in the melody giving us a first-order model. This is shown in Table 2.3(b).

The fixed-order Markov models shown in Table 2.3 exhibit some behaviours that we would also expect of our putative German child in the process of becoming enculturated in the music of their culture. For example, the zeroth-order model shows typical features of the tonal hierarchy (Krumhansl, 1990) such that the tonic, mediant, and dominant (scale degrees 0, 4, and 7 respectively) appear most frequently, followed by the remaining diatonic tones (scale degrees 2, 5, 9, and 11), and then the remaining non-diatonic tones (scale degrees 1, 3, 6, 8, and 10). Meanwhile for the first-order model, the most frequent continuation of the leading tone (scale degree 11) is the tonic (scale degree 0) while the most frequent continuation of the submediant (scale degree 9) is the dominant (scale degree 7).

One could go on increasing the order of the model indefinitely, making its predictions increasingly specific to the context. However, given finite input, the longer contexts used by higher-order models will have appeared less frequently in the experience of the model to the extent that many possible continuations will have appeared very infrequently or not at all. If a fixed-order model experiences a symbol that it has never experienced before then it has no option but to assign it a probability of zero, which is obviously too low since the symbol has, in fact, appeared. This is the *zero-frequency problem* (Witten & Bell, 1991). In Table 2.3, for example, note that the flattened supertonic (scale degree = 1) does not appear at all in this corpus and therefore has an estimated zeroth-order probability of zero while, for the first-order model, there are several possible continuations for each context tone that have zero frequency. A more general issue with higher-order models is that while they benefit from generating predictions that are structurally very specific to the context, these predictions can be statistically rather weak if the context has appeared very rarely. Lower-order models suffer

[3]This collection forms part of the Essen folk song collection (Schaffrath, 1995) available at: https://kern.humdrum.org/cgi-bin/browse?l=essen/europa/deutschl/kinder.

in the opposite direction, they are statistically robust but structurally generic, in that they match many different contexts rather than specifically matching the current context.

A solution to these problems is to combine predictions from models of more than one order. The zero-frequency problem can be addressed by using a lower-order model whenever a symbol appears that is not predicted at a given order. For example, when processing written text, a model may never have seen 'u' following 'tranq' (i.e., a fifth-order model) but may have seen 'u' following 'q' (i.e., a first-order model). In machine-learning parlance using lower-order statistics constitutes an *inductive bias* that allows a model to generalize to previously unseen circumstances. A more general intuition is that it might always be advantageous to combine predictions from models of different order so as to benefit from the structural specificity afforded by higher-order models and the statistical power afforded by lower-order models. Methods for achieving this combination of models of different order are collectively referred to as *smoothing* in the literature on statistical language modelling.

Before considering smoothing itself, the requirement to construct and store multiple models with different orders raises the question of how these models can be stored efficiently in memory.

2.2.3 Suffix trees

An obvious solution to the question of predicting at multiple orders is to store multiple transition tables of the kind shown in Table 2.3. However, it is fairly easy to show that this can lead to unmanageable demands on memory in the general case. Imagine that we have a corpus with 24 different pitches (i.e., a two octave range) and we want to store models up to a maximum sequence length of 10 (i.e., an order of 9), then we would need to store more than 66 trillion individual cells. Assuming 2 bytes to store the frequency count in each cell, this would require more than 130 terabytes of space in memory. Considerations such as these motivate the use of a space-efficient data structure for constructing such probabilistic sequence models: the *suffix tree*.

Suffix trees have been referred to as "the fundamental data structure of combinatorial pattern matching" (Farach, 1997). They can be constructed in time linearly proportional to the length of the input sequence and provide a space-efficient representation of sequences that allows for elegant implementation of many fundamental string-matching algorithms (Giegerich & Kurtz, 1997; Ukkonen, 1995). For present purposes, they provide a compact way of storing contexts of different lengths for implementing probabilistic models of multiple orders.

A suffix tree is a *graph* consisting of nodes and edges, the latter representing connections between the former. It is *acyclic* because there is no path through the graph leading from one node back to itself and it is *directed* because edges can only be traversed in one direction from a *parent* node to its *children*. *Branches* are nodes that have children, while *leaves* are nodes that have no children. Each node in a suffix tree represents a sequence of symbols, corresponding to a single symbol extension of the sequence associated with its parent. One special node, the *root node* represents the empty sequence (ε) with zero symbols. The sequence of symbols associated with a node is expressed as the path from the root node that terminates at the node in question. Every suffix of the input sequence is stored in the tree as a path from the root node to a leaf such that for a sequence of length n, there are exactly n leaves. Multiple sequences can be accommodated by adding a unique *sentinel* symbol to the end of each input sequences to mark the end of the sequence, so that every leaf in the suffix tree ends with the sentinel symbol. By default, IDyOM uses $ as its sentinel symbol. A *compressed suffix tree* is one where nodes that have only one child are combined with their children meaning that their label represents a sequence of two or more symbols. The edges of a suffix tree may be associated with frequency counts to enable probabilistic sequence modelling (Bunton, 1996; Larsson, 1996).

Figure 2.4 shows a suffix tree for the string ABRACADABRA with compressed nodes and frequency counts.[4] The suffix tree might be considered an implementation detail and therefore excluded from this presentation. I have included it for two reasons. First, it helps to illustrate the details of the smoothing methods presented below. Second, it helps us understand cognitive processing of sequences by showing, at Marr's computational level, how variable length sequences and multiple-order frequency counts can be represented efficiently. Consideration of the suffix tree representation illustrates how musical sequences can be represented in overlapping form, combining veridical representations of individual sequences with schematic representations of the structural and statistical regularities present across multiple sequences. This is discussed further in §§ 3.1 and 6.1.2.

2.2.4 Smoothing

In the context of probabilistic modelling, smoothing is the combination of predictions of higher-order models with the predictions of lower-order models. During the 1980s and 1990s

[4] For simplicity of exposition, a single word containing obvious repetition is used to illustrate the operation of the model. However, one could substitute the letters in ABRACADABRA for other entities such as pitches (in a melody), words (in a sentence), or body positions (in a motor sequence).

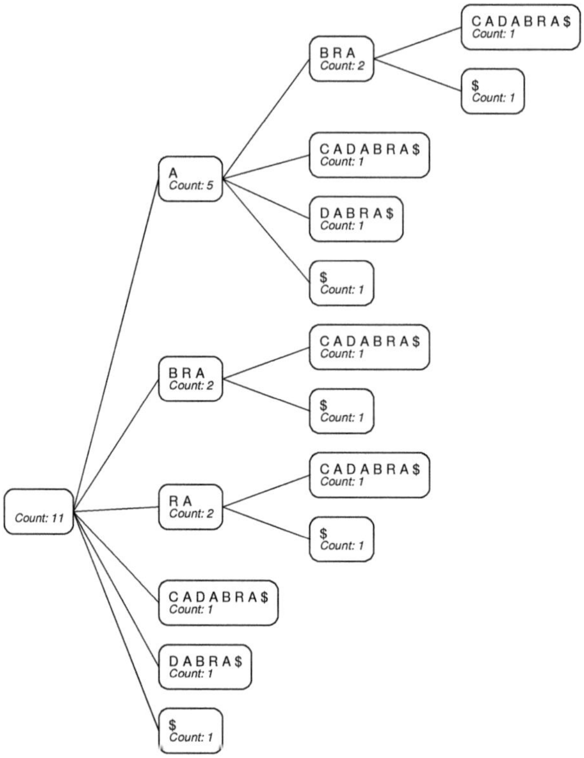

Figure 2.4: A compressed suffix tree for the string ABRACADABRA showing frequency counts for transitions between nodes. Since the sequence has ended, the sentinel symbol is appended to the sequence and added to the suffix tree. The root node represents the empty sequence ε. Note that every path from the root to a distinct leaf ending with $ corresponds to a distinct suffix of the input string: ABRACADABRA, BRACADABRA, RACADABRA, and so on. The counts indicate the number of times the subsequence associated with a given node has appeared in the context of the subsequence associated with its parent node. For example, the count for the root node is equal to the total number of symbols in the sequence.

a range of different techniques were developed in the statistical language modelling literature (Chen & Goodman, 1999; Katz, 1987; Kneser & Ney, 1995) and, apparently independently, in the literature on lossless universal data compression (Bunton, 1996; Cleary & Teahan, 1997; Cleary & Witten, 1984; Howard, 1993; Moffat, 1990). The following presentation is based on the data compression literature but it turns out that the resulting models resemble very closely Kneser-Ney smoothing, the best performing method from the statistical language modelling literature.

Prediction by Partial Match (PPM) is an algorithm developed by Cleary and Witten (1984) for probability estimation in the context of universal lossless data compression. The algorithm is lossless because the original data can be exactly reconstructed (unlike so-called lossy algorithms such as MP3 or JPEG). It is universal because it can be applied to any data. PPM arises from the insight that universal lossless data compression can be decomposed into two subcomponents: first, the construction of a probabilistic model of the data; second, the construction of a compressed representation of the data such that symbols having high probability according to the model are encoded with fewer bits than symbols having low probability (which can be achieved using methods such as arithmetic coding). For present purposes, there is no need to actually produce compressed representations of the input sequence but it is instructive to bear in mind that the probabilities estimated by the model have implications for efficient storage in memory.

2.2.4.1 Backoff smoothing and escape probabilities

In its original formulation, PPM assumes a maximum *global order bound* l which is equivalent to assuming that the suffix tree is limited to paths of depth $l + 1$. PPM uses a technique called *blending*—known in the statistical language modelling literature as *backoff smoothing*—to estimate probabilities from models of different order as follows:

$$p(x \mid \bar{x}) = \begin{cases} \frac{\text{count}(x,\bar{x}) - \beta}{\text{count}(\bar{x}) + \alpha} & \text{if count}(x, \bar{x}) > 0 \\ \frac{\max(\text{count}(x,\bar{x}) - \beta, 0)}{\text{count}(\bar{x}) + \alpha} + \gamma(\bar{x}) \cdot p(x \mid \text{su}(\bar{x})) & \text{if su}(\bar{x}) \neq \varepsilon \\ \frac{1}{|\mathcal{X}|} & \text{otherwise} \end{cases}$$

$$\gamma(\bar{x}) = \frac{t(\bar{x}) \cdot \beta + \alpha}{\text{count}(\bar{x}) + \alpha} \tag{2.7}$$

where $\text{su}(\bar{x})$ returns the suffix of sequence $\bar{x}$ (e.g., su(ABRA) = BRA and su(A) = ε). $\gamma(\bar{x})$ is the proportion of probability mass assigned to novel symbols in the context $\bar{x}$ and provides a weight assigned to probabilities estimated recursively by the lower-order model: $p(x \mid \text{su}(\bar{x}))$. The parameters $\alpha, \beta \in \mathbb{Q}^*$ specify different smoothing methods for combining predictions made by models of different order (Bunton, 1996; Cowans, 2006; Steinruecken et al., 2015).

If a symbol is observed at the highest order (given by the global order bound), backoff smoothing simply uses the maximum likelihood estimate at that order, adjusted according to α and β. If the symbol does not appear, the zero-frequency problem is addressed by reserving a portion of the probability mass for unseen symbols and then occupying this probability mass by recursively seeking an occurrence of the unseen symbol at lower orders. This process is known as *escaping* to a lower-order model and the weight $\gamma(\overline{x})$ is known as the *escape probability*, representing the probability mass that is reserved for unseen symbols at the order corresponding to the length of $\overline{x}$. The recursion terminates either when the symbol appears at a certain order or the zeroth-order model is reached without the symbol having appeared, in which case a uniform distribution over the alphabet is returned. This is known as the order -1 distribution.

As formulated above, the parameters α and β allow different smoothing methods to be specified: α is a *strength* parameter specifying the frequency count for unseen symbols used to determine the escape probability; β is a *discount* parameter that specifies the number of times a symbol needs to appear in a given context in order to be considered novel, adjusting the escape probability accordingly. The optimal values are data dependent but various values have been proposed on heuristic grounds. Five escape methods that have been developed in the data compression literature are expressed below in terms of α and β:

$$
\begin{array}{lll}
A: & \alpha = 1 & \beta = 0 \\
B: & \alpha = 0 & \beta = 1 \\
C: & \alpha = t(\overline{x}) & \beta = 0 \\
D: & \alpha = 0 & \beta = \dfrac{1}{2} \\
AX: & \alpha = t_1(\overline{x}) & \beta = 0
\end{array}
$$

where $t(\overline{x})$ is the *type count*: the number of elements from the alphabet that have appeared with non-zero frequency after context $\overline{x}$ and $t_1(\overline{x})$ is the number of elements that have appeared exactly once. Cleary and Witten (1984) introduced two escape methods, which have become known as A and B. Method A simply assigns a count of one for novel symbols, meaning that the escape probability decreases the more times the context appears. Method B requires a symbol to appear twice in a context before it is considered novel and increases the escape probability as more elements from the alphabet are observed in a given context. Method C introduced by Moffat (1990) was designed to combine the more attractive elements of methods A and B: there is no discounting as in method A but the escape count increases as more symbols from the alphabet are observed as in method B. Method D is a variant of method B introduced by Howard (1993) where the discount is $\frac{1}{2}$ instead of 1. Method AX is inspired by the assumption that the probability of novel symbols appearing is governed by Poisson processes (Moffat et al., 1998; Witten & Bell, 1991).

Figure 2.5 shows how smoothed probabilities are estimated using backoff smoothing with escape method C for a PPM model that has processed the string ABRACADABRA and is predicting the next letter to appear.

2.2.4.2 Exclusion

Exclusion is motivated by the observation that in backoff smoothing, when a symbol appears at a given order, the counts for that symbol in lower-order contexts are not used for estimating the probability of that symbol and yet they still contribute to estimating the probability of other symbols appearing only in lower-order contexts (Cleary & Witten, 1984; Moffat, 1990). Exclusion corrects for this by ignoring these counts when computing the maximum likelihood estimates for other symbols, thereby reclaiming part of the probability mass that was not being used. This has the effect of increasing the weight given to symbols appearing only at lower-orders. Exclusion has been found consistently to improve compression performance (Bunton, 1996; Moffat, 1990). Exclusion is implemented by recomputing the frequency of the context $\text{count}(\overline{x})$ as follows:

$$\text{count}(\overline{x}) = \sum_{x \in \mathcal{X}:\ x \text{ is not excluded}} \text{count}(x, \overline{x}).$$

Note that exclusion is not applied to the transition count, $\text{count}(x, \overline{x})$, or the type count, $t(\overline{x})$. However, it is applied to the order -1 distribution which is redefined as $\frac{1}{|\mathcal{X}|-t(\varepsilon)}$.

Figure 2.5 shows how exclusion is applied in the estimation of smoothed probabilities using backoff smoothing with escape method C for a PPM model that has processed the string ABRACADABRA and is predicting the next letter to appear. Note that the probabilities for A and R, which only appear in the zeroth-order context are substantially increased when exclusion is enabled relative to the probabilities estimated for B, C, and D, which appear in the first-order context. As a result, the sum of the distribution is closer to one than when exclusion is disabled (approximately 0.82 compared with 0.75) illustrating how exclusion reclaims probability mass that is otherwise wasted. Note that although the order -1 model predicts a probability of zero using exclusion, this is never used because all symbols in the alphabet are predicted by the zeroth-order model.

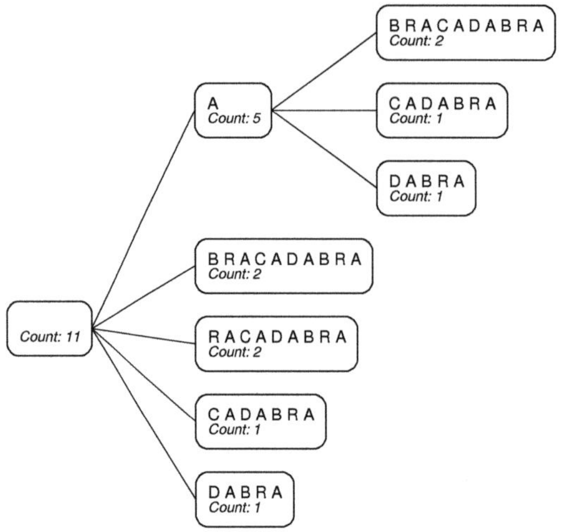

Without exclusion

$A:$	$3/7 \cdot 5/16$	$= 15/112$	0.18
$B:$	$2/7$	$= 2/7$	0.38
$C:$	$1/7$	$= 1/7$	0.19
$D:$	$1/7$	$= 1/7$	0.19
$R:$	$3/7 \cdot 2/16$	$= 3/56$	0.07

With exclusion

$A:$	$3/7 \cdot 5/12$	$= 5/28$	0.22
$B:$	$2/7$	$= 2/7$	0.35
$C:$	$1/7$	$= 1/7$	0.17
$D:$	$1/7$	$= 1/7$	0.17
$R:$	$3/7 \cdot 2/12$	$= 1/14$	0.09

Order = 1				
A	→ B	2	**2/7**	
	→ C	1	**1/7**	
	→ D	1	**1/7**	
	→ *esc*	3	**3/7**	
B	→ R	2	2/3	
	→ *esc*	1	1/3	
C	→ A	1	1/2	
	→ *esc*	1	1/2	
D	→ A	1	1/2	
	→ *esc*	1	1/2	
R	→ A	2	2/3	
	→ *esc*	1	1/3	
Order = 0				
	→ A	5	**5/16**	**5/12**
	→ B	2	2/16	2/12
	→ C	1	1/16	1/12
	→ D	1	1/16	1/12
	→ R	2	**2/16**	**2/12**
	→ *esc*	5	5/16	5/12
Order = -1				
	→ $\mathcal{X}$		1/5	0

Figure 2.5: Probability estimation using backoff smoothing with escape method C and global order bound 1 for a PPM model that has processed the string ABRACADABRA and is predicting the next letter to appear. Each vertical panel of the table corresponds to probabilities estimated at a different order (i.e., 1, 0, and -1). Table columns show the context $\bar{x}$, the symbol x, the frequency count, the estimated probability (including the escape probability), and the estimated probability using excluded frequency counts. Bold font indicates the probabilities and escape probabilities actually used for estimating the probability of a symbol at each order. The corresponding suffix tree is shown with frequency counts for each node. Because the sequence has not yet terminated and prediction is ongoing, the sentinel symbol has not been added to the tree (cf. Figure 2.4). Calculations for estimating probabilities are shown for each symbol in the alphabet $\mathcal{X} = \{A, B, C, D, R\}$ with and without exclusion. Probabilities are given both as exact rational numbers and as real numbers, normalized such that the distribution sums to unity and rounded to two decimal places.

2.2.4.3 Update exclusion

Update exclusion (Moffat, 1990) is a modified strategy for incrementing the frequency counts when a symbol is observed. By default, full counting is used, meaning that the count for a symbol is incremented in every context in which it appears—that is, every suffix of the longest context used. When update exclusion is enabled, the count is updated for the highest-order matching context that predicts the symbol but not lower-order matching suffixes of that context. There are two ways of motivating this strategy. In the data compression literature, update exclusion is motivated on the grounds that it only updates frequency counts in the context that blending actually uses for probability estimation and therefore yields counts that more accurately reflect which events are likely to have been excluded during prediction (see § 2.2.4.2). Update exclusion improves the compression performance of PPM (Bunton, 1996; Moffat, 1990).

In the statistical language modelling literature, update exclusion is typically used in Kneser-Ney smoothing (Chen & Goodman, 1999; Kneser & Ney, 1995), the best-performing smoothing method for natural language corpora. Here, it is motivated through the observation that update exclusion only increments the count for a symbol when that symbol is not predicted in a higher-order context. This should prove advantageous when processing fixed sequences of symbols that tend to appear together in a particular order (i.e., a *collocation*). For example, imagine a word model processing a document that contains many instances of 'United Kingdom' but no instances of 'Kingdom' that are not preceded by 'United'. A first-order model should estimate a high probability for 'Kingdom' when preceded by 'United' but the zeroth-order probability of 'Kingdom' should be low because its high frequency depends on it being part of the fixed sequence 'United Kingdom'. Update exclusion captures this intuition and generalizes it to models of all order.

Update exclusion is implemented by incrementing a distinct frequency count, denoted $\text{count}_{\text{ux}}(x, \overline{x})$, at the corresponding node in the suffix tree only when $\overline{x}$ is the longest matching context for the current input sequence. This implies that x has not appeared in the context of any longer contexts formed by prepending $\overline{x}$ with a single symbol. The number of times this can happen is equal to the number of elements in the alphabet plus one if $\overline{x}$ appears at the beginning of the input sequence. Therefore, the update excluded count is constrained to be less than or equal to the size of the alphabet plus one: $\text{count}_{\text{ux}}(x, \overline{x}) \leq |\mathcal{X}| + 1$.

Figure 2.6 illustrates the use of update excluded counts in the context of estimating smoothed probabilities using backoff smoothing with escape method C for a PPM model that has processed the string ABRACADABRA and is predicting the next letter to appear.

Note that the full count for the root node represents the total number of symbols seen (i.e., the 11 letters in ABRACADABRA) while the update excluded count is equal to the size of the alphabet. Each node stores both full and update excluded counts but in the calculations shown, update excluded counts are used. In this example, update exclusion changes the estimated probabilities for A and R, which only appear in the zeroth-order context. The suffix tree shows that A has a full count of 5 but an update excluded count of 4. This is because although A appears five times in ABRACADABRA, it only appears in four distinct contexts (εA, RA, CA, DA) since RA appears twice. Similarly, B and R have a full count of 2 but an update excluded count of 1 because they always appear in the same contexts (AB and BR respectively). In terms of incrementing the counts during processing, the second time ABRA appears, the highest matching order will be 2 and the first-order update excluded counts for BR and RA, and the zeroth-order counts for A, B and R are not updated. This expresses the inference that ABRA is to be treated as a collocation without updating the frequency counts for lower-order models. In the example, this has the effect of lowering the probability of symbols appearing in the collocation for the lower-order models, which manifests as lower probabilities for A and R, while the probabilities for B, C, and D remain unchanged (see the unnormalized probabilities represented as rational numbers in Figure 2.6).

2.2.4.4 Interpolated smoothing

The original motivation for backoff smoothing was to solve the zero-frequency problem. However, it has been observed that lower-order statistics improve the accuracy of the probability estimates even for symbols that are observed in a higher-order context (Bunton, 1996; Chen & Goodman, 1999). This observation motivated *interpolated smoothing*, which always estimates symbols using a full mixture (a weighted linear combination) of the maximum likelihood estimates at all orders lower than the global order bound:

$$p(x \mid \overline{x}) = \begin{cases} \dfrac{\max(\text{count}(x,\overline{x}) - \beta, 0)}{\text{count}(\overline{x}) + \alpha} + \gamma(\overline{x}) \cdot p(x \mid \text{su}(\overline{x})) & \text{if } \text{su}(\overline{x}) \neq \varepsilon \\ \dfrac{1}{\mathcal{X}+1-t(\varepsilon)} & \text{otherwise} \end{cases}$$

where $\gamma(\overline{x})$ is the escape probability defined in Equation 2.7. While backoff smoothing addresses the zero-frequency problem specifically, interpolated smoothing is motivated by the complementary advantages and disadvantages of high- and low-order models discussed above: the former have high structural specificity to the predictive context but low statistical reliability while the latter have low structural specificity but high statistical reliability.

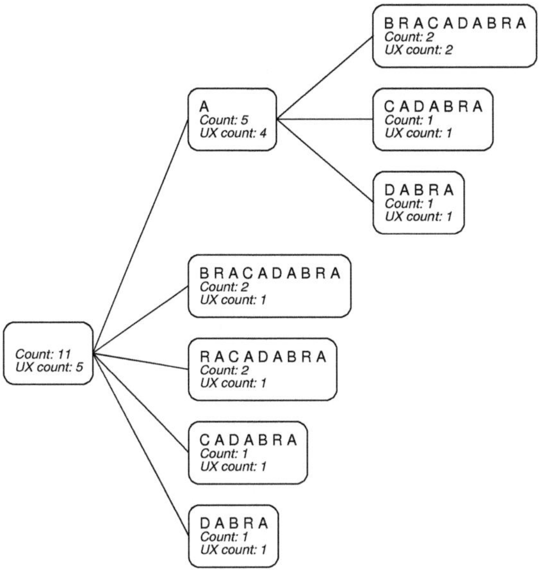

With exclusion and update exclusion

$$
\begin{array}{llll}
A: & 3/7 \cdot 4/10 & = 6/35 & 0.22 \\
B: & 2/7 & = 2/7 & 0.36 \\
C: & 1/7 & = 1/7 & 0.18 \\
D: & 1/7 & = 1/7 & 0.18 \\
R: & 3/7 \cdot 1/10 & = 3/70 & 0.05
\end{array}
$$

Order = 1				
A	→ B	2	**2/7**	
	→ C	1	**1/7**	
	→ D	1	**1/7**	
	→ *esc*	3	**3/7**	
B	→ R	1	1/2	
	→ *esc*	1	1/2	
C	→ A	1	1/2	
	→ *esc*	1	1/2	
D	→ A	1	1/2	
	→ *esc*	1	1/2	
R	→ A	1	1/2	
	→ *esc*	1	1/2	
Order = 0				
	→ A	4	4/13	**4/10**
	→ B	1	1/13	1/10
	→ C	1	1/13	1/10
	→ D	1	1/13	1/10
	→ R	1	1/13	**1/10**
	→ *esc*	5	5/13	5/10
Order = -1				
	→ $\mathcal{X}$		1/5	0

Figure 2.6: Estimation of smoothed probabilities using backoff smoothing with update excluded counts, which are shown in the suffix tree alongside the full counts used in Figure 2.5. Entries shown in bold in the table correspond to the probabilities used in smoothing. Other than using the update excluded counts, the calculations performed are identical to those shown in Figure 2.5 with exclusion enabled.

Interpolated smoothing improves PPM's compression performance (Bunton, 1996) and also improves prediction performance in statistical language modelling (Chen & Goodman, 1999). Just like backoff smoothing, interpolated smoothing can be used with or without exclusion and update exclusion.

With update exclusion enabled, Equation 2.8 corresponds to a generalized form of Kneser-Ney smoothing (Chen & Goodman, 1999; Steinruecken et al., 2015). It can also be considered an approximate sequential construction of a Pitman–Yor process (Goldwater et al., 2005; Teh, 2006) providing a more general-purpose and theoretically grounded motivation for its construction.

Figure 2.7 illustrates the use of interpolated smoothing in the context of a PPM model that has processed the sequence ABRACADABRA and is predicting the next letter to appear.

2.2.4.5 Variable-order modelling

Thus far we have considered models with a fixed global order bound limiting the maximum order used for prediction. However, this constraint can be removed, allowing the model to use the highest-order context in the suffix tree that matches the current context. This produces what is called PPM* in the data compression literature (Bunton, 1996; Cleary & Teahan, 1997), an instance of a *variable-order Markov model* (Begleiter et al., 2004). Rather than always selecting the highest-order matching context, strategies have been developed for selecting lower-order contexts for improved efficiency and potentially also prediction performance.

The simplest strategy, developed by Cleary and Teahan (1997), involves what are known as *deterministic* nodes in the suffix tree, reflecting contexts in which only a single symbol has appeared, $t(\overline{x}) = 1$. If the highest-order matching context is deterministic, rather than using the corresponding node, the lowest-order matching deterministic node is selected as the maximum-order context for probability estimation using smoothing. The intuition here is that the single symbol that has appeared in the chain of deterministic contexts will have occurred at least as many times in the lowest-order context as in higher-order contexts and, therefore, this context will generate the lowest-entropy prediction (Bunton, 1996).

Not using the maximum-order matching context interferes with the principle underlying update exclusion but this can be remedied by using full counts when a lower-order deterministic node is selected as the starting point for smoothing and the update excluded counts for lower-order contexts (Bunton, 1996). When combined with interpolated smoothing and update exclusion, variable-order PPM* results in improved compression performance over the original fixed-order PPM (Bunton, 1996).

Figure 2.8 illustrates the use of variable unbounded orders in the context of a PPM model that has processed the sequence ABRACADABRA and is predicting the next letter to appear. The highest-order matching context is ABRA but since this context is deterministic, a lower-order deterministic context is sought. In this case, the order 3 context (BRA) and the order 2 context (RA) are also deterministic since C is the only symbol that has appeared following

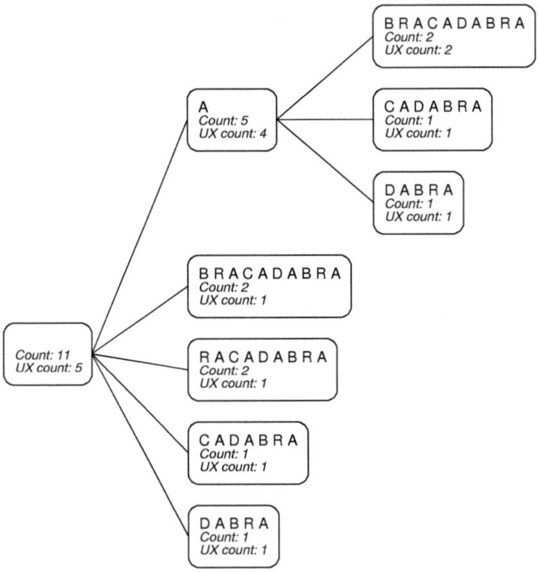

Order = 1				
A	→ B	2	**2/7**	
	→ C	1	**1/7**	
	→ D	1	**1/7**	
	→ *esc*	3	**3/7**	
B	→ R	1	1/2	
	→ *esc*	1	1/2	
C	→ A	1	1/2	
	→ *esc*	1	1/2	
D	→ A	1	1/2	
	→ *esc*	1	1/2	
R	→ A	1	1/2	
	→ *esc*	1	1/2	
Order = 0				
	→ A	4	4/13	**4/10**
	→ B	1	1/13	**1/10**
	→ C	1	1/13	**1/10**
	→ D	1	1/13	**1/10**
	→ R	1	1/13	**1/10**
	→ *esc*	5	5/13	**5/10**
Order = -1				
	→ $\mathcal{X}$		1/5	**1**

With exclusion, update exclusion, and interpolated smoothing

$$
\begin{array}{llllllllll}
A: & 0 & + & 3/7 & \cdot & (4/10 & + & 5/10 & \cdot & 1) & = & 27/70 & 0.19 \\
B: & 2/7 & + & 3/7 & \cdot & (1/10 & + & 5/10 & \cdot & 1) & = & 19/35 & 0.27 \\
C: & 1/7 & + & 3/7 & \cdot & (1/10 & + & 5/10 & \cdot & 1) & = & 2/5 & 0.2 \\
D: & 1/7 & + & 3/7 & \cdot & (1/10 & + & 5/10 & \cdot & 1) & = & 2/5 & 0.2 \\
R: & 0 & + & 3/7 & \cdot & (1/10 & + & 5/10 & \cdot & 1) & = & 9/35 & 0.13
\end{array}
$$

Figure 2.7: Probability estimation using interpolated rather than backoff smoothing. The calculations shown use update excluded counts and exclusion enabled as in Figure 2.6. Entries shown in bold in the table correspond to the probabilities used in smoothing.

these contexts. As we have seen, the first-order context is not deterministic making the second-order context the lowest-order deterministic context, which is used as the maximum-order model for prediction. If the highest-order matching context had not been deterministic, that would have been used instead. In this way, the maximum order used for prediction varies from context to context depending on the longest match in the model's previous experience. It is notable that this higher-order model estimates a much higher probability for C than the models considered so far, which had a global order bound of 1. This results from the deterministic second-order prediction RA→C whereas the first-order context (A) has three possible continuations (B, C, and D).

2.2.5 Long- and short-term models

The modelling techniques introduced in § 2.2.4 assume a single model of a sequence of symbols (corresponding to the square boxes in Figure 2.1). This is appropriate for data compression where a model is constructed dynamically online while the data is compressed. However, in domains such as music and language, it is often fruitful to combine models constructed dynamically in this way from particular input sequences with models trained on a large corpus of sequences in the target domain. This allows a model to generate probability estimates that reflect both the particularities of an individual sequence and the more general structure and statistics of the sequential domain from which it is drawn. Conversely, in natural language processing, a static probabilistic model is typically trained on a large corpus of written or spoken language but it has proved advantageous to combine this model with a so-called cache model constructed dynamically from a portion of the recently processed text (Kuhn & De Mori, 1990).

Similarly in processing the notes making up an unfamiliar piece of music, it is likely that prediction performance will be improved if the model is able to combine knowledge of the syntactic structure of a musical style with dynamically acquired knowledge of repeated patterns indicating motivic or thematic structure within a piece of music. As discussed in Chapter 3, these two sources of knowledge have psychological counterparts in different cognitive processes thought to influence listeners' musical expectations.

Darrell Conklin (1990) introduced methods for computationally simulating these two distinct influences on probabilistic prediction. The *long-term model* is a probabilistic model (i.e., a PPM model) trained on a large corpus of music reflecting general knowledge of the syntax of a musical style while the corresponding *short-term model* is trained dynamically on each piece of music it encounters from an initially empty state to capture incrementally expanding knowledge of the structure of individual pieces. Each model generates a conditional probability distribution for the next event, which must be combined to produce a single distribution. The overall architecture of the long- and short-term models is summarized in

Order = 4			
ABRA	→ C	1	1/2
	→ *esc*	1	1/2

Order = 3			
BRA	→ C	1	1/2
	→ *esc*	1	1/2

Order = 2			
AB	→ R	2	2/3
	→ *esc*	1	1/3
AC	→ A	1	1/2
	→ *esc*	1	1/2
AD	→ A	1	1/2
	→ *esc*	1	1/2
BR	→ A	1	1/2
	→ *esc*	1	1/2
CA	→ D	1	1/2
	→ *esc*	1	1/2
DA	→ B	1	1/2
	→ *esc*	1	1/2
RA	→ C	1	**1/2**
	→ *esc*	1	**1/2**

Order = 1				
A	→ B	2	2/7	**2/6**
	→ C	1	1/7	**1/6**
	→ D	1	1/7	**1/6**
	→ *esc*	3	3/7	**3/6**
B	→ R	1	1/2	
	→ *esc*	1	1/2	
C	→ A	1	1/2	
	→ *esc*	1	1/2	
D	→ A	1	1/2	
	→ *esc*	1	1/2	
R	→ A	1	1/2	
	→ *esc*	1	1/2	

Order = 0				
	→ A	4	4/13	**4/10**
	→ B	1	1/13	**1/10**
	→ C	1	1/13	**1/10**
	→ D	1	1/13	**1/10**
	→ R	1	1/13	**1/10**
	→ *esc*	5	5/13	**5/10**

Order = -1				
	→ $\mathcal{X}$		1/5	**1**

With exclusion, update exclusion, interpolated smoothing, and variable order

$$\begin{array}{llllllllll}
A: & 0 & + \ 1/2 \cdot & (0 & + \ 3/6 \cdot & (4/10 & + \ 5/10 \cdot & 1)) & = \ 9/40 & 0.14 \\
B: & 0 & + \ 1/2 \cdot & (2/6 & + \ 3/6 \cdot & (1/10 & + \ 5/10 \cdot & 1)) & = \ 19/60 & 0.19 \\
C: & 1/2 & + \ 1/2 \cdot & (1/6 & + \ 3/6 \cdot & (1/10 & + \ 5/10 \cdot & 1)) & = \ 11/15 & 0.44 \\
D: & 0 & + \ 1/2 \cdot & (1/6 & + \ 3/6 \cdot & (1/10 & + \ 5/10 \cdot & 1)) & = \ 7/30 & 0.14 \\
R: & 0 & + \ 1/2 \cdot & (0 & + \ 3/6 \cdot & (1/10 & + \ 5/10 \cdot & 1)) & = \ 3/20 & 0.09
\end{array}$$

Figure 2.8: Probability estimation with unbounded, variable maximum order bound (PPM*) added to the model with exclusion, update exclusion, and interpolated smoothing from Figure 2.7. For conciseness and clarity, only the matching context is shown for orders 3 and 4. Note that these contexts are deterministic, so the order 2 model is used as the longest matching context for smoothing. Entries shown in bold in the table correspond to the probabilities used in smoothing.

Figure 2.1 and the ways in which IDyOM can be configured to use the long- and short-term models is summarized in Table 2.1. Here we consider how to combine the distributions generated by the two models.

Let M be a set $\{ltm, stm\}$ containing the LTM and STM, and p_m be the probability distribution estimated by model m in a given context. Conklin (1990) introduced a weighted arithmetic mean to combine the distributions returned by the long- and short-term models, where the weights reflected the entropy of each distribution, such that greater weight is assigned to the model with the lower-entropy distribution:

$$\mathrm{p}(\mathrm{x} \mid \overline{\mathrm{x}}) = \frac{\sum_{m \in M} w_m(\overline{\mathrm{x}})\, \mathrm{p}_m(\mathrm{x} \mid \overline{\mathrm{x}})}{\sum_{m \in M} w_m(\overline{\mathrm{x}})}$$

where the weights w_m correspond to the reciprocal of the relative entropy of the distribution:

$$w_m(\overline{\mathrm{x}}) = \mathrm{H}_{relative}(\overline{\mathrm{x}})^{-b} \tag{2.8}$$

and b is a free parameter that determines the degree to which relative entropy influences the weight ($b = 0$ implies no weighting, higher values imply increasing influence). The relative entropy of a distribution p_m is computed as follows:

$$\mathrm{H}_{relative}(\overline{\mathrm{x}}) = \begin{cases} \mathrm{H}(\overline{\mathrm{x}})/\mathrm{H}_{\max}(\overline{\mathrm{x}}) & \text{if } \mathrm{H}_{\max}(\overline{\mathrm{x}}) > 0 \\ 1 & \text{otherwise} \end{cases}$$

where H and $\mathrm{H}_{\max}$ are defined in Equations 2.3 and 2.4 respectively.

An alternative method for combining distributions introduced by Pearce et al. (2005) is to use a weighted geometric mean:

$$\mathrm{p}(\mathrm{x} \mid \overline{\mathrm{x}}) = \frac{1}{R} \left(\prod_{m \in M} \mathrm{p}_m(\mathrm{x} \mid \overline{\mathrm{x}})^{w_m(\overline{\mathrm{x}})} \right)^{\frac{1}{\sum_{m \in M} w_m(\overline{\mathrm{x}})}}$$

where R is a normalization constant such that the resulting distribution over $\mathcal{X}$ sums to unity and the weights $w_m(\overline{\mathrm{x}})$ are computed using Equation 2.8.

The arithmetic and geometric combination schemes have different properties (Hinton, 2002; Tax et al., 2000). A weighted arithmetic mean produces a combined distribution that is flatter with higher entropy than any of the component distributions. This is useful for suppressing errors in cases where any of the component models erroneously predicts any continuations with very high or very low probability. The weighted geometric mean, on the other hand, yields a distribution that is sharper with higher entropy than any of the component distributions. For a given element of the alphabet it suffices for just one model to correctly assign that element a low estimated probability for it to have a low probability in the combined distribution (regardless of whether another model incorrectly estimates a high probability for that element). This amounts to trusting that one of the individual component models will correctly estimate a low probability for symbols that are actually unlikely to appear in a given context.

In an empirical investigation, Pearce et al. (2005) found that geometric combination produced better prediction performance than arithmetic combination for a corpus of 185 chorale melodies from the collection harmonized by J. S. Bach and the optimal bias value for combining the long-term and short-term models was $b = 2$.

2.2.6 Other modelling techniques

Other techniques have been developed to improve the prediction performance of probabilistic models that are not used here but worth mentioning for completeness. Bunton (1996) developed an improved mechanism for selecting the maximum-order model used for prediction in any given context. This mechanism associates each node in the suffix tree with an information-theoretic measure of prediction performance that is dynamically updated as the sequence is processed. Rather than selecting the highest-order matching non-deterministic context or the lowest-order matching deterministic context, the context with best past performance is selected to provide the highest-order context in smoothing. Information-theoretic context selection improves compression performance by about 3.5%.

Steinruecken et al. (2015) extended PPM to explicitly optimize the generalized smoothing hyper-parameters α and β in Equation 2.8 for different kinds of context varying in terms of order (suffix tree depth) and fanout (or type count, the number of symbols that have followed the context). Optimizing dynamically during compression resulted in compression improvements for text files of about 1.4% over the best available existing PPM implementation.

2.3 Multiple viewpoint representations

The methods for probability estimation introduced in § 2.2 assume a sequence of atomic elements. However, music is inescapably multidimensional. Even when restricted to melody, musical notes have multiple attributes: pitch, onset time, duration, loudness, timbre, articulation, and so on. More subtly but no less importantly, each of these attributes can be usefully modelled at different levels of abstraction both in psychological and computational terms as discussed in § 2.1 (see Figure 2.2 and Table 2.2). If we expand the scope to polyphony, then the picture becomes even more complex; for example, it may be necessary to represent simultaneous sequences of harmonic movement and one or more melodic or contrapuntal lines (as investigated further in Chapter 4).

In the late 1980s, Darrell Conklin, while still an MSc student in Computer Science at the University of Calgary, created a powerful new framework for simultaneous predictive modelling of multiple features of a musical sequence. The framework is called *multiple viewpoints*, a term taken from Ebcioğlu (1988) who observed that the richness of music requires a modeller to observe it from multiple different perspectives or viewpoints. Conklin's design of multiple viewpoints was inspired partly by the work of David Lewin (1987) who developed mathematical methods for deriving abstract musical representations such as pitch interval from more concrete representations such as pitch.

2.3.1 Representing music using multiple viewpoints

The multiple viewpoint framework adopts a representation of music not as a sequence of discrete atomic elements drawn from an alphabet (see § 2.2) but rather as a sequence of discrete events, each consisting of a finite number of attributes τ each of which is associated with a finite alphabet $[\tau]$ from which elements are drawn. The attributes making up an event are known as *basic attributes* while attributes derived from these are known as *derived attributes*. The Cartesian product of the alphabets of n basic attributes $\tau_1, \ldots, \tau_n$ is referred to as the *event space*, ξ:

$$\xi = [\tau_1] \times [\tau_2] \times \ldots \times [\tau_n] \tag{2.9}$$

An *event* $e \in \xi$ is an instantiation of the attributes $\tau_1, \ldots, \tau_n$ and consists of an n-tuple in the event space ξ, which denotes the set of all representable events. Sequences of events are denoted by $e_i^j \in \xi^* = e_i, \ldots, e_j, j \geq i \in \mathbb{Z}^+$ where ξ^* denotes the set of all sequences composed of elements of ξ including the empty sequence ε. The basic event space shown in Table 2.2 consists of three basic attributes describing events (`pitch`, `onset`, and `duration`) as well as four describing the prevailing tonal (`keysig`, `mode`) and metrical (`pulses`, `barlength`) environment in which an event appears (which may have been inferred or encoded). The latter are represented as event attributes for convenience and are used in the derivation of `scale-degree` and `posinbar` respectively.

Every attribute τ has an associated *viewpoint function* which maps sequences of events onto elements in $[\tau]$:

$$\Psi_\tau : \xi^* \rightarrow [\tau]. \tag{2.10}$$

It is useful to distinguish three different classes of viewpoint: basic, derived, and linked.[5] For basic attributes, $\Psi_\tau(e_1^j)$ is simply a projection function, returning the corresponding attribute of the final event in the supplied event sequence e_1^j. For derived attributes, which do not feature in the event space ξ, the viewpoint function computes a more abstract representation derived from one or more basic attributes. Since the function is partial, the result may be undefined (denoted by $\perp$) for a given event sequence. An example of a derived viewpoint is one which represents melodic pitch intervals. Given the basic attribute `pitch` shown in Table 2.2, the derived viewpoint `interval` is defined by the function:

$$\Psi_{\texttt{interval}}(e_1^j) = \begin{cases} \perp & \text{if } j = 1, \\ \Psi_{\texttt{pitch}}(e_j) - \Psi_{\texttt{pitch}}(e_{j-1}) & \text{otherwise.} \end{cases}$$

Table 2.2 lists attributes derived from `pitch` (`interval`, `contour`, `scale-degree`) and `onset` (`ioi`, `ioi-ratio`, `ioi-contour`, `posinbar`), which are illustrated in Figure 2.2.

A linked attribute τ combines two or more primitive (i.e., basic or derived) attributes $\tau_1 \times \ldots \times \tau_n$ such that the alphabet of τ is the Cartesian product of the alphabets of the component attributes: $[\tau] = [\tau_1] \times \ldots \times [\tau_n]$. The viewpoint function is defined as follows:

$$\Psi_\tau(e_1^j) = \begin{cases} \perp & \text{if } \Psi_{\tau_i}(e_1^j) \text{ is undefined for any } i \in \{1, \ldots, n\} \\ \langle \Psi_{\tau_1}(e_1^j), \ldots, \Psi_{\tau_n}(e_1^j) \rangle & \text{otherwise.} \end{cases}$$

Table 2.2 contains two linked viewpoints: `interval⊗scale-degree` derived from `pitch` and `ioi-ratio⊗posinbar⊗barlength` derived from `onset`. These are illustrated in Figure 2.2.

2.3.2 Prediction using multiple viewpoints

Prediction using multiple viewpoints involves choosing a set of *target viewpoints*, specifying the basic attributes to be predicted, and a set of *source viewpoints*, the basic, derived, or linked viewpoints that will be used for prediction of the target attributes. The set of source viewpoints must include at least one viewpoint derived from each target viewpoint τ_b. For

[5]This summary excludes *test viewpoints* and *threaded viewpoints*. See Conklin (1990) and Pearce (2005) for a complete exposition.

each of the source attributes, τ, a representation of the training corpus as sequences in $[\tau]^*$ is obtained by recursively applying $\Psi_\tau()$. This set of sequences is then used to train a probabilistic model as described in § 2.2. For each event in the corpus, the models for each source viewpoint return a distribution over the corresponding viewpoint alphabet $[\tau]$. However, in order to combine the distributions generated by the models for different viewpoints, this must be converted into a distribution over the basic event space ξ. To achieve this, for each target viewpoint, τ_b, the distributions from viewpoint models for attributes derived from τ_b are first mapped from $[\tau]$ onto $[\tau_b]$:

$$\Psi'_\tau : \xi^* \times [\tau] \rightarrow \mathcal{P}([\tau_b])$$

where $\mathcal{P}(S)$ denotes the power set of S. The mapping is, in general, many-to-one since the derived sequence could represent several different sequences of events. Consider as an example `scale-degree` for which the element 0, representing the tonic, could map onto several different pitches in different octaves. As a result, the probability estimates returned by the model for each derived element are divided equally amongst the basic elements onto which they map. To ensure that derived viewpoints such as `interval` allow recovery of a complete distribution over the basic attribute `pitch`, their alphabet is specified prior to prediction of each event such that there is a one-to-one correspondence between $[\tau]$ and the domain of the corresponding basic attribute $[\tau_b]$. If no source viewpoints predict the target viewpoint (e.g., `interval` for the first note in a melody) then a uniform distribution over $[\tau_b]$ is used.

Once the distributions for all source viewpoints have been mapped onto distributions over the corresponding basic alphabet, they are combined using the same techniques described in § 2.2.5 for combining the long-term and short-term models. Pearce et al. (2005) found that geometric combination produced better prediction performance than arithmetic combination for a corpus of 185 chorale melodies from the collection harmonized by J. S. Bach and the optimal bias value was $b = 7$ for viewpoint combination. These parameters are used in the IDyOM simulations reported in this book. As shown in Figure 2.1, the distributions output by the viewpoint models are first combined separately for the long-term and short-term models, whose combined distributions are then combined into a single final distribution for each target viewpoint specified. The overall probability of an event is then computed as the joint probability of the specified individual target attributes, $\tau_1, \ldots, \tau_m$:

$$p(e_i|e_1^{i-1}) = \prod_{l=1}^{m} p_{\tau_l}(e_i|e_1^{i-1})$$

Description	Compositions	Events	Mean composition length
Nova Scotian songs/ballads	152	8,553	56.27
Chorale melodies	185	9,227	49.88
Alsatian folk songs	91	4,496	49.41
Yugoslavian folk songs	119	2,691	22.61
Swiss folk songs	93	4,586	49.31
Austrian folk songs	104	5,306	51.02
German folk songs	213	8,393	39.40
Chinese folk songs	237	11,056	46.65
Total	1,194	54,308	45.48

Table 2.4: The melodic test corpus used for model evaluation. The first subcorpus (Creighton, 1966) is available from http://kern.humdrum.org/cgi-bin/browse?l=users/craig/songs/creighton/nova while the second (Bach, 1875) is available from http://kern.humdrum.org/cgi-bin/browse?l=musedata/bach/chorales. The remaining subcorpora are taken from the Essen folk song collection (Schaffrath, 1995) available from http://kern.humdrum.org/cgi-bin/browse?l=/essen. The final subcorpus consists of 237 melodies randomly selected from the `shanxi` dataset of the Essen folk song collection.

2.4 Performance evaluation

IDyOM's modelling parameters have been introduced in §§ 2.2 and 2.3. The parameter settings used consistently throughout this book are those reported by Pearce (2005), which were selected on the basis of prediction performance on a corpus of melodies. The corpus consisted of 1,194 melodies taken from the eight subcorpora shown in Table 2.4. Prediction performance was assessed in terms of information content, aiming to choose model parameters that minimize the mean information content of sequences in the test corpus:

$$\begin{aligned} \mathrm{H}_m(\mathrm{x}_1^j) &= -\frac{1}{j}\log_2 p_m(\mathrm{x}_1^j) \\ &= -\frac{1}{j}\sum_{i=1}^{j}\log_2 p_m(\mathrm{x}_i \mid \mathrm{x}_1^{i-1}) \end{aligned} \tag{2.11}$$

where a parameterized model m assigns a probability $\mathrm{p}_m(\mathrm{x}_1^j)$ to a sequence x_1^j of symbols drawn from the alphabet $\mathcal{X}$. Given some assumptions, $\mathrm{H}_m(\mathrm{x}_1^j)$ approximates the *cross-entropy* between the distribution estimated by the model and the true generative distribution

Parameter	Short-term model	Long-term model
Exclusion	enabled	enabled
Order bound	unbounded	unbounded
Interpolation	enabled	enabled
Escape method	X	C
Update exclusion	enabled	disabled

Table 2.5: Parameter settings as optimized by Pearce (2005) for prediction performance on the corpora listed in Table 2.4. These parameter settings are used consistently throughout the book.

for the sequence. The model parameters selected, corresponding to IDyOM's defaults, are shown in Table 2.5.[6]

2.5 Comparison with other models

Since IDyOM was originally evaluated on the music corpora described in § 2.4, several other researchers have tested models on exactly the same corpus, allowing for a precise and systematic comparison of model performance. Probabilistic approaches are considered first followed by artificial neural networks.

Langhabel et al. (2017) introduced a mechanism for searching a generalized space of possible features to find feature sets that optimize prediction performance. The underlying model is a conditional random field with long- and short-term models combined using a weighted arithmetic mean as described in § 2.2.5. The resulting optimized long-term model combined `pitch`, `interval`, `contour` and a version of `scale-degree` that distinguishes major and minor keys while the short-term model combined `pitch`, `interval`, and a feature representing the pitch interval from the first note in the piece. As shown in Table 2.6, this model performs better than an IDyOM BOTH+ model configured to predict using the basic viewpoint `pitch` as its source viewpoint but not better than an IDyOM model using the derived viewpoint `interval⊗scale-degree` without any further optimization of viewpoints.

[6]Note that the intention is not to simulate the computational process that generated the melodies in the corpus (see § 9.6 for further discussion). Rather prediction performance is used simply as a principled way of selecting model parameters. Furthermore, there is no guarantee that these parameters optimally simulate perception.

Reference	Method	Cross-entropy	IDyOM	Cross-entropy
Langhabel et al. (2017)	feature optimization	2.395	BOTH+ `interval⊗scale-degree`	2.372
Cherla (2016)	RTDRBM LTM	2.712	LTM `pitch`	2.866
Cherla et al. (2015)	RTDRBM LTM + STM	2.674	BOTH+ `pitch`	2.540
Cherla et al. (2015)	RTDRBM LTM + n-gram STM	2.421	BOTH+ `pitch`	2.540
Lattner et al. (2018)	RTDRBM + RGAE LTM	2.656	LTM+ `pitch, interval`	2.456
Sears et al. (2018)	LSTM	5.583	BOTH+	4.893

Table 2.6: Comparison of IDyOM's performance with other published models of melody prediction. The first column cites the source of the model, the second column gives the modelling technique, the third the cross-entropy. The fourth column gives the corresponding IDyOM configuration and the fifth column the resulting cross-entropy from IDyOM for comparison with the third column.

Srikanth Cherla (2016) developed recurrent neural network models for prediction of melodic pitch. His starting point was the *Restricted Boltzmann Machine* (RBM), a generative model consisting of visible and hidden layers of units which are fully connected with each other, a set of weights for each connection, and biases for the visible and hidden units. The discriminative RBM is able to estimate the conditional probability of a visible event given the preceding visible context, just like IDyOM. The recurrent temporal discriminative RBM (RTDRBM) consists of a sequence of discriminative RBMs whose biases are determined by the RBM at the previous timestep. Cherla showed that the RTDRBM predicted the melodic test corpus slightly better than the corresponding IDyOM LTM (see Table 2.6). Cherla et al. (2015) found that adding a short-term RTDRBM improved performance slightly but not sufficiently to outperform IDyOM in its BOTH or BOTH+ configurations, although this was achieved by adding an IDyOM short-term model (PPM*C) to the long-term RTDRBM.

Lattner et al. (2018) investigated the neural network equivalent of multiple viewpoint prediction by combining Cherla's pitch RTDRBM with a *Recurrent Gated Autoencoder* (RGAE) that learned pitch interval representations of musical sequences. Over the test corpus, the resulting model improved prediction performance over the long-term RTDRBM but did not match the performance of the corresponding IDyOM model configured with `pitch` and `interval` source viewpoints (see Table 2.6).[7]

Finally, Korzeniowski et al. (2018) compared IDyOM to *long short-term memory* (LSTM), a recurrent neural network architecture, on prediction of harmonic sequences. The corpus consisted of 1,116 pieces in a range of styles including the chorales of J. S. Bach, string quartets by Haydn, Mozart, and Beethoven, piano works by Mozart, Beethoven, Chopin, and Scott Joplin, and an assortment of orchestral works. Chords were represented in terms of the scale degree of the bass plus the intervals formed between the bass and the remaining chord tones, ordered in terms of size, with voice doublings removed (see Chapter 4 for a detailed presentation of the representation of harmony within IDyOM). The LSTM is a recurrent neural network whose hidden units depend on both the current input and the hidden unit for the previous timestep, with gates to prevent the error gradient either vanishing or exploding. The results showed that IDyOM convincingly outperformed the LSTM in terms of prediction performance (see Table 2.6). A further analysis comparing model performance with the statistical properties of each sequence, suggested that for IDyOM predictive performance increased primarily with increasing sequence length but decreased as the number of chord types in a sequence increased. For the LSTM, the pattern was markedly different: performance decreased as the proportion of rare events (those with low zeroth-order probability in the corpus) in a sequence increased and when the proportion of adjacent repeated chords decreased.

Overall, these results suggest that IDyOM is competitive with a range of more recent techniques for predictive modelling of musical structure. The overall goal in this book is not to exhaustively compare prediction performance of different models but to understand human perception. To the extent that they implement the statistical learning and probabilistic prediction hypotheses from Chapter 1, any of these models could have been used for the research presented in this book. The results suggest that IDyOM is not necessarily more limited in terms of prediction performance than other models but I will return to this question in Chapter 9.

2.6 Beyond the Markov assumption

The assumption underlying finite-context Markov models such as PPM is that the probability of a symbol depends only on a finite portion of the immediately preceding sequence of

[7] Curiously, Lattner et al. (2018) omitted two of the datasets in the test corpus: Nova Scotian folk songs and Bach chorale melodies.

symbols. As such they fall within a class of syntactic model describing *finite-state* or *regular* languages in which the next state is uniquely determined by the current state and the input symbol (Hopcroft & Ullman, 1979). This constrains the kind of syntactic regularities that can be represented. While IDyOM can learn the regularity with which single symbols follow immediately after particular contexts, it cannot learn so-called *non-adjacent dependencies*, describing regularities between non-adjacent events. More powerful syntactic models such as context-free and context-sensitive grammars are capable of representing such regularities by loosening the constraints on grammatical rewrite rules to allow a non-terminal symbol to be rewritten as any sequence of terminal and non-terminal symbols, rather than a single terminal and a single non-terminal symbol (see Figure 2.9).

Natural languages are generally considered to fall somewhere between context-free and context-sensitive (Joshi & Weir, 1990) and similar arguments have been made for music (Roads & Wieneke, 1979). It is undeniable that many musical styles contain non-adjacent dependencies and prominent theories of music cognition such as Lerdahl & Jackendoff's *Generative Theory of Tonal Music* (see Chapter 1) make substantial use of embedded hierarchical structures reflecting non-adjacent dependencies operating over long timescales. Without a doubt, it is possible to listen to music analytically in this way, consciously paying attention to and appreciating high-level aspects of, for example, thematic and tonal structure.

This raises the question of whether more powerful representations of the syntactic regularities of musical styles are required. In the context of this chapter, it is extremely likely that improved prediction performance could be obtained with models that are capable of capturing less constrained regularities including non-adjacent dependencies. However, the loosening of constraints on the regularities to be learned makes the learning process itself much more difficult, often requiring the imposition of external constraints based on the expert knowledge of the model's developer. This goes against the grain of the present focus on learning the syntactic structure of musical styles. Furthermore, it remains possible that more complex regularities can be acquired to a sufficient extent through the more abstract representations afforded by multiple viewpoints. In particular, *threaded viewpoints* allow representation and prediction of events at non-adjacent temporal locations corresponding, for example, to tactus beats or phrase endings.[8]

It is also worth noting at this point that it is far from clear whether hierarchical representations of syntactic regularities are absolutely necessary for modelling perception of

[8]Threaded viewpoints allow an event to be predicted from a context made up of non-consecutive events. Threaded viewpoints are implemented within IDyOM but are not used in the research presented in this book.

Type 0: unrestricted grammars place no restrictions on their rewrite rules:

$$\alpha \longrightarrow \beta.$$

Type 1: context-sensitive grammars are restricted only in that there must be at least one non-terminal symbol on the left-hand side of the rewrite rule and the right-hand side must contain at least as many symbols as the left-hand side:

$$\beta A \beta' \longrightarrow \beta \alpha \beta'.$$

Type 2: context-free grammars further restrict the left-hand side of their rewrite rules to a single non-terminal symbol:

$$A \longrightarrow \alpha.$$

Type 3: finite-state or regular grammars restrict their rewrite rules further still by allowing only a single terminal symbol, optionally accompanied by a single non-terminal, on the right-hand side:

$$\begin{aligned} A &\longrightarrow aB \\ &\longrightarrow Ba \\ A &\longrightarrow a. \end{aligned}$$

Figure 2.9: Four classes of formal grammar which place increasingly strict constraints on their rewrite rules (Hopcroft & Ullman, 1979). A grammar consists of the tuple (V, T, S, P) where V is a finite set of *non-terminal symbols*; T is a finite set of *terminal symbols*, disjoint from V; $S \in T$ is a distinguished symbol called the *initial symbol*; and P is a set of rewrite rules, representing legal transformations of a sequence of terminal and non-terminal symbols into another such sequence. In describing rewrite rules, $a \in T^*$ denotes a (possibly empty) sequence of terminal symbols, $A, B \in V$ denote non-terminal symbols, $\alpha \in (V \cup T)^+$ denotes a non-empty sequence of terminal and non-terminal symbols and $\beta, \beta' \in (V \cup T)^*$ denote (possibly empty) sequences of terminal and non-terminal symbols.

music. This is because evidence for recognition of non-adjacent dependencies in music perception has proved hard to come by (Bigand & Parncutt, 1999; Cook, 1987; Karno & Konečni, 1992; Marvin & Brinkman, 1999; Tillmann & Bigand, 1996) although some studies have found evidence for a degree of hierarchical processing in non-musicians (Koelsch et al.,

2013; Serafine et al., 1989).[9] Therefore, without denying in principle that listeners might learn more powerful syntactic regularities of musical structure, the IDyOM research programme is guided by a methodologically minimalist approach, examining the extent to which perceptual processing can be simulated using relatively simple syntactic regularities that are learned entirely through exposure rather than committing *a priori* to the assumption that listeners possess cognitive representations of complex syntactic regularities that are difficult or impossible to learn through exposure to a musical style and perhaps need to be posited as innate universals. I will return to this question in Chapter 9 but, as we shall see in the coming chapters, the minimalist approach has proved remarkably profitable in simulating listeners' perception of music.

2.7 Conclusion

This chapter has introduced a computational model of statistical learning and probabilistic prediction that is capable of testing the hypotheses proposed in Chapter 1. The model, known as IDyOM, estimates the probability of every event in a piece of music, given the preceding context and the music to which it has been previously exposed. These probability estimates derive from knowledge about the syntactic structure of a musical style acquired through statistical learning of a large number of musical exemplars in that style. In generating predictions, the model can combine this long-term knowledge with short-term knowledge of regularities within an individual piece of music learned dynamically during processing of the piece. The model is capable of generating predictions for multiple attributes of music including pitch, timing, and harmony and, in doing so, can combine the predictions of models of multiple different musical attributes (e.g., using both pitch interval and scale degree to predict pitch). In estimating probabilities, IDyOM uses smoothing to combine predictions made using multiple different context lengths (or orders), adaptively varying the maximum order for every prediction made, so as to be able to estimate a non-zero probability for every possible continuation and also to benefit from both the structural specificity of higher-order models and the statistical power of lower-order models. The smoothing parameters are optimized by maximizing prediction performance across a range of melodic datasets. Direct comparison with alternative models, including neural networks, suggests that IDyOM is competitive with other models in terms of optimal prediction performance.

[9] Evidence for hierarchical processing in musicians (e.g., Dibben, 1994; Lerdahl & Krumhansl, 2007) remains open to the possibility of deliberate application of explicit musical knowledge.

Optimizing prediction performance provides a useful way of selecting model parameters and ensuring that the model provides a robust and meaningful benchmark for comparison with human performance. However, it does not provide a direct test of the statistical learning and probabilistic prediction hypotheses. For that, we need to compare the responses of the model with human responses to music. The goal of the remaining chapters of this book is to do exactly that, starting in Chapter 3 which focuses on pitch expectation in melody.

3 Predictive Processing of Melody

Chapter 2 introduced a computational model of auditory expectation known as IDyOM, which implements and embodies the statistical learning and probabilistic prediction hypotheses proposed in Chapter 1. As shown in Chapter 2, IDyOM is quite successful at predicting the pitch of a note in a melody. But accurate prediction does not necessarily make for a plausible psychological model, which leads us naturally to ask whether IDyOM provides an accurate simulation of the cognitive and neural processes underlying music perception. The remaining chapters of this book present a programme of empirical investigation that systematically tests the statistical learning and probabilistic prediction hypotheses. A comparison of IDyOM's behaviour with human responses across a range of different musical styles and tasks, tapping into different psychological representations and processes, can shed light on which aspects of music perception can be understood as manifestations of these generalizable psychological processes. In the present chapter, we embark on this empirical enquiry with the melodic pitch expectations of listeners enculturated in Western tonal music. First, however, it is important to characterize and differentiate four distinct psychological influences on expectation and how they relate to the different components of IDyOM.

3.1 Expectation, knowledge, and memory

Huron (2006) distinguished four ways in which musical events can be surprising, each related to a different kind of expectation and corresponding underlying memory system. In the following, I generalize this treatment slightly in differentiating four sources of knowledge—schematic, dynamic, veridical, and sensory—that can influence expectations and take a broader theoretical perspective on the role of consciousness. Following Huron, *schematic knowledge* is acquired prior to listening to a piece of music, while *dynamic knowledge* is acquired during listening to the piece of music itself. *Veridical knowledge* is knowledge about the specific way a familiar piece of music proceeds, resulting from having heard it before.[1] Huron proposes that the different sources of knowledge map onto different memory systems. Schematic knowledge arises from long-term semantic memory, dynamic knowledge from short- and intermediate-term memory, and veridical knowledge from episodic memory. To complete the picture, we should add a fourth category: *sensory knowledge* arising from auditory information stored in echoic memory, a detailed but very short-lived, pre-attentional and modality-specific, low-level sensory representation of the raw auditory input.[2]

To summarise thus far, expectations while listening to a piece of music may be influenced by matching between the current context and the contents of: first, schematic memory, reflecting long-term knowledge of the regularities present in a musical style; second, short-term knowledge acquired dynamically from the preceding portion of the same piece of music; third, veridical knowledge of how the piece of music actually progresses, if the listener has heard it before; and fourth, sensory knowledge, reflecting detailed, unprocessed representations of the last few hundred milliseconds of auditory input.

Huron treated *conscious knowledge* as a separate category. However, it seems plausible, and potentially fruitful both empirically and theoretically, to allow for the possibility that schematic, dynamic, and veridical knowledge can be either conscious or unconscious. For example, a non-musician might have unconscious schematic knowledge about stylistic regularities or norms that they bring to bear on a listening experience while a musician might have comparable (or more extensive) conscious knowledge that can be introspectively interrogated and verbally expressed in terms of concepts from music theory. A non-musician might unconsciously acquire knowledge about repeated motifs within a piece of music while a musician might have conscious access to such knowledge. Regardless of musical training, veridical knowledge resulting from hearing a piece of music several times may or may not be available to consciousness. However, Huron may well be right that once schematic, dynamic, and (perhaps) veridical knowledge becomes consciously accessible through explicit study, instruction, or repeated listening, it is stored in a distinct memory system. Finally, it is worth noting that, even if the knowledge contributing to an expectation is implicit, the expectation itself or its consequences might still be consciously accessible to some degree. For example, a listener might have a conscious expectation for a particular pitch continuation of a melody

[1]Note that this use of the word veridical differs from the sense in which veridicality indicates accurate perceptual representations of an environment (e.g., Tschantz et al., 2020).

[2]Nelson Cowan (1984, 1988) argues for two phases of auditory sensory memory: a primary phase lasting up to 350 milliseconds (ms) reflecting an unanalysed trace of raw sensory input and a secondary phase lasting up to several seconds in which some analysis of the sensory input may be conducted.

but be unable to consciously formulate the schematic, dynamic, or veridical knowledge that underlies that expectation. Alternatively, the expectation itself (i.e., a probabilistic prediction for the different possibilities) might be unconscious but the surprise resulting from an unexpected note might be consciously apprehended.

One final qualification deserves mention. Huron (2006) related veridical expectations for music to episodic memory, on the grounds that veridical knowledge of familiar music starts as an episodic memory of the first time we hear it. He argued, convincingly to my mind, that as we experience and remember a piece of music on more and more occasions, the memory that remains is no longer tied to any one particular spatiotemporal context. In other words, the individual episodic memories merge together into a veridical memory of the piece stripped of the episodic contexts. On this basis, one might equate veridical memory with episodic memory. The problem with this conceptualization is that it fails to distinguish the kind of episodic memory involved in many kinds of recognition memory experiment, where a participant is asked to recall whether or not they heard a particular piece of music earlier in the experimental session, regardless of their prior familiarity with the piece. With this in mind, Crowder (1993) used the term episodic to refer specifically to auditory memory that is located in space and time, which seems faithful to the general conception of episodic memory.[3]

On this view, memory for a familiar piece of music would not be considered episodic unless the memory trace was specific to a particular listening episode, at a particular point in time, and a particular point in space. For example, recognizing the "Ode to Joy" theme on the radio or humming it to oneself (as you might very well be doing right now) would not be considered episodic. An example of a truly episodic memory would be remembering listening to a particular recording while the train was pulling out from York station on your journey from London to Edinburgh last Thursday. This incorporates the knowledge that it was that specific recording you were listening to on that specific carriage, at that point in the journey. Following this line of thought, we can consider episodic musical memory to be the subset of veridical memories that have this specific spatiotemporally located character (without committing to them having the same or different underlying psychological or neural substrates). Veridical memory for music is considered further in §§ 5.6, 6.1.2, and 7.2.3.

[3]Which is not to say that such episodic memories are not prone to various kinds of distortion, as vividly illustrated by Huron (2006) for music.

Knowledge source	Acquisition time	Conscious access	Memory type
Sensory	150–350 ms moving	No	Implicit echoic memory
	window	Yes	Explicit echoic memory
Dynamic	During listening	No	Short-term implicit memory
	episode	Yes	Short-term memory
Schematic	Before listening	No	Long-term implicit memory
	episode	Yes	Semantic memory
Veridical	With repeated	No	Implicit veridical memory
	listening	Yes	Veridical memory

Table 3.1: Different hypothesized sources of knowledge that influence expectations, their time of acquisition, accessibility to consciousness, and hypothesized corresponding memory system. Bharucha (1987) first introduced the distinction between schematic and veridical knowledge. Huron (2006) distinguished between schematic, dynamic, and veridical knowledge. Narmour (1990) used the terms *extra-opus* and *intra-opus* to distinguish between schematic and dynamic expectations respectively. Mckone (1995) introduced a distinction between short-term and long-term implicit memory. Dynamic knowledge might draw on both short-term and intermediate-term memory (the latter operating over timescales from minutes to hours); evidence for intermediate-term memory as distinct from short-term and long-term memory comes primarily from the animal literature (Eichenbaum et al., 1994). The different types of memory listed are not intended necessarily to imply distinct underlying psychological or neural mechanisms (see Brown & Chater, 2007; Crowder, 1993; Wickelgren, 1973). Echoic memory is thought to have two components, the second of which may overlap with short-term memory (Cowan, 1984, 1988); only the earlier component is represented here. The distinction between implicit and explicit echoic memory is largely theoretical.

The resulting characterization of the different sources of knowledge influencing expectation is shown in Table 3.1. According to this characterization, a listener's musical expectations may simultaneously reflect varying contributions—explicit or implicit—from any combination of these four sources of knowledge.

To relate these sources of knowledge to IDyOM's architecture (see Chapter 2 and Table 2.1 in particular), schematic and dynamic knowledge map fairly precisely onto the long-term and short-term models respectively. The short-term model does not distinguish between short- and intermediate-term memory, but simply accumulates knowledge within a listening episode regardless of the time frame (corresponding in the simulations presented here to an individual piece of music). In its BOTH+ configuration (combining both a short-term model and a long-term model; see Table 2.1), IDyOM allows for transfer of knowledge from the short-term model to the long-term model, which occurs for each note as it is processed without any loss of information. An extension to account for memory decay along with experimental evaluation can be found in Chapter 6. IDyOM does not distinguish between conscious and unconscious knowledge, so I will refrain throughout from making any assumptions in this regard. The only exception arises when considering effects of musical training in Chapter 8 which considers the possibility that expectations based on implicit and explicit knowledge might differ in systematic ways.

Regarding veridical knowledge, it is assumed in the IDyOM simulations described below that participants in the empirical studies discussed are listening to a piece of music for the first time, so that veridical knowledge does not influence expectations. Correspondingly, IDyOM will be exposed to each musical stimulus exactly once. However, IDyOM can accumulate knowledge through repeated exposure that is distinguished from schematic knowledge. This knowledge does not correspond to a distinct component within IDyOM's architecture but accumulates within the long-term model in the form of long paths from the root node of a suffix tree to a leaf (see § 2.2.3) that are emphasized by having higher frequency counts as the piece is experienced repeatedly. This conception of veridical memory is discussed further in Chapter 6 (see § 6.1.2). Finally, IDyOM does not provide an explicit account of sensory knowledge but it is compared directly with models of sensory knowledge in Chapter 4 in the context of modelling harmonic expectations; furthermore, as shown in Chapter 6, adding decay to the frequency counts allows aspects of sensory memory to be simulated within the IDyOM framework.

3.2 Melodic pitch expectation

3.2.1 Historical perspective

Pitch expectations have been of interest to music psychologists since at least the 1970s when James Carlsen and his colleagues bemoaned the fact that while there was significant interest in expectation within musical aesthetics and general psychology, the literature on musical expectation was almost exclusively theoretical and speculative (Carlsen et al., 1970). Taking up a proposal for an experiment made by the biophysicist and social scientist John R. Platt (1961), Carlsen presented participants with single interval contexts and asked them to sing the expected continuation. The results of this preliminary study indicated that each context generated not a single expected continuation but a cluster of different pitches which differed between contexts. Later research on musical expectations has also adopted this production paradigm for investigating musical expectations (Carlsen, 1981; Morgan et al., 2019; Schmuckler, 1990; Thompson et al., 1997). This research highlights the fact that any melodic context has a number of possible continuations varying in probability. Melodic pitch expectation is not an all or nothing phenomenon allowing only a single deterministic continuation (strong veridical expectations possibly excepted).

Subsequent research expanded on Carlsen's initial approach in two ways: first, using a wider range of experimental tasks and second, using longer, more realistic musical contexts. Regarding the task, Carlsen et al. (1970) considered asking participants to rate the expectedness of a third tone presented after the two-tone context but rejected this procedure on the grounds that it might not reveal the most expected tone for each participant. The solution to this problem emerged (apparently without any knowledge of Carlsen's work) in the doctoral research of Carol Krumhansl supervised by Roger Shepard at Stanford (Krumhansl & Shepard, 1979), who developed what has become known as the *probe-tone paradigm*. In this paradigm, rather than simply presenting one tone after the context, the context itself is repeated several times with a different final tone that is rated by the participant. Although somewhat time consuming, this paradigm has the advantage of producing expectations for a wide range of possible continuations defined by the experimenter whereas the production paradigm is limited to the continuations actually produced (and producible) by the participants. The probe-tone paradigm can also be easily undertaken by non-musicians without any particular musical training or ability required.

Initial research using the probe-tone paradigm focused specifically on tonal expectations by presenting simple, stereotypical key-defining contexts such as diatonic scales in ascending or descending motion (Krumhansl & Shepard, 1979), triads and cadences (Krumhansl & Kessler, 1982). This research demonstrated the existence of cognitive representations of tonality in which major and minor keys establish a hierarchy of stability such that some pitches are stable, and therefore more expected, than others. However, it did not substantially address the question of expectations generated by sequential melodic structure.

The primary inspiration for research tackling this question was the later work of Leonard Meyer (1973) on melodic processes and that of his former student Eugene Narmour (1977, 1990, 1992), whose Implication Realization (IR) model contains a series of symbolic principles for determining the implications of one melodic interval (the implicative interval) for its successor (the realized interval). This is the so-called bottom-up component of the IR model, which was proposed to represent universal properties of the auditory system in contrast to the top-down component which is culturally dependent (see Chapter 1). These theoretical proposals provided a spur to empirical research that used two-tone contexts (forming an implicative interval) in probe-tone studies where the probe tone completed the realized interval, and all three tones were carefully selected to engage different aspects of Narmour's principles (Cuddy & Lunney, 1995; Krumhansl, 1995b). These studies generally confirmed that listeners' pitch expectations conformed to the predictions of the IR model although the two studies suggested slightly different variants of the principles.

A more important development was to use longer contexts taken from real melodies in the probe-tone paradigm. Schmuckler (1989, Experiment 1) used 10 melodic extracts from Robert Schumann's song cycle *Frauenliebe und Leben* (Op. 42) finding effects of tonality and also Meyer's linear and gap-fill melodic processes. Krumhansl (1995a) reported three studies using a probe-tone paradigm with contexts taken from British folksongs, extracts from Webern's *Lieder* (Op. 3, 4, and 15) and Chinese folksongs. The results showed that listeners' pitch expectations were well described by five quantitative principles derived from Narmour's IR model. Schellenberg (1996) reported a separate analysis of the same data, finding support for a different quantitative formulation of the IR model, containing only four principles. Furthermore, noting significant collinearity between the principles, Schellenberg (1997) produced an even more concise formulation of the IR model containing just two, orthogonal predictors—proximity and pitch reversal—without compromising fit to the data of Krumhansl (1995a) or Cuddy and Lunney (1995). A subsequent set of experiments with Finnish spiritual folk hymns (Krumhansl et al., 1999) and indigenous folk melodies (yoiks) of the Sami people of Northern Scandinavia (Krumhansl et al., 2000) suggested that the model developed by Krumhansl (1995b) provided a better fit to the data than those of Krumhansl (1995a) or Schellenberg (1996, 1997). By contrast, Schellenberg et al. (2002) found the exact opposite to be true in experiments with adults and infants undertaking a probe-tone task using contexts taken from French Canadian folk songs: the two-factor model of Schellenberg (1997) provided a better fit to the data than the models of Schellenberg (1996), Krumhansl et al. (1999), and Krumhansl et al. (2000).

The bottom-up component of the IR model was intended by Narmour to represent a universal and innate mechanism for generating expectations. However, the overwhelming impression of the empirical literature reviewed above is that while some variant of the IR model describes melodic expectations at a general level, the optimal formulation of the bottom-up principles seems to vary between experiments with stimuli from different musical styles and participants with different musical experience (see also Chapter 7). This goes against the grain of what is supposed to be a hardwired, universal system and points towards the need for a more sophisticated model that is able to learn the structure of more extended portions of the melody providing the context for prediction and relate this structure to the longer-term stylistic experience of listeners. IDyOM fulfils exactly this need.

Studies of folk music show that melodies reflect the patterns of the bottom-up principles of the IR model (Thompson & Stainton, 1998). Rather than these musical patterns reflecting conformity to hardwired perceptual expectations, the converse may be true: to the extent that they are universal, these melodic patterns may be related to constraints of performance such as the difficulty of singing large intervals accurately or regression to the mean following large intervals that approach the limits of the vocal range (Russo & Cuddy, 1999; Schellenberg, 1997; Tierney et al., 2011; von Hippel & Huron, 2000). Listeners exposed to such pitch patterns in music then internalize these regularities, through implicit statistical learning, into probabilistic cognitive models of a musical style. According to this account, there is no strict distinction between the patterns of expectation captured by the bottom-up and top-down systems of the IR model. Both can be captured entirely through learning of sequential pitch patterns in music, some of which may represent constraints of performance and, hence, show greater universality across musical styles and cultures.[4]

3.2.2 Pitch expectation as probabilistic prediction

Interestingly, Carlsen and his colleagues (writing in 1970) explicitly discussed an information-theoretic interpretation of expectation, inspired by Platt whose 1961 chapter was in turn inspired by the writings of Leonard Meyer (1957) and Claude Shannon (1948, 1951): "the investigator could stop a musical piece and determine, as Shannon did for language, the precise statistical expectations of an audience, thus actually measuring quantitatively the information conveyed by the next notes" (Platt, 1961, p. 427). However, this probabilistic interpretation of musical expectation based on statistical learning and information theory subsequently fell out of favour. Theoretical and empirical research in music psychology generally, and musical expectation in particular, moved towards symbolic, rule-based models and theoretical frameworks including the later work of Meyer (1973) and Carlsen (1981). This reflected a larger movement in psychology (and related fields such as cognitive science, artificial intelligence, and linguistics) from the late 1950s to the 1980s known as the cognitive revolution away from learning and information theory (which had been popular in the 1940s and 1950s) and towards symbolic, rule-based models of cognition (Gardner, 1987;

[4]Note that bottom-up processes may still play a role in expectation via sensory memory as discussed above and in Chapter 4. See Milne et al. (2015) for evidence that a spectral model of echoic pitch memory can account for the probe-tone ratings of Krumhansl and Kessler (1982) elicited in generic tonality-defining contexts such as scales, chords, and cadences.

Description	Compositions	Notes	Mean length
Canadian ballads	152	8,553	56.27
Chorale melodies	185	9,227	49.88
German folk songs	566	33,087	58.46
Total	903	50,867	56.33

Table 3.2: Composition of the corpus used to train IDyOM's long-term model in much of the research presented in this book.

Miller, 2003). In music cognition, as reviewed in Chapter 1, this movement is characterized particularly by the theoretical work of Eugene Narmour (1977, 1990, 1992) on melodic expectation and Lerdahl and Jackendoff (1983) on generative models of tonal music.

As discussed in Chapter 1, however, the symbolic rule-based approach fails to provide an elegant account of enculturation and one reason that interest in information-theoretic interpretations of expectation waned in the 1960s and 1970s was the lack of sufficiently sophisticated modelling techniques. Using improved computational methods, the model presented in Chapter 2 effectively implements the proposals of Meyer, Platt, and Carlsen; the question to be addressed here is whether it simulates accurately listeners' pitch expectations in comparison with existing rule-based models. The first research to investigate this question conducted reanalyses of data from existing experiments (Pearce, 2005, Chapter 8). In these studies, and those that follow, IDyOM is used in its BOTH+ configuration using a training set for the long-term model composed of Bach chorale melodies (Bach, 1875), German folk songs (Schaffrath, 1995), and Canadian folk songs and ballads (Creighton, 1966) as shown in Table 3.2.[5] This training set was selected to provide a representative selection of Western tonal music balancing corpus size against the time required to train and run IDyOM models using electronically encoded corpora available at the time. Subsequent empirical research has tended to use the same corpus to maintain consistency and facilitate comparisons between experiments.

Before turning to empirical work, it is instructive in terms of interpreting IDyOM's information-theoretic output to consider an example in which its predictions are compared with those of a music analyst. Figure 3.1 applies IDyOM to excerpts from Schubert's Octet for Strings and Winds, which is discussed in detail by Leonard Meyer in his book *Explaining music* (p. 219, Example 121). Since Meyer's analysis pertains to pitch structure, IDyOM is configured to predict pitch only. Referring to the penultimate note in the second bar shown

[5]These corpora are available at: http://kern.ccarh.org/.

(a) bars 1–2

Pitch	Penultimate note		Final note	
(Number)	Probability	IC	Probability	IC
A5 (81)	.013	6.29	.000	13.81
G5 (79)	.186	2.43	.016	5.97
F5 (77)	.128	2.96	.028	5.15
E5 (76)	.189	2.40	.285	1.81
D5 (74)	.344	1.54	.067	3.89
C5 (72)	.106	3.24	**.134**	**2.90**
B4 (71)	.003	8.58	.059	4.08
A4 (69)	.002	8.93	.060	4.06
G4 (67)	**.015**	**6.02**	.289	1.79
Entropy		2.49		2.81

(b) bars 21–22

Penultimate note		Final note	
Probability	IC	Probability	IC
.013	**6.29**	.170	2.56
.186	2.43	**.535**	**0.90**
.128	2.96	.048	4.38
.189	2.40	.034	4.86
.344	1.54	.028	5.16
.106	3.24	.003	8.18
.003	8.58	.003	8.24
.002	8.93	.000	11.41
.015	6.02	.001	10.24
	2.49		2.15

(c) bars 23–24

Position	1	2	3	4	5	6	7	8
Pitch	G5	F5	A4	B4	C5	E5	D5	C5
Probability	.509	.234	.003	.053	.691	.234	.314	.360
IC	0.98	2.10	8.34	4.25	0.53	2.09	1.67	1.47

Figure 3.1: Three excerpts from the fourth movement of Schubert's Octet in F major (D. 803) taken from: (a) bars 1–2; (b) bars 21–22; and (c) bars 23–24. Panels (a) and (b) show probabilities and corresponding information content (IC) and entropy generated by IDyOM for the penultimate and final notes in each excerpt. At each point in processing, IDyOM estimates a probability distribution for the 37 chromatic pitches from B2 (`pitch` = 47) to B5 (`pitch` = 83), most of which have very low probabilities. For purposes of illustration, only the diatonic pitches between G4 and A5 are shown, including those that actually appear in the Octet (highlighted in bold font: G4 and C5 in (a); A5 and G5 in (b)). The entropy of the prediction is computed over the full 37-pitch alphabet. Panel (c) shows the probability and IC for each note in the final two bars of the theme. In all cases, IDyOM was configured to predict pitch with a source viewpoint linking melodic pitch interval and chromatic scale degree (`interval⊗scale-degree`) combining both the short-term and long-term models, the latter trained on the corpora shown in Table 3.2.

in Figure 3.1(a), Meyer writes: "The continuation is triadic—to G—but in the wrong register. The realization therefore is only provisional." IDyOM reflects this analysis, estimating a lower probability for the G4 that actually follows than for the G5 that is anticipated (.015 vs .186). When the theme returns in bars 21–22 shown in Figure 3.1(b), Meyer writes that: "The triadic implications of the motive are satisfactorily realized [...] But instead of the probable

G, A follows—as part of the dominant of D minor (V/II)." IDyOM reflects this analysis, estimating a lower probability for the A5 that actually follows than for the G5 that is, again, anticipated (.013 vs .186). The relatively high probability (.344) assigned by IDyOM to the D5 in this context can be described in terms of another melodic process discussed by Meyer called gap-fill in which a large interval that spans more than one adjacent scale degree (the gap, C5 to E5 in this case) creates an implication for the subsequent melodic movement to fill in the intervening scale degrees skipped over (here D5). The relatively high probability (.189) assigned by IDyOM to the E5 reflects statistical learning of a tendency for small intervals (here a unison, the smallest interval possible).

Meyer adds that: "The poignancy of the A is the result not only of its deviant character and its harmonic context, but of the fact that the larger interval—a sixth rather than a fifth—acts both as a triadic continuation and as a gap implying descending motion toward closure." Again, IDyOM reflects Meyer's analysis: the penultimate A5 in bar 22 allows IDyOM to predict the continuation with greater certainty than it could following the G4 in bar 2 (reflected in the lower entropy of 2.15 compared with 2.81), making the subsequent descent to the G5 at the end of bar 22 (finally making its appearance, resolving the tension introduced by the preceding deviations from anticipated continuation) much more probable than it would have been following the penultimate G4 in bar 2 (.535 vs .016) and indeed more probable than the C5 that actually followed in bar 2 (.535 vs .134). As shown in Figure 3.1(c), IDyOM strongly anticipates the restatement of the G5 on the downbeat of bar 23 while the cadence towards tonal closure in the final two bars is characterized overall by high probability in IDyOM's analysis (average probability = .3).

3.2.3 Simulating listeners' pitch expectations

While the example shown in Figure 3.1 provides a useful illustration, it is important to demonstrate that IDyOM simulates the pitch expectations of a broader range of listeners who are not expert music analysts and does so for a broader range of musical examples. The experiments described below test the hypothesis that a simulated psychological mechanism of probabilistic prediction based on statistical learning can account for pitch expectation, using the rule-based IR theory as a baseline for comparison.

In a first empirical investigation, IDyOM was used to simulate listeners' expectations in the probe-tone study of Cuddy and Lunney (1995) using eight two-tone contexts, accounting for 72% of the variance in listeners' expectedness ratings. This was not significantly different from the 68% accounted for by the two-factor IR model of Schellenberg (1997) consisting of predictors for pitch proximity, pitch reversal, and tonality. The pitch model of Temperley (2007) described in Chapter 1 using parameters determined empirically from the Essen Folk Song Collection (Schaffrath, 1995) accounted for 76% of the variance. The very short two-tone contexts used in this study do not provide a good platform for distinguishing the models. Therefore, a second experiment simulated listeners' expectations in the probe-tone study of Schellenberg (1996, Experiment 1) using eight British folk songs as contexts. In this case, IDyOM accounted for 83% of the variance in listeners' expectedness ratings which was significantly greater than the 75% of variance accounted for by the two-factor model. These longer contexts which better represent the complexities of real music revealed IDyOM's advantage as a model of melodic pitch expectation.

One of the disadvantages of the probe-tone paradigm is that it provides retrospective measures of expectedness only at a particular point during listening to a piece of music. A paradigm which addresses this limitation was introduced originally by Claude Shannon (1951) for estimating the entropy of printed English, further developed by Cover and King (1978), and applied to music by Manzara et al. (1992). Musicians were given a pot of virtual currency and provided with a musical score on a computer screen showing an incomplete melody. By dragging with a mouse, they could change the pitch of the final note; when they thought they had the correct next note, they bet a proportion of their capital on their prediction, continuing to place bets until the correct pitch was predicted. Correct predictions were rewarded and incorrect predictions penalized by incrementing or decrementing the participants' capital sum in proportion to the amount bet. Once the correct next note was predicted, the participant moved on to the next note until they reached the end of the melody. Given 20 pitches to choose from, the information content of a listener's prediction for the note at position n in the melody was estimated as $\log_2 20 - \log_2 S_n$ where S_n was the net capital gained by the listener for predicting that note.

Manzara et al. (1992) had participants perform the task for two of the chorale melodies harmonized by J. S. Bach (BWV 159 and 379). IDyOM accounted for approximately 63% of the variance in the mean information content estimates (reflecting participants' unexpectedness) which was significantly greater than the 13% of variance accounted for by the two-factor IR model. Increasing stimulus complexity still further over Schellenberg's probe-tone experiment highlights even more clearly the better performance of IDyOM over the IR model in simulating listeners' pitch expectations. Figure 3.2 shows a comparison of IDyOM's information content for BWV 379 alongside the average information content of the listeners in Manzara et al.'s experiment.

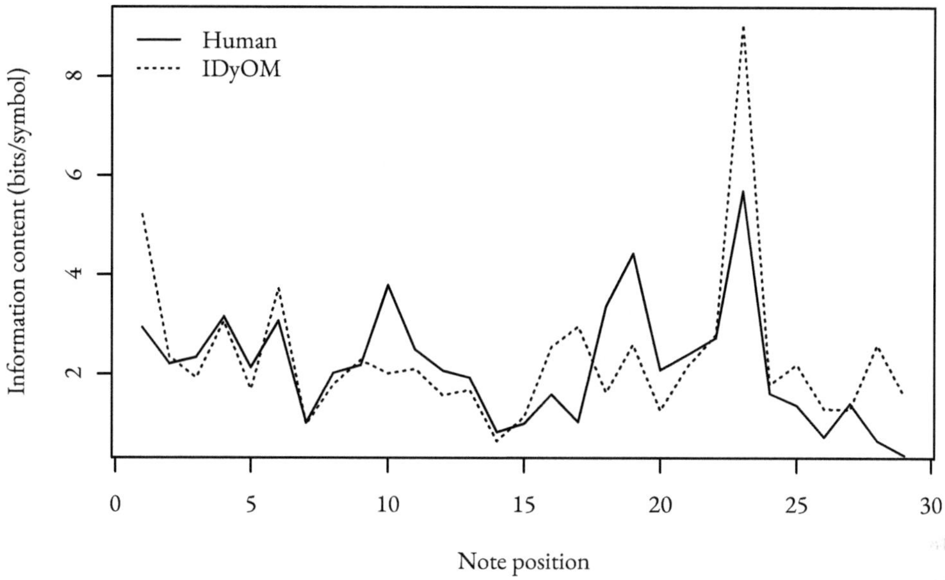

Figure 3.2: Information content generated by IDyOM for the Bach chorale shown in Figure 2.2 (BWV 379) together with mean perceived unexpectedness (Human) from the empirical study reported by Manzara and colleagues (see text for details). IDyOM was configured to predict pitch only with a source viewpoint linking melodic pitch interval and chromatic scale degree (`interval⊗scale-degree`) combining both the short-term and long-term models, the latter trained on the corpora shown in Table 3.2.

The betting paradigm yields specific responses to individual notes throughout an entire melody and rewards accurate estimation of the probability of a continuation, in terms of the amount bet, rather than simply providing a fixed reward for a correct response. Perhaps more importantly, it elicits responses prior to revealing the identity of the note, thereby providing a better prospective measure of expectation generation. However, it requires a degree of musical training and is likely to reflect explicit, reflective cognitive processes rather than direct perceptual processing. A further difficulty with both the betting and the probe-tone paradigms is that the music must be stopped to allow participants to respond. There is evidence that stopping the music in this way can lead to expectations that are specifically related to tonal closure at the end of a melody which are different from those generated during listening to earlier parts of the melody (Aarden, 2003). For this reason, Pearce et al. (2010c) developed a new paradigm, the visually-guided probe-tone paradigm, in which a visual cue is presented that indicates which musical events participants should respond to, allowing data to be collected at multiple points throughout a stimulus without stopping the music.[6]

In order to focus specifically on pitch expectations using this paradigm, without any potentially confounding effects of rhythm, Pearce et al. (2010c) used a set of 28 English hymns in which rhythmic structure had been removed in a musically sensitive way such that all notes had equal duration. IDyOM was used to select one note with high information content and one note with low information content in each stimulus. The results showed greater expectedness for the notes with low information content than those with high information content and IDyOM accounted for 78% of the variance in the mean expectedness ratings which was numerically but not significantly greater than the 73% of variance accounted for by the two-factor IR model. A subsequent probe-tone study (described in more detail in § 3.3) that used 12 isochronous English hymns from the same corpus along with 12 *lieder* by the Austrian composer Franz Schubert showed that IDyOM accounted for approximately 54% of the variance in listeners' expectedness ratings which was significantly greater than the 39% accounted for by the two-factor IR model (Hansen & Pearce, 2014). Since the visually-guided probe-tone paradigm only presents participants with notes that actually appear in melodies, it may be that IDyOM's advantage over the two-factor model lies particularly in simulating perception of the more complex melodic contexts and continuations provided by the Schubert *lieder*.

As discussed above, expectations are thought to be generated largely automatically and implicitly during musical listening. Furthermore, according to the statistical learning hypothesis (see Chapter 1), the cognitive mechanisms underlying musical expectations are acquired implicitly through passive exposure to a musical style. In contrast, all of the paradigms introduced so far for testing musical expectations involve a deliberate response from the listener requiring them to introspect about mechanisms that are thought to be implicit. One way of addressing this disconnect is to have participants respond to a question about a musical event that is unrelated to its contextual predictability and examine differences in reaction time between events that vary in hypothesized expectedness (Bharucha & Stoeckig, 1986). Examples of such tasks include judging the timbre, dissonance, or intonation of particular events where some of the stimuli are manipulated in a task-relevant way (i.e., have altered timbre, dissonance, or intonation). This priming paradigm, as it has become known, rests on the idea that an event that is expected (or primed) given the preceding context will be

[6] Other ways of achieving this include continuous ratings (e.g., Egermann et al., 2013; Gingras et al., 2016) or neuroimaging (e.g., Di Liberto et al., 2020).

Empirical data	Paradigm	Stimuli	N	IDyOM R (R^2)
Cuddy & Lunney (1995)	Probe tone	8 single intervals	24	.85 (.72)
Schellenberg (1996)	Probe tone	8 British folksongs	20	.91 (.83)
Manzara et al. (1992)	Gambling	2 chorale melodies	15	.80 (.63)
Pearce et al. (2010)	Visually-guided probe tone	28 English hymns	40	.89 (.78)
Omigie et al. (2012), Expt. 2, Controls	Visually-guided probe tone	32 English hymns	12	.79 (.62)
Hansen & Pearce (2014)	Probe tone	12 English hymns, 12 Schubert *lieder*	34	.74 (.54)

Table 3.3: A summary of the empirical studies discussed in § 3.2.3 assessing IDyOM's simulations of listeners' melodic pitch expectations by correlating information content with expectedness ratings elicited using a variety of paradigms and musical stimuli.

processed more quickly and elicit faster responses than an unexpected event. Whereas the probe-tone paradigm has been used most prominently for studying melodic expectations (perhaps through its association with the work of Carol Krumhansl), the priming paradigm is more closely associated with research on harmonic expectations following in the tradition of Bharucha and Stoeckig's original study (see Chapter 4 for more on harmonic expectations). However, the priming task has been used successfully to elicit expectations for melodies ending on scale degrees varying in tonal stability (Marmel & Tillmann, 2009; Marmel et al., 2008, 2010).

Does IDyOM also account for expectations when they are elicited implicitly using a priming task? An experiment run by Diana Omigie used the priming paradigm with 58 isochronous English hymns from the same corpus described above (Omigie et al., 2012). Participants were directed using a visual cue to respond to particular notes as in the visually-guided probe-tone paradigm but rather than rating the expectedness of those notes, they judged whether the note was played on a piano (like the preceding notes) or a marimba. Once again, notes were selected that had high and low information content according to IDyOM. As predicted, the results showed that participants responded faster and more accurately to notes with low information content than notes with high information content. This result demonstrates that IDyOM accounts for pitch expectations even when they are elicited implicitly using a paradigm that does not require explicit rating of expectedness.

3.3 Uncertainty for pitch continuations

The results covered in the previous section suggest that information content accurately accounts for listeners' experience of pitch unexpectedness while listening to melodies (see summary in Table 3.3). In other words, pitches that have high information content according to IDyOM tend to be unexpected while those with low information content tend to be expected. But what about the prospective process of generating predictions? According to the theory, expectation generation takes places before the next event actually arrives regardless of how expected or unexpected that event turns out to be. This is a much more difficult phenomenon to study because it is not directly time-locked to a specific event in the way that information content and unexpectedness are. Rather, the prediction emerges from the context and is continually updated as new events appear and are processed. Nonetheless, we can characterize this anticipatory process of prediction in terms of its certainty or uncertainty. A prediction that assigns an equal probability to every possible continuation of a melody is maximally uncertain since every possibility is equally likely (just like the rolling of dice or tossing of coins). On the other hand, a prediction that assigns a very high probability to one particular continuation and very low probabilities to all others is very certain. The information-theoretic quantity entropy (introduced in Chapter 2) precisely captures this intuitive notion of uncertainty and can be computed from the predictive probability distributions estimated by IDyOM.

To illustrate the difference between expectedness and uncertainty, consider the Schubert octet shown in Figure 3.1. In panels (a) and (b), a prospective prediction is generated for the final note of the passage after hearing the penultimate note but before the final note is heard. The prediction consists of an estimation of the probability of each possible continuation. When the final event arrives, it is experienced as being more or less expected according to the prospective prediction. In panel (a), IDyOM's prospective prediction for the final note is relatively uncertain since two potential continuations have a very similar chance of occurring: G_4 with a probability of .289 and E_5 with .285, which between them account for about 57% of the probability mass. In this case, neither of the most probable continuations actually follows, illustrating how music can disconfirm expectations even when the prior expectation is relatively uncertain. In panel (b), by contrast, the prospective prediction for the final note is more certain, with a single continuation accounting for about 54% of the probability mass (resulting in lower entropy: 2.15 vs 2.81). In this case, the relatively certain expectation is confirmed by the highly probable G_5. Had the music continued with a different note, the highly certain expectation would have been disconfirmed, illustrating again how certainty and expectedness do not always go hand in hand.

Perhaps because of the difficulty of pinning it to a particular point in time, prospective uncertainty has received much less attention in the empirical literature than the unexpectedness of concrete events. In their original pilot study, Carlsen et al. (1970, p. 9) reported that "certain two-tone intervals generate a more unified expectancy than others". They did not go into any further detail but I think it is reasonable to interpret the word "unified" here to mean certain. Schmuckler (1989) presented participants with the incomplete contexts used in his probe-tone study and asked them to rate the strength and specificity of their expectations. The strength and specificity ratings were highly correlated with each other and both showed a moderate correlation with a coarse non-probabilistic measure of uncertainty computed as the difference between the highest probe-tone rating and the average rating for each context. It is only recently that detailed probabilistic measures of uncertainty have been assessed in an empirical context.

The question of whether entropy predicts listeners' experience of uncertainty while listening to melodies was addressed in a study run by Niels Chr. Hansen which introduced two measures of predictive uncertainty (Hansen & Pearce, 2014). First, listeners were presented with an incomplete melodic context and asked to state how uncertain they felt about its continuation (*explicit uncertainty*). Second, listeners undertook a probe-tone task with the same contexts followed by nine different probe tones; the mean expectedness ratings for each context were normalized to sum to unity, producing a distribution whose entropy was calculated (*inferred uncertainty*). For comparison, the method introduced by Schmuckler was also investigated as an alternative way of computing inferred uncertainty from IDyOM's probability distributions. The study included musicians and non-musicians and melodic contexts in two musical styles: 12 of the isochronous English hymns used in some of the studies presented in § 3.2.3 (Omigie et al., 2012, 2013; Pearce et al., 2010c) and 12 *lieder* composed by Franz Schubert. The contexts were specially selected such that IDyOM produced a high-entropy prediction for continuation in half of them and a low-entropy prediction in the other half.

The results showed that participants' inferred uncertainty was greater for high-entropy contexts than low-entropy contexts and inferred uncertainty for each context also correlated significantly with entropy across the stimuli. Model comparisons suggested that these effects were driven by the long-term IDyOM component reflecting schematic effects of stylistic exposure to Western music based on model orders of up to three. In most cases, entropy provided a better model of uncertainty than Schmuckler's method (also computed from IDyOM distributions). Although inferred and explicit measures of listeners' uncertainty were correlated, explicit uncertainty only showed significant differences between high- and low-entropy conditions for the English hymns and did not show an overall correlation with entropy. These results suggest that uncertainty is well simulated by the entropy of predictive probability distributions acquired through exposure to Western music via statistical learning but this probabilistic knowledge is largely unavailable to explicit introspection (see Chapter 8 for more on the question of explicit and implicit knowledge).

3.4 Neural measures of predictive processing in melody

Although the priming paradigm provides an implicit measure of expectation, it does so indirectly through the influence of expectation on another task. Perhaps the only way of directly measuring expectations without requiring an overt response is to use physiological measurement, in particular functional neuroimaging methods, which allow recording of cortical responses to unexpected events (rather than less specific effects on skin conductance, for example). Of these, electrophysiological techniques such as electroencephalography (EEG) and magnetoencephalography (MEG) are particularly useful due to their very fine temporal resolution, allowing detailed measurement of responses to individual musical events even for fast tempi or short note durations. While EEG and MEG can be used to understand neural mechanisms, a cognitive neuroscientific approach is adopted here in which these methods are used to understand cognition. This, in turn, presupposes a robust and consistent neural marker for the cognitive process one is interested in.

Violations of harmonic syntax have been studied extensively using EEG and MEG (see Chapter 4) but there has traditionally been less comparable research on melodic expectation, perhaps because it is more difficult to use music theory to construct melodic stimuli that produce such strong violations of expectation. Early EEG studies on melodic expectation used familiar melodies with the pitch of certain notes modified so as to make them less expected while analysis focused on the P300 (Besson & Faïta, 1995; Besson & Macar, 1987; Paller et al., 1992; Verleger, 1990) an endogenous event-related potential (ERP) that is now associated with task-related discrimination of attended stimuli rather than expectation per se (Polich, 2007).

More recent research has focused on the mismatch negativity (MMN), a well-studied neural response to rare events in an otherwise regular auditory scene (Näätänen et al., 2007).

The MMN has a latency of 100–250 ms and generators in both superior temporal gyrus and inferior frontal gyrus (Deouell, 2007). It is typically associated with the oddball paradigm in which, for example, a sequence of tones with the same pitch (the standard) is periodically interrupted by a tone with a different pitch (the rarely occurring oddball or deviant). However, enhanced MMN responses have also been reported for unexpected events in both melody (Brattico et al., 2006) and rhythm (Vuust et al., 2009). Other studies have investigated the early anterior negativity (EAN; or sometimes, when it is right lateralized, the ERAN) reported first in studies of harmonic expectation violation where it peaks at a latency of around 180–200 ms (Koelsch et al., 2000). In studies of melodic expectation, the EAN has been observed in response to out-of-key notes within unfamiliar melodies (Miranda & Ullman, 2007), and both in- and out-of-key notes ending simple five-note melodies (Koelsch & Jentschke, 2010). The EAN often occurs earlier for violations of melodic compared with harmonic expectation, with a latency of 100–150 ms, making it difficult to distinguish from the N1 response (Koelsch & Jentschke, 2010). The neural generators of the EAN are thought to include superior temporal gyrus and inferior frontal gyrus, the latter generally stronger than for the MMN (Koelsch, 2009). These studies used categorical stimulus conditions consisting of predictable and unpredictable notes defined in terms of music theory; none of them employed a computational model of the cognitive mechanisms underlying melodic expectation to make predictions about neural responses to melodic events varying continuously in predictability.

A study conducted by Diana Omigie generalized these findings to real-world unfamiliar stimuli (though in a familiar style) whose unpredictability had been characterized by IDyOM in terms of the information content of each note (Omigie et al., 2013). Participants listened to 58 isochronous English hymn melodies taken from the same corpus as other studies reviewed above. At a latency of approximately 100–120 ms, the peak amplitude of the EAN decreased significantly between notes selected to have high, medium, and low information content and this effect was slightly right lateralized. Furthermore, across all 1,948 notes in all 58 stimuli, the absolute amplitude of the EAN (binned into 10 categories by decile) showed a convincing correlation with information content, $r = .9$. These results suggest that IDyOM's information content accounts well for a neural marker of expectedness in melodic pitch processing in addition to the implicit and explicit behavioural results reviewed in § 3.2.3. Subsequent research by David Quiroga-Martinez showed greater prediction error, indexed by increased MMN amplitude, in low- vs high-entropy melodic contexts (indicated by IDyOM) but only for pitch-related deviants and not for intensity or timbral deviants

(Quiroga-Martinez et al., 2019). This is consistent with lower-entropy predictions focusing probability mass on specific notes resulting in greater prediction error when those strongly expected continuations do not appear (see also § 8.2.3).

3.5 Optimizing cognitive representations

While instrumental music often lacks explicit referential meaning (see Chapter 5), it can still be unusually rich in representational content as a cultural communication signal, possessing complex time-varying structure in many auditory dimensions. As discussed in Chapter 2, a piece of music often contains simultaneous changes in parameters related to melody, harmony, rhythm, voicing, timbre, dynamics, and articulation but music is also rich in terms of its potential for abstract representations in which equivalences are drawn between particular values of particular parameters. Pitch provides a good example. Listeners possess absolute representations of pitch height (Saffran, 2003; Schellenberg & Trehub, 2003) but also more abstract representations such as pitch chroma (Demany et al., 1985; Shepard, 1982) which introduces octave equivalence, for example making middle C (440 Hz) equivalent to the C above (880 Hz) and below (220 Hz) corresponding to doubling and halving of frequency respectively. In addition, listeners are able to derive relative pitch representations in terms of scale degree relative to an inferred tonal centre, drawing an equivalence between notes representing the same scale degree across octaves and tonal centres regardless of their actual pitch (Krumhansl & Kessler, 1982; Krumhansl & Shepard, 1979). Finally, a different kind of relative pitch representation relates a pitch to its melodic predecessor in terms of pitch interval, the signed distance between successive pitches, or, even more abstractly, pitch contour, which only preserves the sign and not the magnitude of the interval (Dowling, 1978, 1994; Dowling & Bartlett, 1981).

How are these different representations organized and selected in cognitive processing of music? Are all representations used simultaneously? If not, what determines which subset of representations is used in a given context? The answers to these questions remain speculative at present. However, a possibility that derives from the theory of probabilistic prediction is that listeners employ representations that are optimally predictive of the music they are currently processing (Huron, 2006). In other words, representations are selected that increase the probability (or equivalently reduce the information content) of the music they are currently processing. This would depend not only on the music itself and the internal predictive model used by the listener based on their previous experience but also on the cognitive representations possessed by that listener (which might themselves depend on experience).

IDyOM provides a useful framework for investigating this hypothesis by virtue of its multiple viewpoint representation scheme which permits the use of different representations in prediction (see Chapter 2). The majority of the results reported in § 3.2.3 used IDyOM models configured to predict `pitch` with a single source viewpoint `interval⊗scale-degree` that links melodic pitch interval with tonal scale degree. However, given a set of candidate viewpoints, IDyOM can be configured to use different combinations of viewpoints allowing an investigation of the fit of each model to listeners' expectations. This provides evidence about which representations listeners are likely to be using when generating pitch expectations. Pearce (2005, Chapter 8) conducted exactly this analysis for the data reported by Cuddy and Lunney (1995), Schellenberg (1996), and Manzara et al. (1992) discussed in § 3.2.3. The results were slightly different for each experiment suggesting adaptation to the stimuli. However, in all three cases support was found for representations related to pitch interval, representations that link pitch structure with rhythmic structure (see also Chapter 4) and, for British folk songs and chorale melodies, representations related to tonality.

For the two chorale melodies used as stimuli by Manzara et al. (1992), the viewpoints selected for optimal fit to the empirical data also reduce information content. This suggests that listeners might select representations that increase prediction performance. In fact, across all 247 viewpoint sets considered in this analysis, there is a strong negative correlation between information content and fit to the empirical data where information content is greater than 2.3 bits/symbol (comprising 45 viewpoint sets) but not lower than this threshold. This provides some evidence that listeners select cognitive representations that maximize the perceived probability of musical events but also that this relationship is subject to other constraints. These might include constraints on the nature and complexity of the cognitive representations that can be derived and the number and kind of representational dimensions that can be attended to concurrently.

3.6 Conclusion

This chapter has introduced a research programme empirically investigating the statistical learning and probabilistic prediction hypotheses presented in Chapter 1 by using the computational model presented in Chapter 2 to simulate listeners' predictive processing of music. This covered both expectedness and predictive uncertainty using explicit, implicit, and neural measures, with a focus throughout on melodic pitch expectations. In the next chapter, the programme of research is extended to temporal and harmonic expectations.

4 Beyond Pitch Expectation: Rhythm and Harmony

The previous chapter covered empirical studies demonstrating that the IDyOM model from Chapter 2 accurately simulates listeners' pitch expectations in a range of melodic contexts. However, there is more to music than melody and more to melody than pitch. The present chapter goes beyond pitch expectation to examine whether the statistical learning and probabilistic prediction hypotheses, as instantiated in IDyOM, can account for listeners' expectations for the timing of musical events and musical harmony. This leads on to the question of how listeners infer the metre and tonality underlying a piece of music.

In § 4.1, IDyOM is used to simulate a series of experiments in which participants responded to melodies whose pitch and temporal structure had been systematically manipulated. This examination of rhythm naturally raises the question of how combined expectations for the timing and pitch of melodic events are generated, using dependent or independent processes of probabilistic prediction. Section 4.2 goes beyond melody to examine expectations for harmonic movement—that is, expectations for sonorities comprising sets of simultaneously sounding notes (vertical slices if you like) in polyphonic music. Once again computational simulations are combined with empirical studies of human listeners to evaluate IDyOM's predictions by comparison with competing models. Finally, predictive processing goes beyond observable properties of the musical surface to latent properties such as tonality and metre that are inferred by listeners based both on their musical experience and the structure of the music (in terms of pitch and timing respectively). The chapter closes in § 4.3 with an investigation of whether these phenomena can be accommodated within the theoretical framework of statistical learning and probabilistic prediction by evaluating an extension of IDyOM to metrical and tonal inference.

4.1 Expectations for event timing

4.1.1 The when as well as the what

The elements of a piece of music are not available to perception all in the same instant as with the appreciation of a painting.[1] Rather we must wait for the music to unfold, revealing its structure as it does so, which leads us naturally to make predictions for *when* musical events will appear as well as *what* those events will be.

While Carl Seashore (1938) suggested that in the first decades of the 20th century, more experimental research had been conducted on rhythm than on any other musical feature, by the end of the century, Eric Clarke (1999) wrote that timing in music had received less attention than pitch. In the third edition of the edited volume in which Clarke's chapter appeared, Henkjan Honing (2013) suggested that research in the 21st century had redressed this balance, accompanied by shifts towards using ecologically valid musical stimuli and investigating the role of musical context. Any theory of music perception would be remiss if it did not address the question of temporal structure as well as pitch structure and the present section focuses on extending the empirical evaluation of the statistical learning and probabilistic prediction hypotheses to temporal expectations.

Many scholars writing on timing in Western music have emphasized the need to distinguish different components of temporal structure including rhythm, metre, tempo, expressive timing, and grouping (Clarke, 1999; Cooper & Meyer, 1960; Honing, 2013; Lerdahl & Jackendoff, 1983; London, 2012). As I will use the terms here, *rhythm* refers to the pattern of event onset times and durations making up a sequence of temporal events, using a discrete, symbolic representation of time, like that conveyed in a musical score. *Metre* is inferred from a rhythm and provides a regularly recurring pattern of pulses at particular temporal locations, potentially including multiple hierarchically embedded levels such that depth in the hierarchy indicates strength of metrical accent. An inferred metre provides a temporal framework relative to which note onsets can be represented and predicted. *Grouping* refers to the aggregation of rhythmic events into sequential segments (or equivalently the identification of boundaries between such segments), which may be hierarchically embedded, such as motifs, phrases, sections, or other aspects of musical form. The focus in this section is on rhythm and metre while grouping is considered in Chapter 6. Reflecting this focus,

[1] This is not to say that the experience of visual art is a static phenomenon, which is certainly not the case, only that the temporal nature of the experience is generally much less constrained by the stimulus than with music.

IDyOM is trained on rhythms as they appear in a score, using a relative rather than absolute representation of time.[2]

There has been discussion in the literature about the extent to which pitch and rhythmic structure are processed in an independent or integrated fashion. Independent processing would mean, for example, that pitch expectations are not affected by the timing of the notes in a melody or, conversely, that temporal expectations are not affected by the pitch of the notes in a melody. Proponents of independent processing have pointed towards evidence that measures of the pitch and temporal stability of the final note of musical phrases combine additively, without interaction, to account for listeners' judgements of phrase completeness (Palmer & Krumhansl, 1987a, 1987b). The alternative account, that pitch and timing are represented and processed jointly, finds support in empirical results showing that listeners can more accurately make judgements about the pitch of notes when their timing is predictable (Jones et al., 1982, 2002).

These contrasting perspectives can be reconciled, given evidence that whether pitch and time are processed independently or jointly depends on several contextual factors. For example, Boltz (1999) reported a series of experimental results suggesting that joint processing is sensitive to coherence within the music (i.e., between the pitch and temporal structure), attention (i.e., directed towards pitch, temporal structure, or both), and veridical familiarity (i.e., previous experience with the stimulus itself). Furthermore, Prince et al. (2009a) found that interactive processing occurred in a pitch comparison task when the melody intervening between the two tones to be compared was atonal, but not when it was tonal. The authors suggested that participants may implicitly attend to pitch structure rather than timing by default due to their experience with Western music where pitch structure is more likely to be an information-bearing parameter.

IDyOM can be configured to predict pitch, timing, or both of these parameters, offering a useful tool for further investigating the conditions under which listeners process these dimensions independently or jointly (see § 4.1.1). The question of how IDyOM can be configured to account for these aspects of temporal expectation is addressed in the next section.

---

[2] For the sake of simplicity, tempo and expressive timing are not considered here but ongoing work is focused on developing models that do incorporate them within an integrated probabilistic framework (Kaplan et al., 2022, 2023).

4.1.2 Temporal expectation

A temporal expectation is a prediction for the timing of some future event. The vast majority of research on temporal expectations in music focuses on the influence of metre. I shall refer to these as *metrical expectations*. London (2012) characterized metre as a highly practised form of entrainment—a synchronization of perceptual, cognitive, and behavioural processes with regularly recurring events in the environment—citing Jeff Pressing (2002, p. 295): "*Musical rhythm arises from the evolved cognitive capacity to form and use predictive models of events*—specifically, predictions of the timing of anticipated future events." The mental synchronization that emerges from this process of temporal prediction obviously includes movement but also perceptual and cognitive processes such as attention and recognition.

A less thoroughly investigated influence on temporal expectation concerns the way in which thematic repetition of rhythmic motifs within a piece of music sets up expectations for an incomplete instance of the motif to be continued. I shall refer to expectations arising in this way as *rhythmic expectations*. This kind of temporal expectation may arise dynamically through perception of repeated rhythmic patterns within a piece of music, as in the case of the rhythmic motifs mentioned above. But generalizing beyond an individual piece of music, rhythmic expectations may also arise schematically (see Table 3.1) reflecting long-term learning of the rhythmic syntax of a musical style, which can be simulated by a model of rhythmic regularities in a stylistically coherent corpus. As noted by Huron (2006), the most obvious examples of rhythmic syntax come from dance styles which are distinguished by characteristic metrical and rhythmic regularities. Note that rhythmic expectations may differ from metrical expectations if, for example, a repeated rhythmic pattern contains events on weak but not strong metrical accents. Indeed, repetition of such contrametrical rhythmic patterns (Kolinski, 1973) is a distinctive feature of many musical cultures from sub-Saharan Africa (e.g., Mali; London et al., 2017) and the Middle East (e.g., Turkish *makam*; Holzapfel, 2015). Beyond music, Patel (2008) suggested that rhythmic expectations reflecting non-periodic aspects of event timing in music may share parallels with speech (see § 7.3.1).

These temporal phenomena are illustrated in Figures 4.1 and 4.2 which show examples of two different dance forms—the *bourrée* and *sarabande*—extracted from the cello suites of J. S. Bach. The *bourrée*—whose opening motif is shown in Figure 4.1—is in duple metre with the downbeat emphasized, has a predictable rhythmic structure and would usually be performed at a lively tempo. The predictable rhythm is illustrated by the fact that IDyOM (configured to predict `onset` using the source viewpoint `ioi-ratio⊗posinbar⊗barlength`) is able to anticipate note timings consistently with relatively low information content (mean = 0.45, standard deviation (SD) = 1.01 across the whole piece). By contrast, the *sarabande*—whose opening motif is shown in Figure 4.2—is in triple metre, emphasizes the second beat rather than the downbeat, and would typically be performed at a stately pace. It also contains unpredictable note onsets, illustrated particularly by the syncopation in the second bar (from note 6 to note 7), which has high information content in the IDyOM simulation. IDyOM finds the seventh note unexpected both because of the unusual preceding inter-onset interval (IOI; and IOI ratio, see below) but also because it appears at a relatively weak metrical position. Dynamic learning of temporal structure within the piece is illustrated by the fact that the next time this rhythmic figure appears in the same metrical position (in the 14th bar) it has significantly reduced information content (5.41 vs 17.32). The temporal unpredictability of the *sarabande* is indicated by the fact that information content is greater and more variable than for the *bourrée* (mean = 1.4, SD = 2.9 across the whole piece). These examples therefore illustrate differences in both metrical and rhythmic syntax between the two dance pieces and also how unpredictable elements can become more predictable through motivic repetition within the piece.

4.1.3 Computational representation and prediction of event timing

Which attributes of musical events do listeners use when generating metrical and rhythmic expectations? And how can these be simulated with IDyOM? Considering first rhythmic expectations, a variety of representations are possible, as discussed in Chapter 2. The absolute onset time of musical events provides a very concrete representation but it does not allow a model or a listener to generalize to rhythms heard earlier or later in the piece or in other pieces. A more abstract representation is the IOI between an event and its predecessor, which allows a rhythm to be identified no matter where in a piece (or indeed any piece) it appears. However, listeners do not appear to represent rhythms only in terms of absolute IOI since they can recognize the same rhythm played at different timescales and regardless of expressive timing variations (Clarke, 1999). An even more abstract representation of the ratio between successive IOIs allows matching of the same rhythm at different temporal scales. In other words, this representation is impervious to augmentation or diminution of the rhythm. Even more abstract representations such as IOI contour (i.e., whether an IOI is larger, smaller, or equal in duration to its predecessor) may be overly indiscriminate and unable to distinguish rhythms that listeners would perceive as being different. Therefore,

(a)

(b)

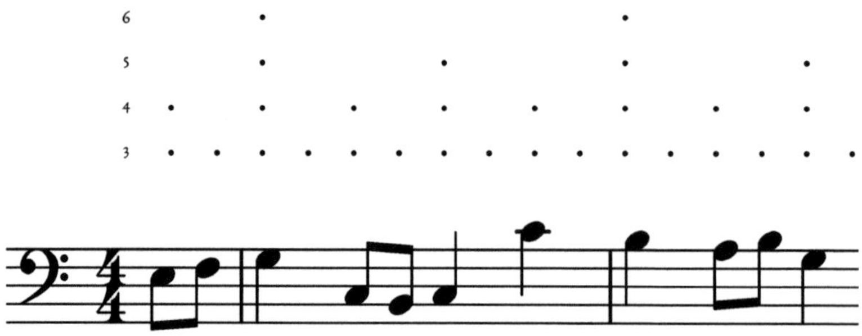

(c)

Note position	1	2	3	4	5	6	7	8	9	10	11
`onset`	72	84	96	120	132	144	168	192	216	228	240
`ioi`	⊥	12	12	24	12	12	24	24	24	12	12
`ioi-ratio`	⊥	⊥	1	2	$\frac{1}{2}$	1	2	1	1	$\frac{1}{2}$	1
`ioi-contour`	⊥	⊥	0	1	-1	0	1	0	0	-1	0
`posinbar`	72	84	0	24	36	48	72	0	24	36	48
`barlength`	96	96	96	96	96	96	96	96	96	96	96
`posinbar⊗barlength`	(72 96)	(84 96)	(0 96)	(24 96)	(36 96)	(48 96)	(72 96)	(0 96)	(24 96)	(36 96)	(48 96)
`ioi-ratio⊗posinbar⊗barlength`	⊥	⊥	(1 0 96)	(2 24 96)	($\frac{1}{2}$ 36 96)	(1 48 96)	(2 72 96)	(1 0 96)	(1 24 96)	($\frac{1}{2}$ 36 96)	(1 48 96)
Information content	4.32	4.32	0.05	1.09	2.37	<0.01	0.63	0.71	0.98	2.78	0.02

Figure 4.1: J. S. Bach, Cello Suite No. 3 (BWV 1009), Bourrée, opening motif, bars 0–2. Panel (a) shows the metrical structure implied by the 4/4 time signature, including metrical levels 6 to 3 where level 4 corresponds to the tactus (with finer grained metrical levels smaller than quaver or eighth note subdivisions omitted). Panel (b) shows the musical score. Panel (c) shows the corresponding viewpoint sequences for temporal viewpoints, using a representation of time in which a crotchet or quarter note corresponds to 24 basic time units. The final row of the table shows the information content of each note given an IDyOM model in the BOTH+ configuration trained on the corpus shown in Table 3.2 predicting `onset` using the source viewpoint `ioi-ratio⊗posinbar⊗barlength`. Note that this viewpoint is undefined (⊥) for the first two notes (on account of `ioi-ratio` being undefined) whose onset is predicted with a uniform distribution.

we can reasonably hypothesize that IOI ratio provides an appropriately balanced level of abstraction for generating rhythmic expectations.

Turning now to metrical expectations, most representations of metre posit an embedded hierarchy of periodic pulses at different temporal scales such that a timepoint is accentuated to the extent that it coincides with pulses at higher levels in the hierarchy with longer periods (Lerdahl & Jackendoff, 1983). Examples of such hierarchies are shown in Figures 4.1 and 4.2 corresponding to time signatures of 4/4 and 3/4 respectively. This representation can be decomposed into two components: the first specifies a time period derived from the inferred metre; the second specifies the probability with which events are expected to fall at different points within that period. Correspondingly, to allow IDyOM to generate metrical expectations, we can posit two simulated perceptual representations: first, perception of a bar-level period (see § 4.3.1 below for how this metrical period can be inferred) represented by the view-

(a)

(b)

(c)

Note position	1	2	3	4	5	6	7	8	9	10
onset	0	24	48	72	90	96	126	132	138	144
ioi	⊥	24	24	24	18	6	30	6	6	6
ioi-ratio	⊥	⊥	1	1	$\frac{3}{4}$	$\frac{1}{3}$	5	$\frac{1}{5}$	1	1
ioi-contour	⊥	⊥	0	0	-1	-1	1	-1	0	0
posinbar	0	24	48	0	18	24	54	60	66	0
barlength	72	72	72	72	72	72	72	72	72	72
ioi-ratio⊗posinbar⊗barlength	⊥	⊥	(1 48 72)	(1 0 72)	($\frac{3}{4}$ 18 72)	($\frac{1}{3}$ 24 72)	(5 54 72)	($\frac{1}{5}$ 60 72)	(1 66 72)	(1 0 72)
Information content	4.32	4.32	0.68	0.30	5.81	<0.01	17.32	15.13	5.14	<0.01

Figure 4.2: J. S. Bach, Cello Suite No. 4 (BWV 1010), Sarabande, opening motif, bars 1–3. Panel (a) shows the metrical structure implied by the 3/4 time signature, including metrical levels 5 to 2 where level 4 corresponds to the tactus (with finer grained metrical levels smaller than semiquaver or 16th note subdivisions omitted). Panel (b) shows the musical score. Panel (c) shows the corresponding viewpoint sequences for temporal viewpoints, using a representation of time in which a crotchet or quarter note corresponds to 24 basic time units. The final row of the table shows the information content of each note given an IDyOM model in the BOTH+ configuration trained on the corpus shown in Table 3.2 predicting `onset` using the source viewpoint `ioi-ratio⊗posinbar⊗barlength`. Note that this viewpoint is undefined (⊥) for the first two notes (on account of `ioi-ratio` being undefined) whose onset is predicted with a uniform distribution.

point `barlength`; second, perception of event onsets relative to the start of the current bar, represented by the viewpoint `posinbar`. The linked viewpoint `posinbar⊗barlength` integrates these two components and allows IDyOM to learn predictive distributions of event timings within bars—i.e., that events are more likely to appear at some temporal offsets from the beginning of the bar (e.g., the downbeat with zero offset) than others. It is important to note that the viewpoint itself does not specify any metrical hierarchy but only the relative representation of time with respect to the bar.[3]

For a zeroth-order long-term model trained on a representative corpus of Western music, this would produce something akin to a traditional metrical hierarchy, illustrated above each notated example in Figures 4.1 and 4.2. However, IDyOM's ability to represent higher-order statistical regularities allows generation of more detailed metrical predictions, depending

[3]Note that `posinbar⊗barlength` does not distinguish simple and compound metres (e.g., 3/4 vs 6/8), which have the same bar length. This can be addressed by adding `pulses` (the numerator of the time signature) to the linked viewpoint but preliminary investigations suggested that this yields only small advantages in prediction performance alongside meaningful increases in computational complexity.

on the preceding context (i.e., the relative onsets of the preceding sequence of notes). Furthermore, these predictions may reflect dynamically or schematically acquired knowledge, via the STM (short-term model) and LTM (long-term model) respectively. This means that sequential patterns of metrical timing can be learned within an individual piece of music or through exposure to music in a given style. Importantly, this includes music with contrametrical rhythmic patterns for which a rhythmic pattern in which note onsets fall on weak metrical accents can be learned by the short-term model (within a piece) or the long-term model (across pieces in a style). Finally, note that while the perceived metre used to infer the hierarchy of accents is assumed, for now, to correspond to the notated time signature, § 4.3.1 extends the approach to metrical inference based on statistical learning and probabilistic prediction.

Prediction of metrical timing doesn't uniquely specify the precise onset time of the next note since a given metrical timepoint (`posinbar`) would be equivalent for the current bar and any future bar. However, predictions for precise onset times can be achieved by linking `posinbar⊗barlength` with `ioi-ratio`. This will affect the prediction if, for example, the `ioi-ratio` model produces a strong expectation for an onset that does not correspond to an expected metrical onset time. Conversely linking `ioi-ratio` with the linked viewpoint `posinbar⊗barlength` adds an element of metrical weight to a prediction based on the IOI ratio. For example, a doubling of IOI might be strongly expected but if that implies that the next event appears on an unlikely metrical position (as predicted by `posinbar⊗barlength`), then it will be perceived as being less probable than other possible ratios. A linked viewpoint combining `posinbar⊗barlength` and `ioi-ratio` therefore enables IDyOM to learn an integrated representation of rhythmic and metrical structure which is analogous to the linked viewpoint representing joint melodic and tonal structure used in Chapter 3. In both cases, the viewpoints provide representational access to structure but the structure itself (e.g., differentiation of scale degrees or metrical positions in terms of probability, dependent on the sequential context) must be learned.

In the following section, IDyOM long-term models (trained on the corpus shown in Table 3.2) are configured to generate temporal predictions with the triple-linked viewpoint `ioi-ratio⊗posinbar⊗barlength` and used to simulate listeners' temporal expectations—both rhythmic and metrical—in experiments using melodies whose temporal and pitch structure have been systematically manipulated.

Figure 4.3: The stimulus from Prince et al. (2009b), Experiments 1 and 2, showing the context and an example probe tone which was rated for goodness of fit. After Prince et al. (2009b, Figure 2).

4.1.4 Simulating listeners' temporal expectations

Jon Prince has conducted a series of experiments examining the simultaneous perception of pitch and temporal structure in musical melody. Here we simulate listeners' responses in these studies using IDyOM models configured to predict pitch only, timing only, and both pitch and timing. IDyOM was configured to use the long-term model only, trained on the corpora shown in Table 3.2. Pitch predictions were made using a single linked viewpoint `interval⊗scale-degree` (see Chapter 3) while temporal predictions were made using a single linked viewpoint `ioi-ratio⊗posinbar⊗barlength`. Three separate models were used, the first making only pitch predictions (for convenience, referred to henceforth as the *pitch model*), the second making only temporal predictions (the *timing model*), and the third predicting both pitch and timing (the *pitch+timing model*). For the purposes of modelling, all stimuli were assumed to have a time signature of 4/4 and the Krumhansl-Schmuckler key-finding algorithm (Krumhansl, 1990, augmented with the improved tone profiles of Temperley, 1999) was used to estimate the key signature for each stimulus.

4.1.4.1 Prince et al. (2009b)

Prince et al. (2009b) conducted two probe-tone experiments for a single two-bar melody with a key signature of C major and a time signature of 4/4 (see Figure 4.3). The probes consisted of all pitches from C4 to B4 placed at eight temporal locations corresponding to quaver subdivisions of the bar, yielding a total of 96 stimuli, which were presented to each participant three times in three separate blocks. One bar of metronome pulses appeared between the context and the bar containing the probe tone. Participants rated how well the probe tone continued the context on a seven-point scale. In Experiment 1, 19 musicians participated while in Experiment 2, 22 musicians participated and attention was directed specifically towards temporal structure, given the finding that pitch structure dominated expectations in Experiment 1. For the IDyOM simulation, the bar of metronome pulses between the context

	Experiment 1	Experiment 2
pitch model	**-.86**	-.16
timing model	-.09	**-.80**
pitch+timing model	-.48	-.78

Table 4.1: Correlation coefficients (df = 94) between IDyOM's information content estimates and average goodness-of-fit ratings in the probe-tone experiments of Prince et al. (2009b). Rows show results for IDyOM models configured to make pitch predictions, temporal predictions, and both pitch and temporal predictions. In Experiment 1, participants' attention was not directed, while in Experiment 2, participants were directed to attend specifically to temporal structure when making their ratings. The strongest correlation in each column is highlighted in bold.

and the probe tone was omitted. Table 4.1 shows the correlations between IDyOM estimates of the information content of the probe tone and the average goodness-of-fit rating. In Experiment 1, where there was no direction of participants' attention, listeners' expectations show a strong negative correlation with the pitch model but not the timing model.[4] By contrast, in Experiment 2 where participants were directed to attend to temporal structure, expectations show a strong correlation with the timing model but not the pitch model. The pitch+timing model yields a lower correlation than the pitch model in Experiment 1 and the timing model in Experiment 2, suggesting relatively little influence of the unattended dimensions, as simulated by IDyOM.

These results suggest that Experiment 1 elicited pitch expectations that are well simulated by the pitch model alone without the timing model. Meanwhile, Experiment 2, which contained an explicit instruction to attend to temporal structure, elicited primarily temporal expectations that are well simulated by the timing model alone. Because participants were never instructed explicitly to attend to both pitch and temporal structure, it was not possible to investigate whether pitch and timing are processed in a dependent or independent manner, as discussed in § 4.1.1. Subsequent research by Prince (2011) addressed this question directly.

4.1.4.2 Prince (2011)

Prince (2011) extended the experimental paradigm from single tones to entire melodies and directed attention explicitly towards pitch structure, temporal structure, or both. Forty-eight seed melodies were taken from a sightsinging manual and systematically manipulated independently in terms of pitch and temporal structure, resulting in 768 stimuli. Participants

[4]The correlation is negative because goodness of fit reflects participants' expectedness while information content reflects IDyOM's unexpectedness.

	Pitch	Temporal	Both
pitch model	**-.91**	-.49	-.76
timing model	-.37	-.84	-.60
pitch+timing model	-.82	**-.98**	**-.93**

Table 4.2: Correlation coefficients (df = 14) between IDyOM's information content and average goodness-of-fit ratings in the experiments of Prince (2011). Rows show results for IDyOM models configured to predict pitch, timing, and both pitch and timing respectively. Columns show results when participants were directed to attend to pitch structure, temporal structure, or both pitch and temporal structure respectively. The strongest correlation in each column is highlighted in bold.

listened to subsets of the stimuli in three blocks of 48 stimuli each, in which they were asked to focus, respectively, on pitch structure, temporal structure, or both (in counterbalanced order). After listening to each stimulus, participants rated its melodic goodness on a seven-point scale. In Experiment 1, 24 musicians participated while in Experiment 2, 24 non-musicians participated but here both groups are analysed together given the very high correlation of $r = .97$ between the average ratings of musicians and non-musicians across conditions.

For the IDyOM simulation, information content was averaged over the notes in each stimulus. Table 4.2 shows the correlations between IDyOM estimates of the mean information content of the stimulus and the average goodness-of-fit rating for each condition of attentional direction (pitch, temporal, both). When attention was directed towards pitch structure, the pitch model but not the timing model yielded a strong correlation with expectedness. Conversely, when attention was directed towards temporal structure, the timing model but not the pitch model produced a strong correlation. In this case, however, the pitch+timing model yielded an even stronger correlation suggesting that listeners were unable to ignore pitch completely. When attention was directed to both pitch and temporal structure, the pitch+timing model yielded a stronger correlation with listeners' expectations than either the pitch or the timing models.

A multiple regression analysis of the mean goodness-of-fit ratings in the Both attentional condition using information content estimated by the pitch and timing models as predictors accounted for 94% of the variance in the data with significant effects of the pitch and timing models but no significant interaction between the two. This demonstrates that the pitch and timing models make independent contributions in accounting strongly for the empirical data, suggesting non-interactive psychological processing of pitch and temporal structure.

4.1.4.3 Prince and Loo (2017)

Prince and Loo (2017) extended the experiments of Prince (2011) by creating a further experimental condition in which the manipulations of pitch and temporal structure were only introduced during the second section of the melody (the *last* condition) as compared with throughout both sections of the melody (the *full* condition). This allowed an investigation of how expected listeners found the material in the second section of each stimulus given the material presented in the first section (instead of the overall goodness judgements of Prince, 2011). Twenty-four participants with varying amounts of musical training judged on a seven-point scale how expected they found the second section of the melody having listened to the first section. Once again, across three blocks participants were asked to focus their ratings on pitch structure, temporal structure, or both. For the IDyOM simulation, the whole stimulus was predicted from beginning to end, including both the first and second sections, and then information content was averaged over notes in the second section of each stimulus in both the full and last conditions, to simulate the instructions given to the participants.

Table 4.3 shows the correlations between IDyOM estimates of the mean information content of the stimuli and the average goodness of fit rating for each condition of attentional direction (pitch, temporal, or both) and structural manipulation (last phrase, full stimulus). When attention was directed towards pitch structure, for both last and full conditions, the pitch model produced strong correlations with expectedness ratings while the timing model did not. Conversely when attention was directed towards temporal structure, the pitch model produced weak correlations while the timing model produced strong correlations. Again, the pitch+timing model yielded a closer fit to these data when attention was directed towards temporal structure, suggesting that participants were not able to ignore pitch structure completely. Finally, when attention was directed towards both pitch and temporal structure, the pitch+timing model produced stronger correlations than either the pitch or timing models.

The results reported by Prince and Loo (2017) showed that listeners found the last condition more unpredictable than the full condition, presumably because the first section of the stimulus set up expectations that were violated in the second section of the last condition but fulfilled in the full condition. IDyOM replicated this pattern of perception, also finding stimuli in the last condition to be more unpredictable (i.e., with higher information content) than stimuli in the full condition. A multiple regression analysis of the mean goodness-of-fit ratings in the *Both* attentional condition and *Last* stimulus condition using information

	Pitch		Rhythm		Both	
	last	full	last	full	last	full
pitch model	**-.91**	**-.84**	-.45	-.51	-.79	-.71
timing model	-.29	-.43	-.85	-.82	-.54	-.57
pitch+timing model	-.71	-.80	**-.97**	**-.98**	**-.87**	**-.85**

Table 4.3: Correlation coefficients (df = 14) between IDyOM's information content and average goodness-of-fit ratings in the experiments of Prince and Loo (2017). Rows show results for IDyOM models configured to predict pitch, timing, and both pitch and timing respectively. Columns show results when participants were directed to attend to pitch structure, temporal structure, and both pitch and temporal structure, respectively, and also when structure was manipulated across the full stimulus (full) or only the last phrase (last).

content estimated by the pitch and timing models as predictors accounted for 95% of the variance in the data with significant effects of the pitch and timing models but no significant interaction between the two. This demonstrates that the empirical data can be accounted for in terms of independent contributions of the pitch and timing models, with no evidence to suggest that participants in this experiment processed pitch and temporal structure in a dependent manner.

Overall, these results show that a trained IDyOM model accurately simulates the temporal expectations of listeners with enculturation in Western music across a range of melodic contexts. Furthermore, the combination of pitch and timing predictions within IDyOM provided the most accurate simulations of expectations when attention was directed to both pitch and rhythmic structure. It is interesting that pitch and temporal expectations appeared to represent independent rather than overlapping psychological processes. This seems likely to reflect the nature of these particular stimuli which were deliberately designed to vary pitch and rhythmic structure orthogonally. It is plausible that interactive effects of pitch and rhythmic structure on expectation would be generated for stimuli in which pitch and temporal structure had greater colinearity. The results also indicate that pitch expectations tend to dominate over temporal expectations and continue to exert influence even when attention is directed to rhythmic structure only. This seems likely to reflect an effect of experience with Western music in which pitch is often a more prominent carrier of information than rhythm, but might vary for listeners with experience of different musical cultures.

4.2 Expectations for harmonic movement

4.2.1 Harmony and harmonic movement

In polyphonic music, more than one note may sound at a time forming sonorities (or chords), which are characterized by their constituent simultaneously sounding notes. Sequences of changing chords constitute a phenomenon known as harmonic movement.[5] Harmonic movement is often presented as *the* quintessential parameter of musical syntax and, although this might not be true of all (or even most) musical styles cross-culturally, it forms an important part of many Western musical styles, with which most existing empirical research on music cognition has been conducted (as a matter of fact rather than principle). According to the present thesis, a listener enculturated in Western tonal music possesses an internal model of regularities in the sequential harmonic structure of this music and can use this model to generate expectations for forthcoming sonorities. One of the central debates in the empirical literature on harmonic expectation has been the extent to which expectations result from such a cognitive model of harmonic syntax, in accordance with the present position, or whether they result from sensory facilitation of pitch components that are repeated in the very short-term context making up echoic memory (see Chapter 3). Such facilitation depends on enhanced processing of any spectral components that are repeated exactly within the timeframe of echoic memory (see Table 3.1; Cowan, 1984).

A defining study in this debate was conducted by Bigand et al. (2003) who examined reaction times for a judgement of whether the final chord (the target) in a sequence of eight was consonant or dissonant. Half of the chords (called foils) were rendered dissonant by the addition of a tone that was close in pitch to one of the chord tones (a semitone away from the tonic or dominant) and responses to the foils were not analysed. The chord sequences were constructed to vary orthogonally cognitive and sensory effects on expectation for the target chord. The cognitive manipulation contrasted a syntactically likely target (tonic chord) with a less likely target (subdominant chord). The stimuli were constructed such that the same final two chords (e.g., G-C) functioned as a likely target in one condition (i.e., V-I when the preceding chords implied a key of C major) and as an unlikely target in the other condition (i.e., I-IV when the preceding chords implied a key of G). The sensory manipulation was achieved by controlling whether or not the subdominant chord (IV) appeared in the context; its appearance was hypothesized to increase the expectedness of the

[5] Harmony can also be implied by music and inferred by listeners when some of the constituent tones of a chord are missing or temporally displaced so that they don't sound simultaneously. This is true even for melody.

syntactically unlikely target (IV) through sensory facilitation. As hypothesized, the results showed slower reaction times for unlikely compared with likely targets, demonstrating a cognitive effect of learned harmonic syntax. However, the hypothesized effect of sensory facilitation was notably absent except for subtle differences when the stimuli were played at very fast tempi in a separate experiment (13 chords per second compared with 1.5 chords per second in the original experiment). This suggests that sensory facilitation does exist but is heavily outweighed in strength by cognitively generated syntactic expectations.

However, Bigand et al. (2014) showed that many other experimental results, some of which had originally been taken to indicate cognitive generation of syntactic expectations, could be accounted for in terms of sensory effects simulated by the *periodicity pitch* (PP) model of echoic memory created by Leman (2000). The PP model simulates several different components of auditory perception. First, peripheral auditory processing is modelled including filtering effects of the outer and middle ear, frequency decomposition by the basilar membrane, and transduction of the signal by inner-hair cells into rate-code firing patterns of frequency-tuned neurons in the auditory nerve. Second, this primary image is transformed into a PP image representing the instantaneous strengths of pitch periodicities, in the range 80–1250 Hz, that are present in the firing patterns. Finally, this PP image is smoothed over time with a constant chosen to reflect different components of sensory memory. As a model of sensory expectation, a *local* PP image with a short time constant (e.g., 0.1 s) is correlated with a *global* PP image with a longer time constant (e.g., 1.5 s) to simulate the similarity of current auditory input to the contents of echoic memory.

Collins et al. (2014) introduced a model designed to integrate low-level sensory and high-level cognitive components. The model combines four components using multiple regression to fit linear coefficients on empirical data from seven reaction time studies of melodic and harmonic expectation. The first component is Leman's PP model. The second component is a temporally smoothed representation of the extent to which each of the 12 pitch classes is present in the pitch periodicity representation of Leman's model. The third component is higher level, the tonal space model of Janata et al. (2002) which maps the local and global images from Leman's model onto a toroidal representation of pitch, capturing tonal proximity between pitches. The mapping was achieved by training a Self-Organising Map (SOM) on a specially composed melody that modulates through all 24 major and minor keys. Expectations are computed as the correlation between tonal space representations of the local and global images. The fourth component is a closure predictor, assuming high values when the final event of a stimulus is strongly predicted by the tonal space model (i.e., with a strong correlation between the local and global tonal spaces exceeding that for the penultimate event). The closure predictor was motivated by the observation that without it, the model underpredicts the tonic in several experimental situations.

While this model has advantages in combining simulations of low-level peripheral auditory processing with higher-level representations of tonality, it does not explicitly model sequential dependencies between melodic or harmonic events. Therefore it cannot learn the syntactic structure of musical styles through exposure as listeners do during stylistic enculturation, according to the statistical learning hypothesis outlined in Chapter 1. As a result, the jury is still out on whether expectations are governed primarily by sensory or cognitive factors and, for this reason, the results presented below include comparison of IDyOM—a much more strongly cognitive model of musical syntax than that of Janata et al. (2002)—with the more sensory models discussed above.

4.2.2 Probabilistic prediction of harmonic movement

As with melody and rhythm, individual chords and chord sequences may be characterized in terms of multiple representations, reflecting both the generative principles underlying their composition and the perceptual sensitivities of enculturated listeners. Figure 4.4 shows a simple harmonic sequence along with the representations afforded by the harmony viewpoints implemented in IDyOM (see Harrison, 2019; Sears et al., 2018). Given these representations, IDyOM can learn and predict harmonic movement in exactly the same way as melodic or rhythmic sequences. In particular, an IDyOM LTM trained on a corpus of polyphonic Western music constitutes a strongly cognitive model of harmonic expectation, reflecting sequential harmonic regularities learned through long-term exposure to Western music.

To illustrate the application of IDyOM to harmony, it is useful to draw on the concept of *cadence*, which dates back to the French composer and music theorist Jean-Philippe Rameau (1722). The cadence is a prominent example of harmonic expectation in Western tonal music that creates a sense of resolution and closure indicating the end of a musical phrase. In a cadence, the harmonic and melodic context produce strong expectations that are fulfilled to varying degrees to create this sense of closure. Inspired by William Caplin (2004), David Sears conducted a series of experimental investigations that compared five kinds of cadence in the music of Mozart and Haydn (Sears et al., 2014, 2018, 2019), examples of which (taken from Mozart's keyboard sonatas) are shown in Figure 4.5. Authentic cadences create and fulfil expectations for phrase-final movement from the dominant (or fifth: V) scale degree

Note names	G G B D	B G D G	C G C E	D F♯ A C	G D G B
Octave	3 4 4 5	3 4 5 5	4 4 5 5	4 4 4 5	3 4 4 4
Inversion	root	first	root	root	root
Bass	G	B	C	D	G
Root	G	G	C	D	G
Chord quality	major	major	major	dominant 7th	major
Tonal function	I	I^6	IV	V^7	I
`pi-chord`	(55 67 71 74)	(59 67 74 79)	(60 67 72 76)	(62 66 69 72)	(55 62 67 71)
`pi-chord-type`	(0 12 16 19)	(0 8 15 20)	(0 7 12 16)	(0 4 7 10)	(0 7 12 16)
`pc-chord`	(7 (2 11))	(11 (2 7))	(0 (4 7))	(2 (0 6 9))	(7 (2 11))
`pc-chord-rel-prev-bass`	⊥	(4 (0 7))	(1 (5 8))	(2 (0 6 9))	(5 (0 9))
`pc-set`	(2 7 11)	(2 7 11)	(0 4 7)	(0 2 6 9)	(2 7 11)
`pc-set-type`	(0 4 7)	(0 4 7)	(0 4 7)	(0 3 6 8)	(0 4 7)
`pc-set-rel-bass`	(0 4 7)	(0 3 8)	(0 4 7)	(0 4 7 10)	(0 4 7)
`pc-set-rel-prev-bass`	⊥	(0 4 7)	(1 5 8)	(0 2 6 9)	(0 5 9)
`pc-set-rel-root`	(0 4 7)	(0 4 7)	(0 4 7)	(0 4 7 10)	(0 4 7)
`bass-pc`	7	11	0	2	7
`bass-int`	⊥	4	1	2	5
`bass-pc-rel-root`	0	4	0	0	0
`root-pc`	7	7	0	2	7
`root-int`	⊥	0	5	2	5
`sd-chord`	(0 4 7)	(4 0 7)	(5 0 9)	(7 2 5 11)	(0 4 7)
`sd-set`	(0 4 7)	(0 4 7)	(0 5 9)	(2 5 7 11)	(0 4 7)
`bass-sd`	0	4	5	7	0
`root-sd`	0	0	5	7	0

Figure 4.4: A chord sequence (adapted from Koelsch et al., 2002, Figure 1, transposed into G major) with annotations illustrating harmony viewpoints. The top panel describes the chord sequence in traditional musical terms while the remaining panels show viewpoint sequences representing various features of harmonic movement, with degree of abstraction generally increasing from the upper panels to the lower panels. The panels separate viewpoints into representations of chromatic pitch ('pi', in which constituent notes are represented by midi pitches in ascending order), pitch class ('pc', invariant to octave but not inversion), pitch class sets ('pc-set', invariant to octave and inversion), bass notes ('bass', the lowest pitch in the chord), root notes ('root', inferred using the algorithm of Parncutt, 1988), and scale degrees ('sd', relative to the tonal centre of the prevailing key, G major). All but those in the lowest panel are taken from Harrison (2019) who introduced the 'type' terminology (`pi-chord-type`, `pc-chord-type`, `pc-set-type`) for capturing transpositional invariance by representing the chord relative to the bass note (note that `pc-chord-type` is omitted as it is equivalent to `pc-set-rel-bass`). Note that `pc-set-rel-bass` and `sd-chord` were referred to as `vintcc` and `csdc` by Sears et al. (2018, 2019). ⊥ indicates that the viewpoint is undefined for a given chord (e.g., due to there being no preceding chord). Note the correspondence between the descriptors in the top panel and the viewpoints in the lower panel: for example, `pi-chord` corresponds to note names and octave, `root-pc` to root, `pc-set-rel-bass` to chord quality, and `root-sd` to tonal function.

(a) Perfect Authentic Cadence (PAC): root position dominant to final root position tonic chord (V-I) with tonic scale degree in the melody ($\hat{1}$).

(b) Imperfect Authentic Cadence (IAC): root position dominant to root position tonic chord (V-I) with the third scale degree in the melody ($\hat{3}$).

(c) Half Cadence (HC): phrase-final root position dominant chord (?-V) with second, fifth, or seventh scale degrees in the melody ($\hat{2}$, $\hat{5}$, $\hat{7}$).

(d) Deceptive Cadence (DC): dominant to non-tonic chord (V-?, typically vi) with a stable scale degree in the melody ($\hat{1}$ or $\hat{3}$).

(e) Evaded Cadence (EC): dominant to non-tonic chord (V-?) with a non-tonic scale degree in the melody (typically $\hat{5}$).

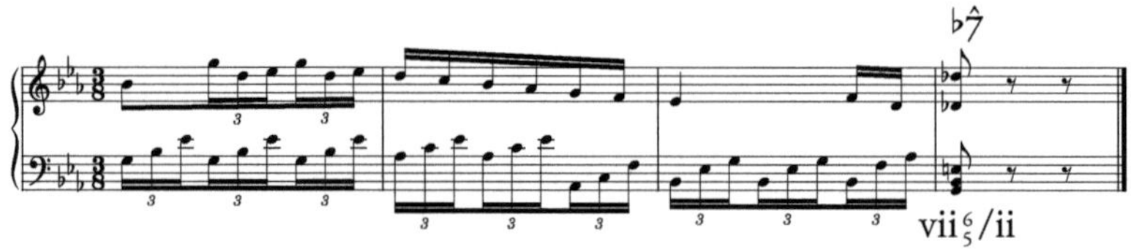

Figure 4.5: Cadence types from W. A. Mozart's keyboard sonatas. The final chords are annotated with their diatonic function and scale degree of the melody note. Melodic dissonances such as suspensions or appoggiature have been removed. A circumflex above a number indicates a melodic scale degree (i.e., the scale degree of the highest chord note in these examples) to distinguish these from the root scale degree of the chord, expressed in Roman numerals. (a): K. 281, i, mm. 5-8; (b): K. 311, i, mm. 1-4; (c): K. 333, iii, mm. 60–64; (d): K. 457, i, mm. 42–48; (e): K. 281, ii, mm. 96–99. After Sears et al. (2019, Figure 2).

to the tonic (first: I). The *perfect authentic cadence* (PAC) fulfils these expectations both melodically and harmonically while the *imperfect authentic cadence* (IAC) retains the V-I harmony but deviates melodically, arriving on the third scale degree rather than the tonic. The imperfect authentic cadence disrupts the anticipated tonic arrival slightly in the melody, but there are two cadence types that introduce much stronger deviations. The *deceptive cadence* (DC) moves from the dominant harmony to a non-tonic chord but retains a stable scale degree (usually the first or third) in the melody. The *evaded cadence* (EC), meanwhile, having introduced a dominant context that implies movement to the tonic, confounds those implications entirely, being characterized by a melodic leap to a higher scale degree, often the fifth. Finally, the *half cadence* (HC) is distinct in moving to a phrase-final dominant harmony, involving less specific expectations for closure and weaker fulfilment of those expectations than authentic cadences (PAC and IAC) but not to the extent of the cadential deviations (DC and EC).[6]

The patterns of expectation set up by these different cadence types depend on learning of harmonic regularities in Western music. Figure 4.6 shows the output of an IDyOM long-term model predicting the final cadential chord in 40 extracts from Mozart keyboard sonatas, eight in each of the five cadence categories (PAC, IAC, HC, DC, EC), including those shown in Figure 4.5. For each extract IDyOM was trained with leave-one-out cross-validation on the remaining 39 extracts. Results are shown for harmonic predictions, using the `root-sd` viewpoint, and melodic predictions, using the `interval⊗scale-degree` viewpoint from Chapter 3. The `root-sd` viewpoint represents the root of each chord as a scale degree (i.e., relative to the tonic of the prevailing key) allowing IDyOM to learn sequential harmonic relationships that underlie the anticipatory differences between the cadence types. IDyOM's response to these cadence types illustrates the fulfilment of harmonic expectations by the authentic cadences relative to the other cadence types and the dramatic violation of melodic expectations associated with the evaded cadence.

4.2.3 Simulating listeners' harmonic expectations

David Sears conducted an empirical study of listeners' harmonic expectations including both implicit and explicit measures (i.e., reaction times and ratings respectively) that compared the

[6]This cadential taxonomy and the corresponding musical examples provide a convenient illustrative starting point for examining IDyOM as a model of harmonic expectation. However, it is worth noting that the examples were chosen to represent clear examples of each cadence type; actual practice is much more varied. There is no suggestion that the cadential taxonomy applies more widely to musical styles beyond those reflected in the corpora considered here.

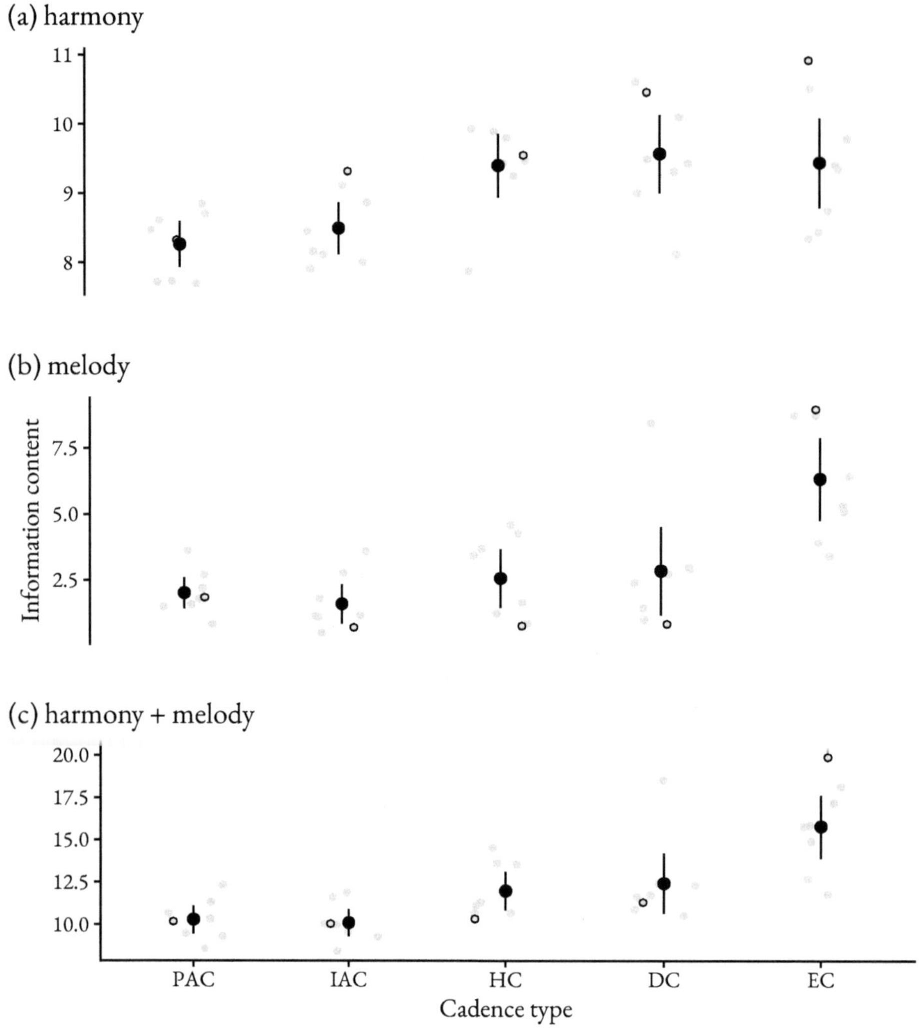

Figure 4.6: Information content for the final cadential chord in 40 extracts from Mozart keyboard sonatas (Sears et al., 2019) generated by an IDyOM long-term model trained using leave-one-out cross-validation on the remaining 39 extracts. The x-axis distinguishes five cadence types, with each type represented by eight extracts, each of which is plotted as a point. Points plotted with a bold outline correspond to the extracts shown in Figure 4.5. Panel (a) shows harmonic predictions generated by IDyOM configured with the source viewpoint `root-sd`. Panel (b) shows melodic predictions generated by IDyOM with the source viewpoint `interval⊗scale-degree`. Panel (c) shows the output of a composite model that computes the joint probability of the harmonic and melodic models. Black points represent mean information content and error bars represent 95% confidence intervals. The different scales for information content reflect the different alphabet sizes for the different models. PAC = perfect authentic cadence; IAC = imperfect authentic cadence; HC = half cadence; DC = deceptive cadence; EC = evaded cadence. Note that Sears et al. (2019) used EV for the evaded cadence because EC is often used to denote an elided cadence; here EC refers to evaded cadences for consistency with the other chord types.

Model	Fit-ratings (Expt. 1)	RTs (Expt. 2)
Leman (2000)	.15	-.18
Janata et al. (2002)	.16	-.24
IDyOM		
- melody	**-.64**	**.63**
- harmony	**-.72**	**.50**
- melody + harmony	**-.79**	**.67**

Table 4.4: Correlations between expectation models and listeners' expectations in Experiment 1 (goodness-of-fit ratings) and 2 (Reaction Times, RTs) of Sears et al. (2019). The PP model (Leman, 2000) was configured with half lives of 0.1 s for the local image and 4 s for the global image. Adapted from Sears et al. (2019, Table 2). Significant correlations are highlighted in bold font.

performance of sensory (Leman, 2000), sensory/cognitive (Collins et al., 2014), and cognitive (IDyOM) models (Sears et al., 2019). In a first experiment, 40 participants listened to 40 excerpts from Mozart's keyboard sonatas ending on one of five different types of cadence (perfect authentic, imperfect authentic, half, deceptive, and evaded) as illustrated in Figures 4.5 and 4.6. They judged prospectively the strength and specificity of their expectations for continuation of a version of the stimulus with the final event truncated. Subsequently, they listened to a nontruncated version and rated how well the actual continuation matched their expectations. A second experiment presented 40 participants with the nontruncated stimuli in two versions where the final tone and chord were either in tune or out of tune (40 cents sharp), measuring reaction times on a mistuning detection task.

IDyOM was used to simulate the experimental data in three configurations, all trained on a corpus of 50 movements from Haydn's string quartets. The first configuration simulated harmonic expectations using the viewpoint `pc-set-rel-bass` in which each chord is represented as the set of absolute intervals in semitones modulo 12 between each of the upper parts and the lowest instrumental part (the bass). The second configuration simulated melodic expectations using the viewpoints `interval` and `scale-degree⊗pitch`. The source viewpoints for the melodic (`interval`, `pitch⊗scale-degree`) and harmonic models (`pc-set-rel-bass`) were selected from a larger set to optimize prediction performance on the nontruncated stimuli including all events (not just the final cadence). The third configuration combined these simulations of melodic and harmonic expectations for cadences by computing the joint probability of their melodic and harmonic components.

For all three configurations, IDyOM's information content estimates for the final event showed strong negative correlations with the goodness-of-fit ratings in Experiment 1 and positive correlations with the reaction times in Experiment 2, as shown in Table 4.4. The fit was stronger for the combined melodic and harmonic configuration (simulating the joint probability of the melody and harmony) than for either the melodic or harmonic models run in isolation. By contrast, a model comparison revealed that neither the sensory PP model (Leman, 2000) nor the tonal space model (Janata et al., 2002) showed significant correlations with the empirical data in either experiment.

While this model comparison produced relatively unambiguous findings, it is possible that results for cadences might not generalize to harmonic expectations generated in a broader range of musical contexts. In other words, cadences might elicit harmonic expectations governed by cognitive models of stylistic syntax but this might be less true of harmonic expectations elicited in non-cadential musical passages. To investigate this question, Peter Harrison ran a study in which 50 participants listened to 300 sequences of eight chords each taken from songs randomly sampled from the McGill Billboard corpus (Burgoyne, 2011), consisting of chord transcriptions for 739 pop songs from the Billboard magazine's US *Hot 100* chart from 1958 to 1991 (Harrison, 2019; Harrison & Pearce, 2018). Chord sequences were presented in a piano timbre at a tempo of 60 beats per minute (1 s per chord) without metrical cues, with bass notes in the octave below middle C and non-bass notes in the octave above middle C. Participants rated the surprisingness of the sixth chord in each eight-chord sequence in a harmonic version of the visually-cued probe tone paradigm (see § 3.2.3).

In a computational simulation, long-term and short-term models were configured with a set of 11 viewpoints from Figure 4.4 (`pc-chord`, `pc-chord-rel-prev-bass`, `pc-set`, `pc-set-rel-bass`, `pc-set-rel-prev-bass`, `pc-set-rel-root`, `bass-pc`, `bass-int`, `bass-pc-rel-root`, `root-pc`, `root-int`). The long-term models were trained using cross-validation on the Billboard corpus. Unusually, the weights for combining the viewpoints were derived not using the entropy-weighting scheme presented in § 2.2.5 but rather through maximum-likelihood estimation for optimal prediction performance across the Billboard corpus. This can be likened to the viewpoint selection used in § 3.5, representing another way of implementing Huron's (2006) proposal that listeners select perceptual representations that maximize prediction performance.

In the analysis of the empirical data, IDyOM was compared with the sensory models of Leman (2000), Milne et al. (2011), and Collins et al. (2014). As shown in Table 4.5, the model comparison demonstrates that IDyOM provides the most convincing simulation of

Model	Surprisal ratings
Leman (2000)	.036
Leman (2000) optimized	**.33**
Milne et al. (2011)	-.169
Collins et al. (2014)	-.007
IDyOM	**.67**

Table 4.5: Pearson correlation coefficients between four models of harmonic expectation and the mean surprisal ratings of 50 participants in the experiment reported by Harrison and Pearce (2018) and Harrison (2019). Significant correlations are indicated in bold font. Two versions of the PP model of Leman (2000) are shown: the first uses default half-life parameters (local: 0.1 s; global: 1.5 s) while the second uses parameters optimized on the empirical data (local: 0.5 s; global: 4 s). IDyOM was run using an implementation in *R* (Harrison et al., 2020) configured with an unbounded order; see text for other configuration details. The results for Leman (2000) optimized is taken from Harrison (2019, Table 4.6) while the remaining results are from Harrison and Pearce (2018, Table 1).

listeners' expectations for harmonic continuations in this experiment, though the sensory PP model also accounts for significant variance when optimized on the empirical data.[7]

In a follow-up EEG experiment using the same 300 stimuli (Goldman et al., 2021), the same models were used to predict four event-related potential (ERP) responses often associated with unpredictable auditory or musical events: the early right anterior negativity or ERAN (150–300 ms; Koelsch et al., 2000), the P3a (250–450 ms; Polich, 2007), the N5 (450–600 ms; Koelsch et al., 2000), and the P600 (600–900 ms; Patel et al., 1998). In this case, IDyOM was configured with the short-term model using a single source viewpoint (`pi-chord`) and the long-term model using four source viewpoints: `pi-chord`, `pc-set-rel-bass`, `pc-set-rel-prev-bass`, and `sd-chord` (see Table 4.4). Following Harrison and Pearce (2018), the viewpoint weights were optimized for prediction performance on each chord. ERP responses were analysed for the second to the sixth chord and a measure of predictability derived from each model was entered into a regression analysis with the amplitude of each respective ERP component as the dependent variable. IDyOM and the PP model (with default half lives: 0.1 s local; 1.5 s global) were significant predictors of the P3a amplitude.[8]

[7] See Harrison (2019, Chapter 4) for a more detailed analysis in which expectations also appear to be influenced by measures of consonance, including interference between partials (Hutchinson & Knopoff, 1978) and harmonicity/periodicity (Harrison & Pearce, 2020; Milne, 2013).

[8] Surprisingly none of the models significantly predicted any of the other ERP components, perhaps reflecting differences in experimental design from previous experiments. For example, the ERAN and N5 are typically elicited in experiments in which a small set of hand-constructed chord sequences are presented with a stylistically

These behavioural and EEG results suggest a strong effect of cognitive processes on harmonic expectation supplemented by an additional weaker effect of sensory processes. Further evidence to support this conclusion comes from an experiment run by Vincent Cheung et al. (2023) comparing IDyOM with the PP model (Leman, 2000). The stimuli were 30 chord sequences once again selected from the McGill Billboard Corpus of commercially successful pop songs. Compared with the stimuli from Harrison and Pearce (2018) and Goldman et al. (2021), the stimuli were much longer, ranging from 30 to 38 chords. Furthermore, the stimuli were selected from the corpus specifically to ensure wide coverage across a factorial space combining high and low information content with high and low entropy, according to an IDyOM model trained using cross-validation on the Billboard pop corpus. In this case, IDyOM was used in its BOTH configuration to predict chord symbols representing the root scale degree, chord quality, and inversion as encoded in the McGill Billboard Corpus using the format of Harte et al. (2005). All stimuli were transposed to C major and there was an interval of 2.4 s between the onset of each successive chord. A sample of 25 participants gave continuous ratings of perceived surprise for each chord in each of the 30 stimuli. The results showed that IDyOM and the PP model (with half lives optimized on the empirical data: 0.5 s local; 4 s global) each explained significant unique portions of the variance in surprisingness ratings, with IDyOM accounting for at least twice as much variance as the PP model.[9]

Overall, these results show that strongly cognitive models such as IDyOM capture important aspects of harmonic expectation, demonstrating that the underlying principles of statistical learning and probabilistic prediction generalize beyond melody to perception of polyphonic music. An important question for the future is to understand whether sensory and cognitive processing influence harmonic expectations independently (in parallel or in sequence) or in an integrated manner. Furthermore, it remains possible that some auditory contexts might elicit a greater effect of sensory processes while others evoke a stronger cognitive effect, perhaps also depending on the auditory experience of the listener (see Chapter 8). Much also remains to be understood about psychological representations of harmony and the extent to which they are influenced by musical enculturation.

---

appropriate or inappropriate final chord, with each individual stimulus repeated many times (Koelsch et al., 2000, 2007; Steinbeis et al., 2006). Goldman et al. (2021), on the other hand, sampled the stimuli randomly from a corpus of pop music, presented each stimulus only once, and looked for correlations between ERP amplitude and model-derived measures of expectation violation. PP model prediction was little affected by choice of half life.

[9] Approximately three times as much for musicians and two times as much for non-musicians. Differences between musicians and non-musicians in this experiment are discussed further in Chapter 8.

4.3 Inferring latent musical attributes: Metre and tonality

The models used in Chapter 3 and § 4.1 to simulate expectations for the pitch and timing of musical notes operate on surface representations such as `interval` and `ioi-ratio` but also more abstract representations such as `scale-degree` and `posinbar⊗barlength` which assume a latent tonality and metre respectively. Neither tonality nor metre exist as physical properties of music that are directly available to perception. Rather these latent attributes must be inferred while listening to a musical passage, depending both on the structure of the musical stimulus and the prior musical experience of the listener. Because evidence suggests that tonality and metre perception are learned through exposure and vary between musical cultures, they constitute excellent candidates for research on enculturation, as discussed further in Chapter 7. However, the IDyOM simulations conducted so far lack an account of how perception of metre and tonality are acquired through exposure to music. Can perceptual inference of metre and tonality be simulated within the same theoretical framework of statistical learning and probabilistic prediction?

4.3.1 Metrical inference

As noted in § 4.1, metre is a hierarchically embedded pattern of periodically recurring accents that is not present in the stimulus but is inferred and aligned with a piece of music during listening. Metrical structures vary between musical cultures but once enculturated, listeners automatically infer metre for stylistically familiar music and use it to coordinate their perception of and movements with music.

Metre is an important factor in determining temporal expectations. Palmer and Krumhansl (1990) used a probe-tone task to assess expectations for event timing in stimuli consisting of non-pitched rhythms implying different metres (2/4, 3/4, 4/4, 6/8) and a final tone whose onset varied systematically in terms of metrical accent. Ratings reflected the hierarchical structure of the metre and also, like tonal hierarchies (Krumhansl, 1990), the statistical distribution of onsets in music, leading Palmer and Krumhansl to suggest that listeners' metrical expectations reflect learned temporal distributions, though the mechanisms of learning were not specified in any detail. Broadly speaking, computational models of metre perception have focused either on traditional rule-based approaches (Lee, 1991; Lerdahl & Jackendoff, 1983; Parncutt, 1994; Povel & Essens, 1985; Temperley, 2001) which are not sensitive to experience, or coupled adaptive oscillator models (Large & Kolen, 1994; McAuley, 1995) and, more recently, neural resonance models consisting of gradient-frequency networks of coupled oscillators (Large et al., 2015), which map neatly onto low-level processes of oscillatory neural resonance in the brain but naturally emphasize an explanation of metre in terms of stimulus structure rather than long-term probabilistic learning (van der Weij et al., 2022).

An alternative approach adopts a Bayesian perspective, by defining a generative model that estimates the likelihood of a rhythm given a metre and uses Bayes' rule to combine this with a prior distribution over the set of possible metres to infer the posterior probability of a metre given a rhythm. The most prominent example is the model proposed by David Temperley (2007), inspired by an earlier rule-based model (Temperley, 2001) which it outperforms in predicting Western time signatures. This model specifies a generative process for metre that uses distinct distributions for the likelihood of note onsets given a metre at each timepoint in each hierarchical level. For example, distributions are specified that determine whether the supratactus and subtactus metrical levels are binary or ternary, with the parameters taken from the distribution of time signatures in the Essen folk song collection (Schaffrath, 1995).

Bastiaan van der Weij developed an empirical Bayesian framework for inferring metre (van der Weij, 2020; van der Weij et al., 2017) that replaces such distributions with a general-purpose model of probabilistic prediction based on statistical learning.[10] The model estimates the *posterior* distribution over all possible metres, $m \in M$, and the phase ϕ with which they can be aligned to a given rhythm x_1^n of length n as specified by Equation 4.1:

$$\overbrace{p(m, \phi | x_1^n)}^{\text{posterior}} = \frac{\overbrace{p(x_1^n | m, \phi)}^{\text{likelihood}} \overbrace{p(m, \phi)}^{\text{prior}}}{\underbrace{p(x_1^n)}_{\text{evidence}}}. \tag{4.1}$$

The *likelihood*—the probability of the rhythm x_1^n given a metre and phase—is generated by IDyOM long-term models trained selectively on rhythms with a particular metre; the *prior* distribution over metre and phase is empirically derived from the distribution in the data; and the *evidence* (reflecting the marginal probability of the rhythm) is a constant for purposes of metrical inference that simply ensures the distribution sums to unity. Equation 4.1 can be expanded to compute the posterior as a recursive combination of the per-event conditional probabilities estimated by IDyOM as shown in Equation 4.2 (where the evidence has been

[10] The framework is available in the Jackdaw software library (van der Weij, 2020): https://github.com/experiencedlisteners/jackdaw. The metrical inference model is available at: https://github.com/experiencedlisteners/jackdaw-tutorial/blob/master/examples/rhythm.lisp.

omitted for clarity):

$$\overbrace{p(m,\phi|\,\mathrm{x}_1^n)}^{\text{per-event posterior}} \propto \begin{cases} \overbrace{p(\mathrm{x}_n\,|m,\phi,\mathrm{x}_1^{n-1})}^{\text{per-event likelihood}}\,\overbrace{p(m,\phi|\,\mathrm{x}_1^{n-1})}^{\text{updated prior}} & \text{if } n > 1 \\ p(\mathrm{x}_n\,|m,\phi)p(m,\phi) & \text{otherwise.} \end{cases} \tag{4.2}$$

The prior used for the first event is computed as the probability of the metre, derived empirically from the corpus, weighted by the duration of the metrical period (bar length), T_m, to avoid metres with long periods being disadvantaged due to the greater number of possible phase alignments that they allow relative to metres with shorter periods:

$$p(m,\phi) = \frac{p(m)}{\sum^m T_m p(m)},$$
$$p(m) = \frac{N_m}{N}$$

where N_m is the frequency of occurrence of m and N the total number of rhythms in the training data. The prior is updated on each step to reflect the posterior distribution for the prefix of the context (e.g., the prior for the last event at index n is the posterior for all events up to $n-1$).

The per-event-likelihood in Equation 4.2 is generated by IDyOM using the `posinbar` viewpoint, which represents the time of an event relative to the start of the bar (i.e., the temporal interval between the start of the bar and the onset of the event; see § 4.1). By virtue of IDyOM's statistical modelling framework both the per-event likelihood and the per-event prior are also conditional on the preceding rhythmic context. The model is also naturally dynamic—the metrical interpretation can vary incrementally event-by-event during online processing, taking into account the previous rhythmic context. The major difference with the model of Temperley (2007) is that a general-purpose model of sequential statistical learning and probabilistic prediction is used rather than zeroth-order or first-order distributions constructed specifically for the task of inferring metre.

Table 4.6 illustrates the performance of the metrical inference model in predicting time signatures in several collections of folk songs. As shown in Figure 4.7, van der Weij et al. (2017) found that the model using metrical inference improves prediction performance for German folk songs, indicated by lower average information content, over an IDyOM model predicting using `ioi`. In other words, having inferred a metre allows the model to make more accurate temporal predictions. These results were replicated by van der Weij (2020, Chapter 7) who also compared the metrical inference model, for which the learned probability of

Corpus	Time signatures	Songs	Accuracy
Canadian	4	149	.96
German	14	4,900	.85
Chinese	10	1,817	.82
Dutch	14	2,269	.74
		9,501	.82

Table 4.6: Classification performance of the metrical inference model (proportion of correctly classified time signatures). The Canadian corpus consists of 152 folk songs and ballads from Nova Scotia (Creighton, 1966). The German and Chinese corpora consist of 5,332 and 2,230 songs respectively from the Essen folk song collection (Schaffrath, 1995). The Dutch corpus consists of 2,903 songs from the Meertens Tune Collections (Kranenburg et al., 2014). Songs with no time signature or multiple time signatures were removed as were those whose time signature was unique in the corpus. The figures in the table reflect these removals.

a note onset is conditional upon the preceding pattern of note onsets interpreted within a given metre, with a *classical* model, for which the probability of an event depends only on a learned association with metrical salience within a bar, given a meter. The metrical inference model exhibited better prediction performance (lower cross-entropy) than the classical model for corpora of European folk songs (German and Dutch) and, especially, for Turkish *makam* music.[11] The metrical inference model also produced plausible simulations of enculturation effects when trained and tested on the music of different cultures; these results are considered in Chapter 7. Of course, time signatures do not necessarily equate to human perception of metre and these computational simulations must be complemented by simulations of empirical data on human metrical inference.

4.3.2 Tonal Inference

Like metre in the temporal domain, tonal structure in the pitch domain is not explicit in music but is inferred by enculturated listeners and differs in both kind and significance between musical cultures. In Western music, tonality provides a scale of importance for the pitches that may appear such that more important pitches tend to be more predictable or stable by virtue of appearing more frequently in salient positions. This differentiation of

[11]A limitation of this approach is that metrical periods are derived from Western time signatures but a related probabilistic approach has been successfully used for style discrimination in Turkish *makam* music, using the corresponding concept of *usul* (Holzapfel, 2015). In Turkish *makam* music, an *usul* is a rhythmic pattern that defines a cyclic period of a certain number of beats in which some beats are more accented than others but with less emphasis on hierarchy, symmetry, and isochronous periodicity than in the Western conception of metre.

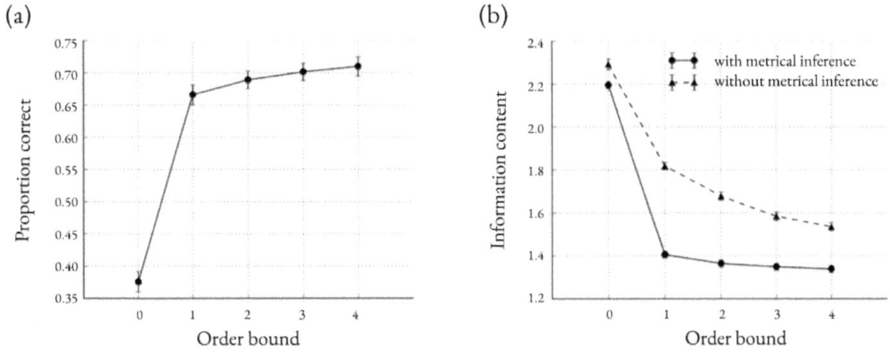

Figure 4.7: (a) Classification performance (proportion of correctly classified time signatures) for metrical inference models varying in order bound. (b) Prediction performance (average information content) for metrical inference models varying in order bound compared with IDyOM models predicting onset using `ioi` without metrical inference. Performance was assessed by 10-fold cross-validation on 4,966 German folk songs from the Essen folk song collection (Schaffrath, 1995). Points represent averages across the 10 folds; error bars represent SD. Reproduced from van der Weij et al. (2017, Figure 3) created by Bastiaan van der Weij and licensed under CC-BY 4.0.

pitches by tonal stability helps listeners generate accurate expectations by relating a new piece of music to patterns learned from previously heard music. The key defines the most stable pitch (e.g., C), while mode (e.g., major) defines the relative stability of the other pitches. Together they specify a *tonal centre* (C in this case), relative to which other pitches are represented (as scale degrees), and a *tonal hierarchy*, which distinguishes the set of possible pitches in terms of tonal stability (Krumhansl, 1990). More stable pitches tend to appear more frequently, more prominently (e.g., with longer duration) and in more salient positions (e.g., phrase endings, metrically strong beats). For example, a piece in the key of C major has C as its most stable pitch (the tonal centre or tonic, not always the most frequent pitch) followed by G, then E (the fifth and third scale degrees, making up the *tonic triad*), then the other pitches in the C major scale (A, B, D, F), then the non-scale tones (C#, D#, F#, G#, A#). An enculturated listener will be able to predict which pitches are more likely to appear based on an implicit schema reflecting an inferred tonality that is derived through statistical learning. In Chapter 3 the tonality was not inferred by IDyOM but taken from the score (though the effect of tonality on IDyOM's pitch predictions was acquired through statistical learning); here the approach is extended to model tonal inference directly.

Computational models of key finding go back to the work of Christopher Longuet-Higgins and Mark Steedman in Edinburgh in the early 1970s (Longuet-Higgins & Steedman,

1971). However, their rule-based model based on scale membership and importance of the tones in a melody leaves open the question of how rules about diatonic scales are internalized by listeners. The Krumhansl-Schmuckler key-finding algorithm (Krumhansl, 1990) estimates tonality by finding the maximum correlation between the duration-weighted distribution of tones in a musical stimulus and reference distributions for the 24 major and minor keys derived from the tone profiles empirically established by Krumhansl and Kessler (1982) for musicians listening to major and minor key contexts.

The use of the Krumhansl-Kessler tone profiles is somewhat unsatisfying from a psychological perspective—since they reflect average ratings from an experiment, it is not clear how they would be derived as psychological representations in the minds of listeners. Nonetheless, this basic idea has dominated research on key finding ever since. Subsequent work has sought to improve performance by: using different tone profiles for comparison (Aarden, 2003; Sapp, 2011; Temperley, 2001), including profiles empirically derived from large musical corpora (Aarden, 2003); introducing segment-based key finding with a mechanism for handling modulations of key within a piece (Temperley, 2001); and using Euclidean distance from a corpus-derived reference profile rather than correlation (Albrecht & Shanahan, 2013). Methods that involve empirical estimation of the reference profile from a corpus (Aarden, 2003; Albrecht & Shanahan, 2013) are especially interesting from a perceptual perspective as they specify a psychological mechanism to account for the acquisition of the reference profiles, suggesting that tonal hierarchies are acquired through statistical learning from distributions of pitches in the music to which a listener is exposed.

As noted by Temperley (2007), a weakness of these approaches is that they treat key finding as an isolated task without specifying how it is integrated in a more general understanding of the perception of music. To address this, Temperley introduced a Bayesian probabilistic model that estimates the posterior probability of a key given the likelihood of a melodic pitch sequence given the key. The generative model used to estimate the likelihood defined a series of distributions that control the selection of a key and, subsequently, the generation of a pitch sequence given that key. Like the key finding algorithms discussed above, the model includes reference tone profiles for the 12 pitch classes in each possible key but adds Gaussian distributions for a central pitch, the pitch range of a melody and melodic pitch interval size, so as to be able to generate specific pitches (see Chapter 1). In the tests reported by Temperley (2007), the rule-based and probabilistic models performed comparably well in predicting the notated keys of common practice music which, as he notes, reflects the fact that they are quite similar in their design.

Corpus	Key signatures	Songs	Accuracy
Canadian	14	151	.83
German	16	5,329	.86
Chinese	12	2,229	.92
Dutch	18	2,903	.91
		10,612	.88

Table 4.7: Classification performance of the tonal inference model (proportion of correctly classified key signatures). The model was trained and tested on each corpus using leave-one-out cross-validation. Songs with no key signature or multiple key signatures were removed as were those whose key signature was unique in the corpus. See Table 4.6 for details of the corpora.

Taking Temperley's argument one step further, we can ask if the Bayesian framework for modelling metrical inference discussed in § 4.3.1 can be extended to simulate tonal inference. The approach, developed by Bastiaan van der Weij, shares many similarities with that of Temperley (2007), the major difference being that the generative model is based not on zeroth-order distributions for selected properties of music such as pitch range, pitch proximity, and tone profiles nor is it limited to Gaussian distributions or fixed reference tone profiles.[12] Rather, the likelihood of a pitch given a key is estimated by a general-purpose variable-order predictive model that acquires domain knowledge through statistical learning. The tonal inference model specifies a generative process in which, given a tonic and mode, scale degree is generated from a distribution conditional on the tonic and mode (i.e., since scale degree depends on the tonic, different PPM models are trained with the source viewpoint `scale-degree` for major and minor modes), then a pitch distribution is specified given the tonic and scale degree (with probability uniformly distributed over the different possible pitches for each scale degree).[13] Following training, the posterior distribution over tonic and mode is estimated given the likelihood derived from IDyOM and an empirical prior over tonic and mode derived from the corpus.

Table 4.7 illustrates the performance of the tonal inference model in predicting key signatures in several collections of folk songs. While this provides a proof of concept, key signatures are compositional devices serving to direct performers rather than reflections of tonality perception. As with the metrical inference model, it remains important to

[12] For example, the tonal inference model does not assume an identical frequency distribution of pitch classes for all keys of the same mode (major, minor), which has been shown not to be the case in analyses of large corpora of European tonal music (Quinn & White, 2017).

[13] The tonal inference model developed by Bastiaan van der Weij uses Jackdaw and is available at: https://github.com/experiencedlisteners/jackdaw-tutorial/blob/master/examples/tonality.lisp.

conduct careful and detailed empirical experiments that compare the model with human perception of tonality, followed by iterative refinement of the model as necessary. However, the present results provide a computational-level demonstration that the perceptual problem of tonal inference can be addressed in the same way as metrical inference using underlying mechanisms of probabilistic processing that have accounted well for perceptual expectations for pitch (Chapter 3), timing (§ 4.1), and harmony (§ 4.2).

4.4 Conclusion

In this chapter, the approach outlined in Chapter 3 has been extended from simulating listeners' melodic pitch expectations to expectations for the timing of musical events and harmonic movement in polyphonic music. In addition, it has been demonstrated that the approach extends naturally to models simulating perceptual inference of latent tonal and metrical structures that condition musical sequences. It is striking that the same process of probabilistic prediction based on statistical learning appears capable of simulating expectations for such a wide range of properties of musical stimuli. In Chapters 5 and 6, the approach is extended beyond expectation to other psychological mechanisms involved in music perception: complexity perception; emotional and aesthetic experience; memory; grouping structure; and similarity.

5 Emotional and Aesthetic Experience

Chapters 3 and 4 addressed predictive perceptual processing of musical structure. However, music is not simply processed in a functional way—it has a rare capacity to induce powerful states of emotion and pleasure. It turns out that these experiences of music can be related directly to probabilistic prediction based on statistical learning. As noted in Chapter 1, the idea that expectation plays a key role in aesthetic and emotional experience of music goes back at least to Eduard Hanslick and Leonard Meyer, writing 100 years apart in the mid-19th and mid-20th centuries respectively. Today, expectation is widely accepted as one of the central psychological mechanisms involved in the evocation of emotion and pleasure by music. And yet, until recently, there was very little concrete, empirical evidence to link prediction with emotional or aesthetic experience. I believe this disconnect between theoretical and empirical enquiry was due, in part, to the lack of an operational model for quantitatively characterizing the expectedness of musical continuations in particular contexts, which meant that research was limited to characterizing the expectedness of stimulus conditions in terms of subjective experience.

This chapter presents research that uses the IDyOM model from Chapter 2 to provide concrete quantitative characterizations of musical expectation in attempting to empirically investigate relationships between probabilistic prediction based on statistical learning and emotional and aesthetic experience of music. Before coming to this empirical research, it is important to establish a theoretical position on how music communicates emotion and evokes aesthetic experiences; this provides the backdrop against which the empirical results can be appropriately interpreted and evaluated. Therefore, in § 5.1, I consider what is communicated by music before examining how music can communicate and induce emotional states (§ 5.2) and how it can evoke aesthetic experience (§ 5.3) from both philosophical and empirical perspectives. Section 5.4 then presents empirical research using IDyOM to investigate the influence of expectation on emotional experience of music, perception of musical complexity, and aesthetic appreciation of music. This is followed by a consideration of psychological and neural mechanisms in § 5.5 and aesthetic experience of familiar music in § 5.6.

5.1 Music as a communicative medium

The idea that music supports expression and communicates meaningfully to an audience (though not necessarily with explicit agency or purpose) has a long history and might, in fact, be considered one of the defining properties that distinguishes music from other kinds of organized sound. But it has proved difficult to pin down exactly what it is that music communicates. As discussed in Chapter 1, Leonard Meyer (1956) considered two philosophical traditions of thinking on this question: *formalism* and *referentialism*. Referentialists believed that musical communication is primarily a process of designating non-musical entities (natural scenes, animals, characters, psychological states, and so on) while formalists believed that music communicates information in terms of relations between purely musical entities, which are apprehended intellectually by a listener familiar with the corresponding stylistic conventions. As a formalist, Hanslick (1854) argued that the aesthetic function of music is not to induce emotion: music cannot represent definite feelings (which have objects) because it cannot represent the objects that support those feelings; it can represent dynamic changes in intensity but not as properties of specific emotions since other phenomena also share such dynamic changes.

Meyer (1956) argued that this debate had been founded on a false assumption that referentialism is associated exclusively with emotional *expressionism*—the idea that aesthetic experience of music has to do with the expression and recognition of emotional states. Meyer pointed out that the fact that referentialists tended to be expressionists does not render expressionism incompatible with formalism in principle. Equally, the fact that prominent formalists such as Hanslick emphatically denied the role of emotion in musical aesthetics (though not that music can induce emotion) does not necessarily render formalism incompatible with expressionism. Hodges (2017) distinguished these different perspectives along two orthogonal dimensions: first, *absolutism* (meaning is found within the music) contrasted with referentialism (meaning is found in extra-musical referents); and second, formalism (meaning arises from appreciation of musical relationships) contrasted with expressionism (meaning arises from the appreciation of the emotion expressed by the music). As shown in Table 5.1, this allows an accommodation between the two traditional positions of absolute formalism and referential expressionism, allowing for Meyer's absolute expressionism, while

	Absolutism	Referentialism
Formalism	Hanslick	—
Expressionism	Meyer	Wagner

Table 5.1: Distinctions between different philosophical positions about the way in which music communicates meaning, inspired by Hodges (2017, Table 7.3).

also accounting for the conflict between these positions in terms of the impossibility of formal referentialism. Meyer sought to develop an absolutist account of emotional expression in music founded on the psychological process of expectation which creates patterns of tension and resolution in a listener's experience of music. These patterns, in turn, generate affective states such that violations of expectation are negatively valenced, indicating predictive failure, while subsequent confirmation of those expectations communicates positively valenced states of resolution (Meyer, 1956, p. 27). Meyer argued that these mechanisms can communicate a range of emotional states including doubt, confusion, apprehension, fear, anxiety, disappointment, hope, relief, and pleasure.

The philosophical position taken in the empirical work described in § 5.4 below is essentially one of absolute expressionism. However, a modern pluralist perspective allows for multiple ways (encompassing both formalism and referentialism) in which music can evoke emotional and aesthetic experiences. Therefore, §§ 5.2 and 5.3 take a broader perspective on the psychological mechanisms involved in emotional and aesthetic experience of music, so as to paint the background against which the empirical studies using IDyOM can be properly interpreted.

5.2 How does music evoke emotions?

There is an important distinction to be made between the recognition of an emotion expressed by music and the induction of an emotional state through listening to music. Listeners are often able to recognize the emotional connotations of instrumental music but do not necessarily experience the emotion that is recognized. Meyer includes both phenomena in his discourse but here we are principally interested in emotional experience rather than recognition of the emotions expressed by music. Although he doesn't explicitly discuss it in these terms, Meyer's characterization of the debate between formalism and referentialism leaves the door open for an explicitly plural theory of emotion induction by music that accommodates the potential involvement of multiple psychological mechanisms.

Such a theory has been developed most comprehensively by Patrik Juslin and Daniel Västfjall (Juslin, 2013, 2019; Juslin & Västfjäll, 2008) who distinguished a number of different mechanisms other than expectation.[1] For present purposes, we are interested in mechanisms that enable emotional experience to be related to specific structural properties of music in a way that generalizes across listeners. However, many of the mechanisms covered by Juslin and Västfjäll are specific to particular individuals and not even critically dependent on music being the triggering stimulus. For example, repetitive pairing of a piece of music with another affective event can lead to a conditioned emotional response to the music. Furthermore, music may evoke episodic memories, triggering explicit recall of specific events from the past and the emotional state associated with them. These mechanisms involve veridical memory representations of music (see Chapter 3) and seem to be completely agnostic to the nature of the music that evokes the emotion.

Other mechanisms do involve musical structure but only to the extent that this structure matches the structure of non-musical phenomena. For example, music may evoke emotions through mimicking the vocal or behavioural expression of emotional states (e.g., fear or tiredness) or it may activate imagery for non-musical phenomena through associations between those phenomena and the structure of the music (which may be concrete or abstract—e.g., the storm in the fourth movement of Beethoven's sixth symphony compared with the more abstract pastoral imagery of the first movement). This is the kind of mechanism invoked by the 19th-century referentialists but it seems highly unlikely that these mechanisms can be generalized to cover every possible kind of musical structure (not to mention all possible non-musical referents).

Finally, there are mechanisms that are related to processing time-varying acoustic and structural properties of a musical stimulus. These include affective responses resulting from auditory processing of sudden, loud, dissonant, or rapidly pulsing sounds (an obvious example being the surprise in the second movement of Haydn's Symphony No. 94) and entrainment of physiological or behavioural processes (e.g., respiration or dancing) to the pulse of music. This category would also include expectations based on sensory memory (see Chapters 3 and 4). These mechanisms evoke rather stereotyped responses to specific characteristics of sounds in general rather than the flexible culture dependent effects of expectations derived from learned stylistic syntax.

There would seem to be no doubt that when listening to a piece of music more than one of these psychological mechanisms may be invoked simultaneously. Indeed, this may

[1] For empirical support, see Juslin et al. (2015, 2022).

be one of the reasons that music proves to be so rich and expressive but also difficult to characterize in precise communicative terms. However, the considerations outlined above suggest that expectation is the mechanism of greatest interest to the present enquiry, since it can be most concretely linked to perception of culture-specific aspects of musical structure.[2] Therefore, for present purposes we are interested specifically in the influence of expectations on emotional experience. To be even more precise, following the statistical learning and probabilistic prediction hypotheses from Chapter 1, we are interested in the influence of expectations derived from learned cognitive representations of musical structure, whether these reflect the syntactic structure of a musical style or the internal structure of an individual piece. As we shall see in the next section, such expectations also play an important role in the aesthetic experience of music.

5.3 Aesthetic experience

5.3.1 What is aesthetic experience?

It is important to be clear about what is meant by aesthetic experience and how it relates to emotional states that might be evoked by a piece of music. This is thorny philosophical territory that has attracted interest and debate across many centuries of scholarly thinking. My position here takes inspiration from a sketch written by Jerrold Levinson (2009) which presents a lucid formulation of aesthetic appreciation and how it relates to the concepts of aesthetic attitude, aesthetic attention, aesthetic satisfaction, and aesthetic experience. For present purposes, it also has the advantage of being conceived with music specifically in mind, though it is in no way specific or limited to music.

According to Levinson, *appreciation* means experiencing something in a way that is intrinsically pleasurable, which distinguishes it from merely perceiving music without finding it pleasurable or finding music pleasurable without experiencing it at all or experiencing it only incidentally (e.g., as background music). *Aesthetic appreciation* is appreciation of the form and content of music rather than appreciation arising from any other source, including the external uses or consequences of listening to music. I will come to what is meant by form and content and the distinction between them below. Levinson is careful here not to exclude musical form and content that has an (intended) external function which might

[2]This is not to say that the other mechanisms are any less important or interesting in a broader sense, nor that they wouldn't exhibit cross-cultural variation.

be, for example, religious, patriotic, celebratory, or memorial. As discussed below, while aesthetic appreciation is pleasurable, the experience that feeds it might not be.

Aesthetic appreciation comprises two components: *aesthetic experience* and *aesthetic estimation*. An *aesthetic experience* is one that involves *aesthetic attention* to and *aesthetic satisfaction* from a piece of music. This leaves three components to understand: aesthetic attention, satisfaction, and estimation. If one is to appreciate an experience of the form and content of music, attention must first be directed, to some extent, towards that form and content. *Aesthetic attention* can be characterized either in terms of the directing of attention towards the form and content of music directly or the aiming to achieve a fully aesthetic experience of the music, or both of these. My own interpretation of aesthetic attention is that it doesn't necessarily require conscious intellectual focus on the music but rather that one is listening to the music in such a way that one can perceive (though not necessarily verbalize) its form and content. *Aesthetic satisfaction* is the pleasure that derives from directing aesthetic attention to music.[3] I would suggest that aesthetic satisfaction is distinct from the emotional content of the music and its emotional impact, such that satisfaction can arise from music that has negatively valenced content or impact on the listener. In this context, one can understand the *aesthetic attitude* (Dickie, 1964; Kemp, 1999) as being a set of circumstances, both psychological and contextual, that encourage attention to be focused on the form and content of the music such that an aesthetic experience becomes more likely, without making the adoption of an aesthetic attitude a necessary component of the experience itself. Finally, the satisfaction resulting from experience of the content and form may lead to a positive *estimation* (i.e., an explicit judgement) of the music, though it need not always do so.

This is an epistemic vision of aesthetic experience rather than a phenomenological one (defining it in terms of how it arises rather than in terms of properties of the experience itself) and therefore avoids prescribing aesthetic experience in terms of psychological states that may not generalize across individuals, cultural domains, and societies (Iseminger, 2003). At the same time, it does not deny that aesthetic experience has an absorbing and evocative phenomenological dimension as emphasized by authors such as Shusterman (1997). While it focuses on the form and content of the aesthetic object (music in this case), thus distinguishing aesthetic experience from non-aesthetic pleasurable sensory experiences, it is conceived in broad enough terms to accommodate aspects of content beyond the strictly formal such as imaginary movement, gesture, and expression as well as form and content that does have

[3]Levinson acknowledges that aesthetic attention can lead to experiences that are neutral or dissatisfying but finds it convenient to associate, by default, aesthetic experience with aesthetic satisfaction. I follow suit here.

external function. The sketch is also agnostic as to whether an aesthetic experience has intrinsic or instrumental value, or both (Budd, 1995; Carroll, 2000). Finally, it does not rely on any pre-existing theory about which aspects of form and content make up the focus of aesthetic experience, allowing it to account for aesthetic experience of natural phenomena as well as aesthetic experiences across cultures and historical periods.

Given the above characterization of aesthetic experience, questions naturally arise about which aspects of form and content provide the focus of that experience, with an understanding that the answer is likely to depend on the cultural experience of the listener (and, therefore, may differ across musical cultures and over time). Regarding the formal content of music, Levinson (1997) has promoted a view known as *concatenationism* which has been contrasted with the *architectonicism* of writers such as Peter Kivy (2001). According to the architectonicist perspective, aesthetic experience derives from appreciation of the large-scale tonal and thematic structure of a piece of music. The concatenationist perspective, by contrast, posits that aesthetic experience derives from the perception of limited portions of the music, held in memory during the listening process, and concatenated together in a chain of connected experiences, rather than overall perception of the global form of the music. Of course, these positions are not mutually exclusive (Levinson, 2006) but it should be clear from the preceding chapters that concatenationism more closely characterizes the kind of musical listening simulated by IDyOM.

Turning to musical content, Levinson (2009) suggests that small-scale musical form (i.e., changes in melody, rhythm, harmony, dynamics, and timbre) give rise to imagined movement (e.g., soaring, swelling, shrinking, or subsiding), which in turn gives rise to mimicry of human behavioural gestures (e.g., sighing or scowling) which in turn gives rise to music's ability to express psychological, in particular affective, states (e.g., sorrow or serenity). This expressive aspect of content is echoed by the contagion mechanism described by Juslin and Västfjäll (2008). Levinson also identifies three non-expressive aspects of musical content: first, non-expressive higher-order perceptual qualities such as delicacy that do not express a psychological state; second, narrative-dramatic content, going beyond expressivity as in narrative musical genres such as opera, though the narrative need not always be explicit; third, representational content, whereby music evokes imagery for some person, object, or event, echoed by the visual imagery mechanism of Juslin and Västfjäll (2008). Note, however, that nothing in Levinson's account requires that aesthetic appreciation involves feeling a specific emotion induced by the music or even the recognition of an emotion expressed by the music (though it may do). The pleasure that characterizes aesthetic appreciation results from experiencing the form and content of the music, not a specific emotional state that results from listening to the music. Conversely, it has been suggested that aesthetic appreciation could itself induce particular kinds of emotional state, which Juslin (2013) terms *aesthetic emotions*.

As noted above, the focus here is exclusively on the absolute content of music but it is useful to situate the empirical work presented below within a broader context. Doing so clarifies both the significance of the absolute expressionist approach but also the work that ultimately needs to be done to weave it into a fully-fledged psychological theory of musical aesthetics at the same level of detail. In Levinson's explicitly plural sketch, the focus on form can be motivated as providing the primary content of music from which other aspects of content are derived. In Juslin and Västfjäll's survey of psychological mechanisms, expectation based on learning of absolute properties of music can be motivated as the mechanism that is most specific to the structure of music and, therefore, open to the influence of enculturation.[4]

For an absolutist, the focus on form in aesthetic experience is self-evident: In one of his last essays, Peter Kivy (2017) arrived at the conclusion that the essential power of absolute music lies in internal repetition within a piece of music. According to the statistical learning hypothesis from Chapter 1, such repetition allows a listener to develop a coherent internal cognitive model of the thematic structure of an individual piece of music (see also § 6.3.3). According to the same hypothesis, it is repetition of musical structure, across many musical pieces, that allows a listener to develop, over the long-term, a coherent internal model of the syntax of a musical style. This in turn allows an individual to appreciate, during listening, the extent to which the music conforms to or departs from conventional stylistic structures. Putting these mechanisms together, music may confirm or deny expectations derived from a long-term model of stylistic syntax and a short-term model of thematic structure. These specific aspects of musical form will be investigated in § 5.4 with regard to the aesthetic satisfaction they evoke. Having set the philosophical scene, the next section considers empirical research on aesthetic experiences of music.

5.3.2 Empirical aesthetics of music

An interest in musical aesthetics is evident in the very beginnings of experimental psychology. Hermann von Helmholtz's *On the sensations of tone as a physiological basis for the theory*

[4]Although note that the underlying mechanism of statistical learning responsible for acquiring internal models of musical structure generalizes across musical cultures.

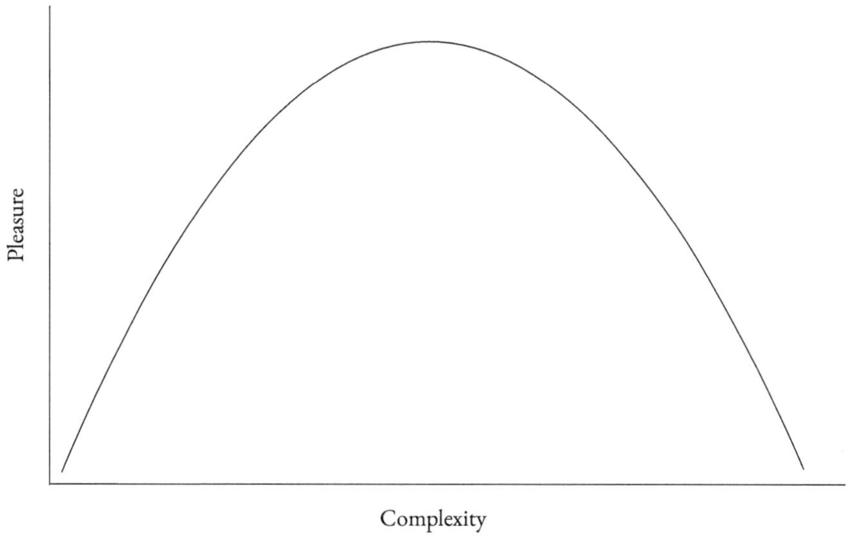

Figure 5.1: The hypothesized inverted-U shaped relationship between complexity and pleasure (or Wundt effect).

of music (1863) was inspired in part by Hanslick's formalist aesthetic. In the third part of his book, Helmholtz tentatively associated the aesthetic qualities of musical notes and scales with the physical properties of tones and their processing in the peripheral auditory system, relating some of his observations specifically to an implicit psychological process of expectation: "[The listener] is startled by a false cadence and feels its unexpectedness but is not at all necessarily conscious of the reason of its unexpectedness" (von Helmholtz, 1863, p. 369). Wilhelm Wundt, a founder of modern experimental psychology and one-time assistant of Helmholtz in Heidelberg, developed a more psychological approach to the study of aesthetics. For example, Wundt introspected about how his own sensations of pleasure, tension, and excitement varied with the tempo of a metronome pulse (see Miller & Buckhout, 1973). Elsewhere, he demonstrated that physiological arousal is related to stimulus complexity and argued that aesthetic pleasure is maximal at intermediate degrees of optimal complexity (Wundt, 1874). This so-called *Wundt effect* produces an inverted-U shape when complexity is plotted on the abscissa against pleasure on the ordinate (see Figure 5.1).

Daniel Berlyne (1971) further developed this relationship between complexity and aesthetic experience, proposing two complementary physiological mechanisms linking arousal and pleasure in different directions and with different thresholds. Together, these mechanisms generate an inverted-U shaped function linking the 'arousal potential' of a stimulus with its 'hedonic value' such that intermediate degrees of arousal correspond to maximum pleasure. As noted by Huron (2006), this inverted-U shaped relationship accounts for both the mere exposure effect (Gordon & Holyoak, 1983; Zajonc, 1968), whereby preference increases with increasing exposure, and a boredom effect (Cantor, 1968), whereby preference decreases with increasing exposure, by positing different initial levels of stimulus-related arousal. Another way of characterizing the Wundt effect distinguishes between subjective and objective complexity. While pleasure is greatest at intermediate levels of subjective complexity, there might be individual differences in the relationship between subjective complexity and objective complexity depending on factors such as familiarity, expertise, and psychological state (Heyduk, 1975; Orr & Ohlsson, 2005). This potentially allows for changes in these factors to alter the relationship between subjective pleasure and objective measures of complexity (see § 8.3).

In general, there have been two experimental approaches to addressing these phenomena in the empirical aesthetics of music. The first, drawing on traditional experimental psychology, creates artificial stimuli in which certain stimulus features are systematically manipulated while others are carefully controlled. Paul Vitz (1972), for example, found inverted-U shaped relationships between preference and the frequency or intensity of individual sine tones. Following an earlier study by Vitz (1966), Crozier (1974) established that complexity and interestingness increase linearly with stimulus uncertainty, while pleasantness and beauty show an inverted-U shaped relationship with uncertainty. The stimuli were computer generated tone sequences varying in the number of possible pitches, durations, and loudness levels across five levels of uncertainty quantified in terms of entropy (e.g., for simple sequences, every tone had the same loudness and duration but one of two randomly selected pitches while for complex sequences each tone was randomly assigned one of 18 pitches, eight durations, and four loudness levels). Using a more musically sensitive approach, Heyduk (1975) composed four pieces for piano in which complexity was manipulated by varying the number of chords and degree of rhythmic syncopation while keeping other factors constant. Ratings of perceived complexity increased across the four pieces confirming the experimental manipulation while liking showed an inverted-U shaped relationship with increasing complexity.

One can argue about how convincingly these studies corroborate an inverted-U shaped relationship between complexity and aesthetic pleasure (see Chmiel & Schubert, 2017, for a comprehensive review). For example, the relationship reported by Heyduk relies critically on the fourth piece being liked slightly less than the third one; the relationship between intensity and preference reported by Vitz (1972) depends crucially on the quietest sine tones being preferred slightly less than the next quietest of six different levels. Nonetheless, the experimental approach allows for confirmatory hypothesis testing and provides clearly interpretable results that permit relations to be drawn between specific stimulus properties and aesthetic experience. On the other hand, the approach is extremely limited in terms of external and ecological validity, since the stimuli do not resemble real music and therefore the results may not generalize to aesthetic experience of real music. In many areas of psychology, ecological validity is not of paramount importance but in the aesthetic experience of cultural phenomena such as music, the nature of the stimulus gains greater significance. If the stimuli do not resemble real music, how can we be sure that the results generalize to aesthetic experiences of real music?

The second approach ensures ecological validity by using actual music as stimuli, in some cases selected by individual participants to have particular aesthetic effects, or otherwise rated by listeners in terms of such effects. For example, Orr and Ohlsson (2001, 2005) used specially composed jazz and bluegrass improvisations to examine relationships between subjective complexity and liking in non-musicians and musicians specializing in jazz and bluegrass. The results showed a Wundt effect only for non-musicians listening to bluegrass suggesting effects both of musical style and musical training. However, it is difficult to understand the psychological mechanisms underlying these effects in any detail since complexity was assessed only using subjective ratings. Another good example of this approach is provided by the well-known studies of music-evoked chills from Robert Zatorre's lab (e.g., Blood & Zatorre, 2001; Salimpoor et al., 2011). Participants provided the pieces of music that most consistently elicited chills for them personally, and these pieces were used as stimuli in neuroimaging experiments.

While this approach ensures ecological validity and (at least when the stimuli are self-selected by participants) genuine aesthetic experiences, the problem is that it tells us very little about the stimulus properties (and how they are processed psychologically) that give rise to these experiences. Computational modelling provides a potential solution that allows a fruitful combination of ecological validity with quantitative control over stimulus properties and the hypothesized psychological mechanisms involved in processing them. An example of this approach is the use of continuous acoustic features derived from the audio signal (e.g., brightness and spectral complexity) as predictors of voxel time series derived from fMRI during listening, enabling the identification of brain regions responsive to those features (Alluri et al., 2012; Burunat et al., 2016). However, such acoustic features are rather different both in kind and in scale from the kind of syntactic stylistic knowledge upon which melodic, rhythmic, and harmonic expectations are based, according to the statistical learning and probabilistic prediction hypotheses. The next section uses IDyOM as a hypothesis-driven computational model of perceptual processing of music to bridge the knowledge gap between musical structure and aesthetic musical experiences.

5.4 Simulating listeners' emotional and aesthetic experience of music

5.4.1 Expectation and emotional experience

In spite of its widespread theoretical influence, there was until recently rather little concrete experimental evidence that specifically supported Meyer's proposed relationship between expectation and emotional experience of music. Koelsch et al. (2008a) played participants five-chord sequences in which the final chord was either stylistically conventional (a tonic chord) or unconventional (a Neapolitan sixth chord) asking them to rate how pleasant they found the final chord. Unconventional chords were experienced as significantly more unpleasant than conventional chords. Using more ecologically valid stimuli, Steinbeis et al. (2006) recorded skin conductance and subjective tension as measures of physiological arousal while participants listened to excerpts from six of J. S. Bach's chorales in which the final chord was stylistically unconventional as a continuation of the context (i.e., distant on the circle of fifths from both the home key and the immediately preceding tonality) as well as versions that were manipulated to make the final chord more conventional (i.e., a return to the tonic) or even less conventional (a Neapolitan sixth). Continuations that were less stylistically conventional elicited greater tension and emotionality ratings, accompanied by increases in skin conductance. Koelsch et al. (2008b) replicated the effects of skin conductance for excerpts from 25 classical and romantic piano sonatas (some of which were expressively performed).

The lack of any quantitative measure of stylistic conventions makes aspects of the results of these studies difficult to interpret. For example, Steinbeis et al. (2006) found a significant difference in skin conductance between conventional and unconventional continuations but not between unconventional and very unconventional continuations whereas Koelsch et al. (2008b) found a significant difference between very unconventional and unconventional continuations but not between unconventional and conventional continuations. Without an operational measure of stylistic convention, it is impossible to interpret these differences between stimulus categories across experiments. More importantly, since these studies relied on the music-theoretic expertise of their authors to construct stimuli differing in stylistic conventions for continuation, they provide very little detail about the psychological mechanisms involved. Finally, these studies involve intensive repeated listening over a short period of time to a small number of artificially recomposed musical fragments. For example, the EEG experiment of Steinbeis et al. involved presenting a single phrase from six different chorales, each in three versions differing by a single chord, each repeated six times: a total of 108 stimuli, consisting of 18 repetitions of each phrase in one version or another. While such repetition is necessary to achieve acceptable signal-to-noise ratio in the analysis of the experimental manipulation, it is far removed from the conditions under which music is usually appreciated.

To address issues such as these, Hauke Egermann ran an experiment to examine whether changes in melodic predictability assessed by IDyOM are associated with characteristic changes in emotional state (Egermann et al., 2013). A collection of six monophonic works for solo flute were assembled covering styles ranging from the Baroque to 20th-century non-tonal music: *Acht Stücke für Flöte allein: VI.* by Paul Hindemith, *Un joueur de flûte berce les ruines* by Francis Poulenc, *Density 21.5* by Edgar Varèse, *Syrinx* by Claude Debussy, and the second and third movements from J. S. Bach's *Partita in A minor for solo flute*. IDyOM was used to estimate the information content of each note in each of the stimuli as a measure of its unpredictability. IDyOM was configured to use both the long-term and short-term models with the long-term model trained on the corpus shown in Table 3.2. Since the stimuli range broadly from firmly tonal to distinctly non-tonal, IDyOM was configured to predict pitch using chromatic pitch (`pitch`) as a source viewpoint. A tonal representation such as scale degree would not be appropriate for a piece such as Density 21.5 and, although it might be appropriate to varying degrees for other pieces in the stimulus set, a common pitch representation was used for consistency in the analysis across pieces.

To ensure ecological validity, continuous subjective and physiological measures of emotional state were collected while 50 participants listened to a live performance of four of the stimuli, having first listened to recordings of the pieces by Hindemith and Poulenc to familiarize themselves with the experimental task. Half of the participants gave continuous ratings of expectedness while the other half gave continuous ratings of their emotional state on a two-dimensional scale of valence (positive, negative) and arousal (calm, excited). Continuous physiological recordings of skin conductance, heart rate, and respiration rate were

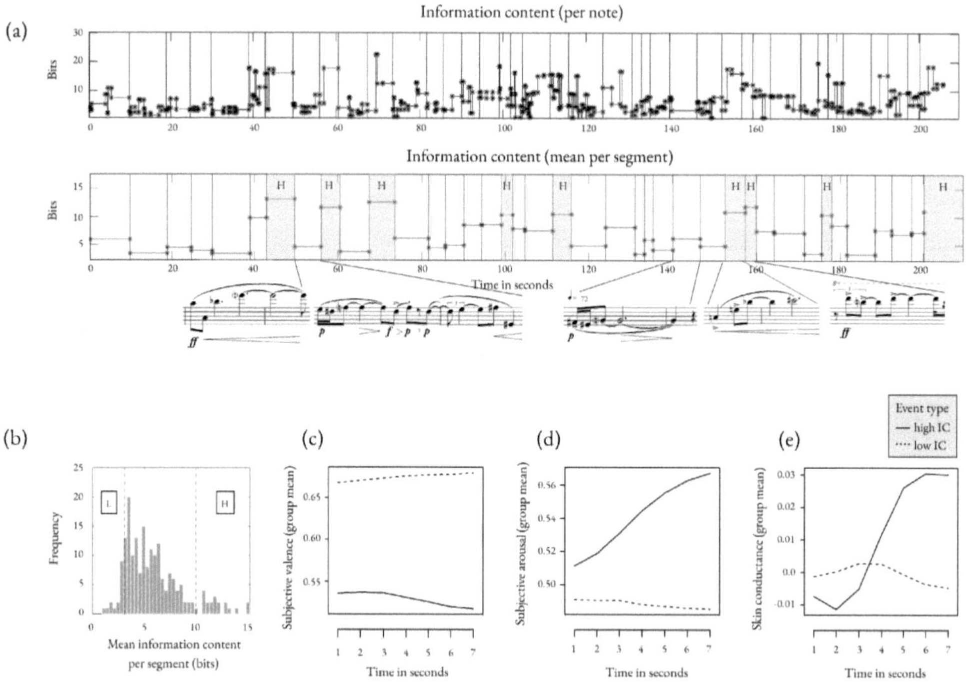

Figure 5.2: Analysis and results from Egermann et al. (2013). (a) An example of the analysis strategy for one of the stimuli, *Density 21.5* by Edgar Varèse. The upper panel shows note-by-note information content values resulting from the IDyOM simulation. The second panel shows information content averaged over segments identified through musicological analysis with segments labelled as high (H) or low (L) information content (being the most unpredictable stimulus, this piece only contains high information content segments). Score extracts are shown for selected segments for purposes of illustration. (b) Histogram of mean information content for all segments across all pieces illustrating the selection of high (H) and low (L) information content segments as the highest and lowest decile of the distribution respectively. (c) Mean subjective valence as a function of time for high and low information content segments (solid and dashed lines respectively). (d) Mean subjective arousal as a function of time for high and low information content segments. (e) Mean skin conductance as a function of time for high and low information content segments. Adapted from Egermann et al. (2013, Figures 1, 2, 4, and 5), created by Hauke Egermann and reproduced with permission from Springer Nature.

taken from all participants. To accommodate any lag between predictable and unpredictable events in the music and the continuous recordings of emotional state (both subjective and physiological), the analysis was conducted at the level of musical motifs identified by a trained music theorist rather than individual notes (see Figure 5.2). The mean information content of each segment was used to select highly predictable and highly unpredictable segments.

Linear mixed-effects analyses showed a significant increase in arousal and decrease in valence for high information content segments compared with low information content segments coupled with a significant increase in skin conductance. These results, summarized in Figure 5.2, demonstrate that unpredictable melodic events, identified using a computational simulation of probabilistic prediction based on long- and short-term statistical learning, induce increases in subjective and physiological arousal. The effect of valence was much weaker and the negative trend may reflect the fact that many of the unpredictable high information content events came from Density 21.5, which would have been stylistically unfamiliar to most of the participants (just as it was to IDyOM trained on a corpus of tonal melodies). Importantly, these results can be generalized naturally to real-life musical experiences across a range of Western musical styles and historical periods.

These empirical results shed light on the relationship between objectively simulated computational measures of predictability and continuously varying emotional state while listening to music. While they provide an account of arousal, they do not give any insight into more specifically aesthetic experiences such as tension and resolution. Addressing this, a pair of studies investigated the relationship between unpredictability, performance timing, and tension using as a stimulus the *Prélude non mesuré No. 7* by the French Baroque composer Louis Couperin (Gingras et al., 2016). Being unmeasured (i.e., without explicit timing information encoded in the score), this piece of music offers great scope to the performer for expressive timing and being performed on a harpsichord limits expression largely to timing rather than dynamics. IDyOM was used to characterize the structure of the piece in terms of the unpredictability (information content) and uncertainty (entropy) associated with each note. IDyOM was configured with its long-term component only, trained on the corpora shown in Table 3.2 but also allowed to learn dynamically while processing the prelude (the LTM+ configuration, see Table 2.1). IDyOM made pitch predictions using a single source viewpoint linking pitch interval and scale degree (`interval⊗scale-degree`).

The first study run by Bruno Gingras investigated variations in performance timing by 12 professional harpsichordists, in terms of the inter-onset interval (IOI) between notes. The results of a time-series analysis indicated that information content and entropy were predictive of continuous changes in performance timing such that performers slowed down in anticipation of high information content events, and timing was more variable across performers around points of high information content and entropy. In a second study, Meghan Goodchild collected continuous tension judgements for the 12 performances of the piece from 10 harpsichordists, 20 non-harpsichordist musicians, and 20 non-musicians. Consistently for all groups, the time-series analysis showed that while information content and entropy were predictive of tension, they did not account for any additional variance above and beyond that already accounted for by performance timing (IOI). This paints a picture of a communicative process in which music's compositional structure (its *form* in Levinson's terms) as captured by IDyOM is interpreted by performers and communicated directly in terms of expressive timing to listeners, who experience it in terms of variations in tension and resolution.

This research is interesting because it encompasses compositional form, expressive performance, and aesthetic experience. However, it covers only one aspect of aesthetic experience—tension—without touching on aesthetic appreciation per se and extends to only a single piece of music. We do not know if these results would generalize to other musical works in different styles, played on different instruments. In an attempt to understand aesthetic appreciation more broadly, we can examine the pleasure arising from aesthetic experience and how it relates to the underlying stylistic predictability of musical structure. This can be achieved across a broader range of music by expanding the focus from individual notes to the predictability of whole pieces of music. In this case, it is important to evaluate whether aggregate information-theoretic measures of unpredictability (i.e., information content averaged across the notes in a composition) do meaningfully simulate a perceived aesthetic property of the composition. In the next section we focus on perceived complexity which, as discussed in § 5.3.2, has a theoretically and empirically established relationship with aesthetic experience.

5.4.2 Complexity

Complexity is a core concept in empirical aesthetics research across modalities with theoretical proposals suggesting that optimal aesthetic appreciation is associated with intermediate degrees of complexity (Berlyne, 1971). If we hypothesize that unpredictable musical passages will be perceived as more complex, an inverted-U shaped relationship is consistent with Meyer's proposal that both violation of expectations and their subsequent fulfilment are important components of aesthetic experience. Therefore, pieces of music with intermediate degrees of unpredictability on average should be experienced as most pleasurable. However, the link between predictability and perceived complexity should not be assumed, it must be empirically demonstrated.

Ana Clemente composed a collection of 200 musical stimuli whose aesthetic properties were systematically varied (Clemente et al., 2020). The collection includes a set of 50 stimuli varying continuously in complexity (number and variability of melodic and rhythmic elements) while controlling other formal properties such as length (two bars, 4 s), style

(Western tonal), timbre (piano), dynamics (flat), balance (balanced, in terms of distribution of notes towards the beginning, middle, or end of the melody), symmetry (symmetrical, such that the second half of each stimulus is a mirror image of the first half), and contour (moderately smooth in both rhythm and melody). Forty-three European non-musicians gave ratings of perceived complexity for this stimulus set and their ratings were simulated by four models: 1) the Free Lossless Audio Codec (FLAC) compression algorithm applied to the raw audio files; 2) a feature-based model called EV_4 (Eerola, 2016), comprising a linearly weighted combination of four features: pitch proximity, tonal ambiguity, rhythmic variability, and pitch variability; 3) a feature-based model derived by taking features reflecting the compositional variation of complexity in the stimuli (the number and variability of pitches and durations), optimizing them using principal components analysis and fitting the two resulting principal components (reflecting: first, event density and pitch variability; and second, rhythmic variability) to the empirical complexity ratings; 4) IDyOM configured to use the short-term model only predicting pitch using a single viewpoint linking pitch interval with scale degree (`interval⊗scale-degree`) and onset using the single viewpoint IOI-ratio (`ioi-ratio`), then using the summed information content of each event in a stimulus as a measure of its complexity.[5] The results of the analysis showed that IDyOM, without any prior optimization for these stimuli, fit the complexity ratings better than the other models, albeit not significantly in the case of the feature-based model that was fitted to the stimuli and complexity ratings.

The manipulations of complexity in Clemente et al. (2020) were all stylistically congruent. However, a study run by Sarah Sauvé examined whether IDyOM's predictability estimates extend to variations of complexity covering much wider variation of stylistic congruence (Sauvé & Pearce, 2019). The study was causal in that the melodic, rhythmic, and harmonic structure of a two-bar stimulus for three voices was recomposed in order to generate systematic variation across eight levels of information content, according to IDyOM, separately for each parameter of interest (melody, rhythm, harmony) while holding the information content of other parameters as constant as possible. Twenty-eight participants with a moderate level of musical training provided complexity ratings for the stimuli, which were accurately simulated by a combination of melody and harmony information content, accounting for 54% of the variance in the ratings. By comparison, the feature-based EV_3

[5]The use of summation here rather than averaging reflects the fact that the stimuli varied widely in number of notes and this influenced complexity perception. The use of the short-term model reflects the stylistic homogeneity of the stimuli.

IDyOM configuration	mean IC	sum IC	mean entropy	sum entropy	sum IC * mean entropy
STM	.58	.68	.67	.63	**.8**
LTM	.12	.62	.32	.57	.73
BOTH	.14	.63	.29	.57	.73

Table 5.2: Regression coefficients between perceived complexity for 40 drum rhythms (Senn et al., 2023) and information-theoretic measures resulting from IDyOM simulations. Each row shows a different IDyOM configuration: short-term model (STM); long-term model (LTM); and both STM and LTM (BOTH). Each column shows different predictor variables used: mean information content (mean IC); summed information content (sum IC); mean entropy; summed entropy; and an optimal combination of these variables determined by forward stepwise selection, which retained as predictors summed information content, mean entropy, and the interaction of these two predictors. The short-term model provided the most accurate simulation of perceived complexity, accounting for approximately 61% of the variance ($R = .8$, $R^2 = .64$, $R^2_{adj} = .61$) with summed information content and mean entropy combined as predictors (see first row, final column, highlighted in bold).

model consisting of predictors for pitch proximity, tonal ambiguity, and rhythmic variability (Eerola, 2016), accounted for a lower proportion of the variance (22%).

The relationship between information content and complexity also holds for a very different aspect of music. Senn et al. (2023) collected measures of perceived complexity from 220 participants for 40 four-bar drum patterns taken from rock, pop, funk, and soul tracks (including, for example, Nine Inch Nails, The Police, The Meters, and Jamiroquai) selected from a larger collection (the Lucerne Groove Corpus; Senn et al., 2018) to have consistent instrumentation and cover a broad range of complexity. IDyOM was applied to this stimulus set (in quantized form) in three configurations including the short-term model, the long-term model, and both combined. The long-term model was trained on the remaining 211 rhythms from the Lucerne Groove Corpus (Senn et al., 2018) and 1,215 rhythms from the Loop Loft corpus (Hosken et al., 2021). The simulation results are shown in Table 5.2. The optimal IDyOM configuration combined summed information content with mean entropy from the short-term model to account for 61% of the variance in perceived complexity.[6]

Together, these results show that IDyOM accounts rather well for perceived complexity of music, suggesting that complexity perception can be simulated accurately in terms of

[6]The IDyOM simulations described here were run on mechanically timed versions of the drum rhythms, whereas the versions used as stimuli by Senn et al. (2023) contain expressive timing. Tom Kaplan has shown that combining a predictive model of expressive timing (PIPPET: Cannon, 2021; Kaplan et al., 2022) with IDyOM produces an improved simulation of perceived complexity (Kaplan, 2024, Chapter 6).

predictability resulting from an underlying process of probabilistic prediction based on statistical learning.

5.4.3 Aesthetic experience

Having established that information content and entropy accurately simulate perceived complexity, is there any evidence that IDyOM can account for hypothesized relationships between complexity (operationalized as information content and entropy) and the aesthetic experience of music? The results of Egermann et al. (2013) demonstrate that greater predictability is associated with more positively valenced emotional states (see also Sauvé et al., 2018). But as argued above, aesthetic appreciation is to be distinguished from the emotional state induced by listening to music. To address whether expectation-related complexity influences aesthetic experience, Ben Gold ran a study in which 44 participants rated how much they liked 55 30 s excerpts from melodies in a wide range of styles (Gold et al., 2019). The stimuli included Baroque, Classical, Romantic, and 20th-century art music (e.g., Telemann, Mozart, Schumann, Berg), jazz (e.g., Chick Corea, Dizzy Gillespie, Duke Ellington, Herbie Hancock), exercises for flute, and traditional music from China, Ireland, Japan, and the USA (performed on a synthesized flute or clarinet). The stimuli were selected to represent a wide range of IDyOM-based levels of unpredictability (information content) and uncertainty (entropy).

IDyOM was configured to predict using both long-term and short-term models using a chromatic pitch viewpoint (`pitch`) and an IOI-ratio viewpoint (`ioi-ratio`). Rather than taking simple averages as in Sauvé and Pearce (2019) or sums as in Clemente et al. (2020), duration-weighted means of information content and entropy were computed across the notes in each melody to incorporate a measure of duration-based salience on unpredictability (see, e.g., the Krumhansl-Schmuckler key-finding algorithm; Krumhansl, 1990). The results showed significant quadratic relationships between liking and both information content and entropy. There was also an interaction between information content and entropy such that the inverted-U shaped relationship between information content and liking was clear for low-entropy stimuli but for high-entropy stimuli it broke down at the low predictability end of the scale where there was no difference in liking between high and medium information content pieces. In other words, high predictive uncertainty (entropy) was associated with reduced pleasure for highly unpredictable pieces.

These findings suggest that the inverted-U shaped relationship posited between predictability and pleasure is modulated by the uncertainty of the prediction. However, since

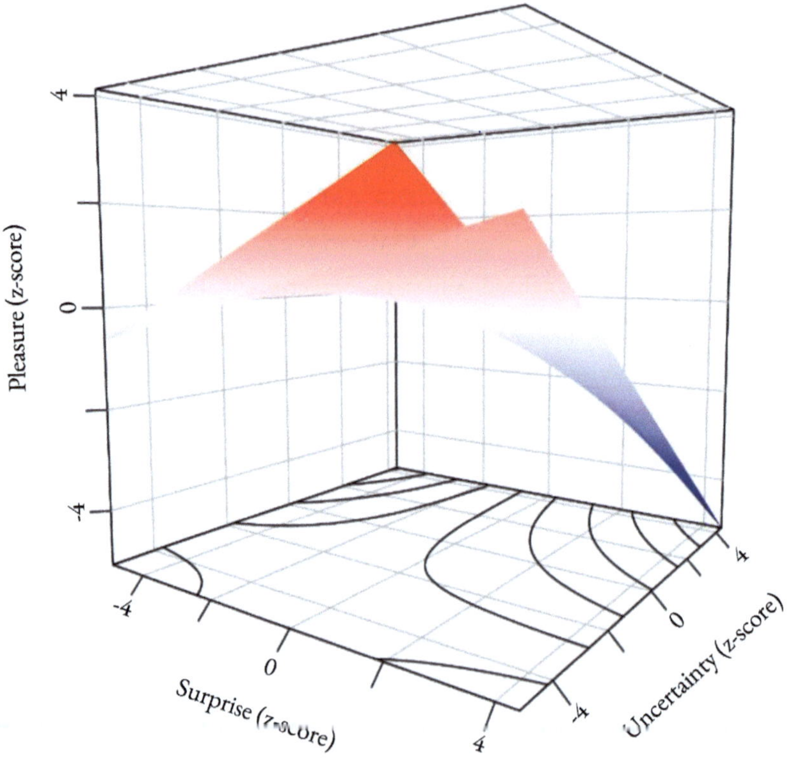

Figure 5.3: The influence of information content (surprise) and entropy (uncertainty) on pleasure while listening to music. Mean continuous pleasure ratings while listening to chord sequences extracted from popular music in the US Billboard Hot 100 chart between 1958 and 1991 (Cheung et al., 2019, Figure 2c). Created by Vincent Cheung and reproduced with permission from Elsevier.

this study used naturalistic melodic stimuli, it is possible that the space of relations between information content and entropy was not fully covered. This was addressed in an investigation of the relationship between harmonic predictability and pleasure run by Vincent Cheung using stimuli selected by Peter Harrison (Cheung et al., 2019). In this experiment, 30 harmonic sequences were selected from the McGill Billboard Corpus representing the US Billboard Hot 100 chart between 1958 and 1991 (see Chapter 4). The stimuli were selected specifically so as to ensure wide coverage across a factorial space combining high and low information content with high and low entropy. To achieve this, IDyOM was configured to predict chord symbols appearing in the corpus using both the short-term and long-term models, the latter trained on the McGill Billboard Corpus. In the experiment, 39 participants gave chord-by-chord pleasure ratings while listening to the chord sequences rendered using jazz guitar, acoustic guitar, and marimba sounds on top of a pop drum pattern. The results showed that pleasure ratings were significantly predicted by the interaction between uncertainty and information content, with an inverted-U shaped effect of both increasing information content and entropy on pleasure. However, as shown in Figure 5.3, there was also an interaction suggesting a modulatory effect of uncertainty on the relationship between predictability and pleasure. When uncertainty was low, increasing unpredictability was associated with greater pleasure but when uncertainty was high increasing unpredictability was associated with lower pleasure. This effect was so strong that the greatest pleasure occurred with maximum uncertainty and minimum surprise, followed by minimum uncertainty and maximum surprise.

These findings for melody and harmony have been replicated in experiments on rhythm perception where syncopation index (Longuet-Higgins & Lee, 1984) has been used as a quantitative measure of rhythmic complexity. Greater syncopation indicates that events fall at unexpected points in time according to the prevailing metrical interpretation of the rhythm. The experimental results show that short drum patterns with intermediate degrees of syncopation elicit greater pleasure and urge to move than drum rhythms with low or high syncopation (Witek et al., 2014). Furthermore, listening to rhythms with moderate syncopation activated reward-related neural structures (ventral striatum, medial orbitofrontal cortex) to a greater extent than highly syncopated rhythms (Matthews et al., 2020).

Taken together, these results suggest that the inverted-U shaped relationship between predictability (as a measure of complexity) and aesthetic appreciation is an oversimplification. While it does seem that intermediate levels of information content are associated with greatest pleasure, this effect is modulated by entropy (or predictive uncertainty). High entropy shifts the apex of the inverted-U towards lower information content while low entropy shifts it towards higher information content. In other words, for stimuli allowing certain predictions, listeners find higher levels of unpredictability most pleasurable. By contrast, high uncertainty leads to lower levels of unpredictability being more pleasurable. This may go some way to explaining the inconsistent pattern of results in empirical studies that have not manipulated or controlled uncertainty.

While these experiments use stimuli that are much more realistic than those used in previous research on musical pleasure that has attempted to quantify musical properties (e.g., Crozier, 1974; Heyduk, 1975; Vitz, 1966, 1972), they still fall short by some distance. In both Gold et al. (2019) and Cheung et al. (2019), the music is generated using a limited set of synthesized instruments without expressive timing, dynamics, or timbre. In the former study, the stimuli lack harmonic accompaniment while in the latter, they lack complex rhythmic and melodic structure. Furthermore, the same stimuli were used for all participants and, for any given participant, there is no guarantee that these stimuli would generate a wide range of pleasure, encompassing both the peaks and troughs of real-world experience. As noted above, a different approach is to allow participants to self-select their own favourite music and identify specific points of optimal pleasure within those tracks. Using such stimuli in an experiment all but guarantees a genuine real-world experience of musical pleasure.[7] This approach was pioneered by Blood and Zatorre (2001) who used the chills or shivers sometimes experienced while listening to music as an objective physiological marker of those pleasurable moments (see de Fleurian & Pearce, 2021, for a systematic review of research on music-evoked chills).

The naturalistic approach has significantly expanded our understanding of the neural basis of music-induced pleasure (see Zatorre, 2024, Chapter 7), but the problem of connecting the experience (and underlying neural processing) to the music remains. Some evidence comes from research by Sloboda (1991) who distributed a questionnaire asking for details of up to three pieces of music which had induced one or more of a supplied list of physical responses in the past five years. For each piece, respondents were asked to give the nature and consistency of the response, and to identify as precisely as possible the specific musical event with which it was associated. Out of 500 British adults approached, 83 participants responded with 165 musical pieces, of which 57 also located the specific musical event. Shivers down the spine were the most frequent response experienced by 90% of participants, followed by laughter, lump in the throat, tears, and goose pimples. Analysis of the musical events specified suggested that shivers and goose pimples (i.e., chills) were associated with unexpected events such as new or unprepared harmonies, sudden dynamic or textural changes, or unexpected melody notes. Subsequent experimental studies have attempted to identify stimulus properties associated with music-evoked chills (Bannister, 2020; Bannister & Eerola, 2018; Goodchild et al., 2019; Grewe et al., 2007; Guhn et al., 2007; Mori, 2022; Solberg & Dibben, 2019) but have focused on acoustic features, notably changes in loudness, timbre, and texture, rather than high-level predictive processes.[8]

[7]Though the effect of the experimental context, especially where neuroimaging is involved, is not to be discounted.

[8]It should be noted that music-evoked chills have been associated with multiple acoustic, musical, and emotional elicitors; they also show significant individual differences and effects of context, meaning that high-level predictive processing is likely to reflect one of several psychological mechanisms involved (de Fleurian & Pearce, 2021).

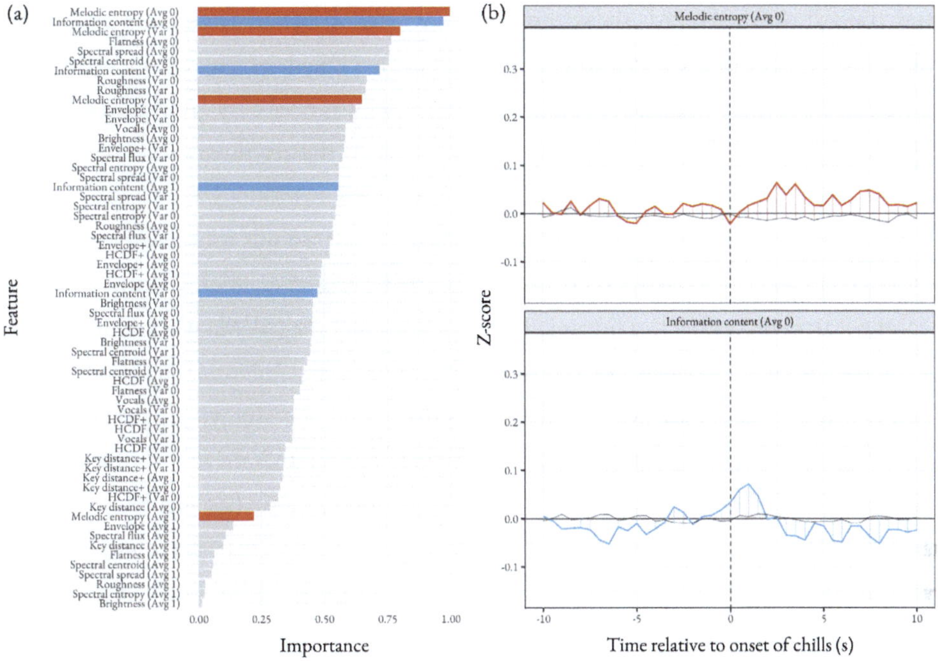

Figure 5.4: Predicting chills moments with low-level acoustic features and high-level informational features. (a) Feature importance for an SVM trained to distinguish chills moments from non-chills moments. Entropy and information content (mean and variance of the zeroth and first derivative; Avg 0, Var 0, Avg 1, Var 1 respectively) are highlighted in red and blue respectively. (b) Time series for mean entropy (top, red) and information content (bottom, blue) relative to chills onset with vertical bars indicating significant differences from baseline segments with no chills reported. Figure created by Rémi de Fleurian and used with permission.

To address this, Rémi de Fleurian conducted a computational analysis of a large collection of musical excerpts supplied online by 2,069 participants along with the precise timing of the chills experienced for each track. After cleaning, this resulted in 1,806 chills moments across 1,019 pieces of music, covering a broad range of artists from Béla Bartók to Amy Winehouse, Nick Drake to Metallica, Bill Evans to Aphex Twin. Drawing on previous experimental research, it was hypothesized that chills moments would be distinguished by both low-level acoustic features (changes in loudness, texture, harmonic content, or tonal strength; increases in roughness, spectral centroid, spectral flux, event density, or frequency range; and presence of vocals) and high-level features related to melodic expectation (information content and entropy derived from an IDyOM simulation). Acoustic features were computed over 500 ms segments. IDyOM was configured to predict pitch using the `interval` viewpoint using both the long-term and short-term models (see Table 2.1), the latter trained on the corpus shown in Table 3.2. Melodies were extracted automatically from every piece of music (Salamon & Gomez, 2012); while this results in a less than perfect rendition of the melody, automatic melody extraction was necessary given the large number of audio files. For all features, the mean and variance of the zeroth and first derivative (i.e., the time series itself and the differenced time series) were computed.

In the analysis, 500 ms segments falling within 1 s of a chills moment were assigned a positive label and all other segments were assigned a negative label. Taking a supervised machine learning approach, a linear support vector machine (SVM) was trained using five-fold cross-validation to distinguish positive from negative segments. The resulting binary classifier achieved above chance classification performance and an investigation of the extent to which each feature contributed to its performance showed that mean information content and entropy were the two most important features by some margin (see Figure 5.4) while three other measures of information content and entropy also fell within the 10 most important features. The other five features in the top 10 were mean spectral flatness, spectral spread, and spectral centroid followed by the variance of roughness and its derivative. More detailed analysis of the 20 s window around each chills moment showed that chills were associated with an increase in information content (relative to non-chills moments) starting about a second before the onset of the chill and peaking about a second afterwards. Chills moments were also associated with generally higher entropy (again relative to non-chills moments) both before and after the chill but also a sharp decrease in entropy precisely at the onset of the chill.

5.5 Psychological and neural mechanisms

The results reviewed in the previous section suggest that pleasure can arise when expectations are challenged while listening to music, especially when predictions can be made with high certainty or precision. On the other hand, in uncertain predictive contexts, expectation confirmation is associated with greater pleasure. But this does not tell us *how* pleasure arises from predictive processing of music. Several psychological mechanisms have been proposed.

5.5.1 Tension and resolution

Perhaps the simplest hypothesized mechanism is the association of pleasure with optimal levels of physiological arousal resulting from intermediate degrees of so-called *collative variables* such as complexity (Berlyne, 1971). Too little arousal is experienced as dull or boring while too much is experienced as overwhelming or agitating. This makes a certain amount of sense at the level of whole pieces of music but, taken literally, it implies that listeners would prefer music in which every note had the same moderate degree of predictability. In contrast, Meyer (1956) argued that pleasure arises from the combination of tension, arising from an event that was unexpected in a context that implied a different continuation, *and* a subsequent resolution of that tension when the implied continuation is subsequently introduced. This is potentially compatible with the Wundt effect since music that includes a balance of both predictable and unpredictable events, creating dynamic patterns of tension and resolution, will have moderate levels of predictability overall. Furthermore, it can be hypothesized that pleasure associated with tension and resolution would depend on the certainty of the context. In highly uncertain contexts associated with high arousal, background tension would be high, leading to a preference for the resolution (and lower arousal) associated with expected events while in contexts affording certain predictions, the background state of arousal might be too low, leading to a preference for tension (and higher arousal) associated with unexpected events. This could account for the interactions between uncertainty and unexpectedness observed in the experiments described above (Cheung et al., 2019; Gold et al., 2019).

5.5.2 Cognitive fluency

An alternative proposed mechanism is that the pleasure associated with predictive processing of music is related to perceptual or cognitive fluency rather than arousal (Huron, 2006; Reber et al., 2004). According to this account, which Huron (2006) calls the *prediction effect*, predictable stimuli are processed more fluently and the resulting positive affect is misattributed as a direct effect of the stimulus. This hypothesized mechanism predicts that pleasure would decrease as a monotonic function of increasing predictability. However, Reber et al. (2004) suggested that the prediction effect may be modulated by the extent to which fluency is expected. For a simple stimulus, the source of fluency is readily apparent while for intermediate degrees of complexity, any fluency is unexpected and, therefore, especially pleasurable. It is not clear how this can explain interactive effects of uncertainty and unexpectedness on pleasure (e.g., pleasure arising from surprising events in a context that provides certain predictions; Cheung et al., 2019).

In addition to the prediction effect, Huron (2006) proposes a distinct mechanism to account for pleasure associated with notably high unexpectedness: *contrastive valence*. The negative valence associated with predictive failure is contrasted with a subsequent neutral appraisal that the surprise is not threatening in the way that other surprises can be (e.g., discovering you are being unexpectedly watched or noticing that a bus is coming when you are about to cross the road). This conscious appraisal may be positively valenced but even if it is only neutral, the contrast with the negative valence of the fast and obligatory prediction response is misattributed as a positive property of the stimulus.

5.5.3 Learning progress, information seeking, and curiosity

Another possibility that is consistent with the empirical data is that the pleasure associated with predictive processing of music arises not from fluency of processing or optimal arousal but rather from opportunity for learning. The learning progress hypothesis states that:

> the brain, seen as a predictive machine constantly trying to anticipate what will happen next, is intrinsically motivated to pursue activities in which predictions are improving, ie, where uncertainty is decreasing and learning is actually happening.
>
> (Oudeyer et al., 2016, p. 265)

There are several ideas to unpack here. First of all, acquiring information is intrinsically pleasurable (Biederman & Vessel, 2006) and the intrinsic pleasure that arises from learning is associated with the trait of curiosity, which can be characterized as information seeking (Gottlieb & Oudeyer, 2018; Kidd & Hayden, 2015; Loewenstein, 1994). Consider a curious agent learning about their environment through experience by incrementally constructing an internal predictive model and seeking to improve the model by reducing uncertainty and prediction error (Oudeyer et al., 2016). Such an agent will avoid environments where uncertainty is low, since the structure of such environments will be learned quickly, and offer no opportunity for subsequent learning. It will also avoid environments where uncertainty is irreducibly high, since such environments also offer no opportunity for sustained learning. Environments with reducible uncertainty on the other hand (or learning progress niches) offer opportunities for sustained learning, indicated by reduced uncertainty (i.e., lower entropy) and prediction error (i.e., lower information content). The curious agent will seek out and dwell in such environments. Given an established model of such an environment, events that have high probability for the agent offer little opportunity for learning since they are consistent with the agent's current model of the environment (which is why they have low information content); learning occurs when unexpected events arise and cause the agent to update its internal model of the environment. The greater the prediction error, the greater the information gain. And the greater the information gain, the greater the pleasure.

In isolation, the information gain resulting from prediction error would suggest that pleasure increases monotonically as a function of information content, which wouldn't account for the empirically observed inverted-U shaped relationship between information content and pleasure (Cheung et al., 2019; Gold et al., 2019). However, prediction error is also an indicator of a mismatch between a learned model and the current environment, which should be penalized to motivate learning of predictively accurate models. If faced with sustained irreducible prediction error (i.e., many successive high information content events), our hypothetical curious agent could fulfil this motivation by either replacing its predictive model with another or choosing another environment (or both). The combination of increasing reward arising from information gain and decreasing reward arising from predictive failure could lead to an inverted-U shaped relationship between prediction error and pleasure. Furthermore, it seems reasonable to hypothesize that in more uncertain environments, an agent would focus on establishing a coherent model by reducing the penalty associated with errors of sensory prediction rather than the reward associated with information gain given an established model. In contexts affording certain predictions, on the other hand, where it is more straightforward to establish a coherent model of the environment, an agent might be expected to prioritize the information gain that arises from prediction error over any penalty associated with prediction failure. This would have the effect of shifting preference towards predictable events in uncertain contexts and unpredictable events in certain contexts, as observed in the experiments described above (Cheung et al., 2019; Gold et al., 2019).

Outside music perception, evidence for learning progress comes in the form of experiments showing that infants exhibit better learning of mappings between visual objects and sounds as well as greater object exploration following experience of the objects in question that violated their expectations—regarding solidity and spatiotemporal continuity—for physical object properties (Stahl & Feigenson, 2015). However, infants look away from visual events (presented in a sequence) that have very high or very low probability (Kidd et al., 2012) suggesting that they direct attention towards events with intermediate levels of information content. Furthermore, experiments in which adults freely explore game-like environments containing learnable structure show that their choices are spontaneously guided by learning progress and perceptual novelty (Poli et al., 2022; Ten et al., 2021).

5.5.4 Reward prediction

It has been proposed that *reward prediction* may be central to the experience of pleasure in music (Salimpoor et al., 2015; Zatorre, 2024). Although this mechanism involves prediction, it is important to distinguish it carefully from the psychological process of *sensory prediction* that is the focus of this book. Sensory prediction involves generating expectations for the properties of forthcoming sensory events. In the present case, this means prediction of the properties of auditory events such as notes and chords. Reward prediction is the prediction of forthcoming reward, which may in turn be associated with properties of those events. Errors of reward prediction can be positive or negative, respectively reflecting whether the actual reward was greater or lower than predicted. Pleasurable experiences of music should occur for positive reward prediction errors; in other words, when the reward experienced is greater than predicted.

This account has its origins in animal studies of midbrain dopaminergic neurons in the ventral tegmental area (VTA) and substantia nigra which project to the ventral striatum and (for the VTA) prefrontal cortex (Schultz, 2016; Schultz et al., 1997). These dopaminergic neurons show a rapid phasic response to primary rewards (e.g., a squirt of fruit juice) but, following conditioning, stop responding to the reward itself and start responding instead to the conditioned auditory or visual stimulus that cues the reward in proportion to the reliability of the cue. However, if the reward is varied so that it is greater (i.e., more or less juice) than presented during conditioning, the dopaminergic neurons show increased firing; if the reward is lower they show decreased firing. In other words, they appear to be coding reward prediction error. Reward prediction error can drive learning, motivating an organism to develop an improved predictive model of the sensory cues or actions that are associated with greater reward. In humans, primary and secondary rewards (e.g., food and money respectively) produce activation of a modality independent reward system consisting of the ventral striatum along with ventromedial prefrontal cortex, anterior insula, and amygdala (Sescousse et al., 2013). Reward prediction has been investigated in humans with a monetary incentive delay task in which an initial cue signals how much money can be won on any given trial, followed by a short delay to allow reward predictions to be formed, and then a task requiring a response with a button click to brief presentation of a visual stimulus. The difficulty of the task is calibrated to ensure that the reward is not always obtained. During the prediction period, activation in the ventral striatum increases in proportion to the reward predicted on a given trial (Knutson et al., 2001) while the reward itself activates medial prefrontal cortex (Knutson et al., 2003).

Music-evoked pleasure activates the same network of brain regions as primary and secondary rewards suggesting a common neural substrate (Mas-Herrero et al., 2021). The role of dopamine in music-evoked pleasure is demonstrated by evidence from ligand-specific positron emission tomography (PET) imaging showing dopaminergic activation in dorsal striatum during anticipation of chills moments in familiar music and in ventral striatum during experience of the pleasurable chill itself (Salimpoor et al., 2011). Further causal evidence comes from studies showing that pharmacological manipulation of dopaminergic function with a dopamine precursor (levodopa) and antagonist (risperidone) respectively increased and decreased subjective pleasure, physiological arousal, and incentive salience (how much money participants would pay) for music (Ferreri et al., 2019) whereas for an opioid agonist (oxycodone) and antagonist (naltrexone) this was not the case (Mas-Herrero et al., 2023).

5.5.5 Outlook

The learning progress hypothesis fits most neatly with the emphasis in the present work on statistical learning and probabilistic prediction. However, the empirical results presented in this chapter are potentially compatible with any of these four proposed mechanisms. Furthermore, the mechanisms are not necessarily mutually exclusive and may operate individually or in tandem to different degrees for different pieces of music, individuals, and listening situations. Future research must try to disentangle this web of possible psychological and neural mechanisms underlying the observed relationships between probabilistic prediction and aesthetic experience of music.

5.6 Aesthetic experience of familiar music

In this chapter (and the book more generally), it is assumed that a listener is hearing an unfamiliar piece of music for the first time. However, experiences of music very often reflect repeated listening to the same familiar pieces. How is it that music provides such enduring appeal, with pleasure persisting, and even increasing, across hundreds or thousands of hearings? This has become known in the literature as Wittgenstein's puzzle (Dowling & Harwood, 1986, p. 220; Wittgenstein, 1966, p. 29).

Leonard Meyer (1961) considered several potential solutions to the puzzle. Because music is complex and memory is imperfect, each rehearing is in some ways a first hearing because aspects of the experience will not have been encoded or stored perfectly (or at all) on previous hearings to enable accurate retrieval on repeated listening (see also Salimpoor et al., 2015). As we have seen, music is a multidimensional phenomenon whose complex structure extends and evolves in time. It can be perceived from multiple perspectives or viewpoints, each highlighting a different component or feature. This complexity might prevent listeners from being able to construct and maintain in memory completely accurate mental representations of music on first, or even subsequent, hearings. This effect would be multiplied in the case of a new performance of a familiar piece which can create a very different interpretation. In fact, going further, some pieces of music may possess such great multidimensional complexity that it is impossible to perceive them from all possible viewpoints at the same time, even for a listener who is highly familiar with the piece. In this case, even an experienced listener would be unable to accurately predict the entirety of the music because by focusing on one dimension (e.g., harmonic structure) they necessarily fail to attend to another dimension (e.g., rhythmic structure). This might also account for the sense of awe one experiences when listening to certain pieces of music, akin to the experience of certain architectural works, in which there is too much complex structure along too many dimensions on such a massive scale to be able to take it all in at once.

Meyer also suggests a different mechanism that relates directly to schematic expectations based on statistical learning. Changes in a listener's schematic expectations between repeated hearings mean that a piece of music is never really heard in precisely the same way. Because schematic expectations depend on experience and the listener will have experienced more music in between hearings, their schematic expectations will be different and this will change the experience of the piece. While this effect may be subtle, Meyer suggests that it will be stronger for listeners with limited experience of music (both in absolute terms and in breadth of stylistic experience) and also when the intervening musical experience is stylistically distant from the previous experience of the listener.

Another possibility, not considered explicitly by Meyer (1961) relates to the distinction between schematic and veridical knowledge which may operate as separate knowledge stores underlying expectation (see § 3.1). If schematic expectations, acquired implicitly over many years of musical listening, are unconscious and obligatory, then a listener could continue to experience schematic unpredictability even while listening to music that is highly familiar in a veridical sense (Bharucha, 1994; Huron, 2006). In fact, the tension between schematic expectations and veridical expectations may even prove to be a source of pleasure in its own right. As discussed in § 7.2.3, there is evidence that schematic knowledge of music is not easily overwritten by conflicting veridical knowledge.

As a counterweight to these perspectives, Meyer closes his essay by suggesting that our tolerance for repeated listening might actually not be as strong and pervasive as is often thought. Composers are driven to create new kinds of musical structure, performers seek out new works and modes of interpretation, listeners are intrinsically motivated to discover new musical pieces and performances:

> it is partly because familiar works and accustomed styles become exhausted that new compositions are needed and new styles are developed. And these, by changing us—our ears, our minds, and our habit responses—are paradoxically able to redeem and revitalize the very works and styles they were created to replace.
>
> (Meyer, 1961, p. 267)

In other words, some intolerance of veridical repetition is a driving force for the cultural evolution of new musical styles, which is where the present enquiry began in Chapter 1. Through mechanisms of prediction based on statistical learning, these new styles can create new aesthetic experiences while also, according to Meyer, refreshing aesthetic experience of familiar music.

5.7 Conclusion

In this chapter, IDyOM has been used to test a hypothesized role of expectation in emotional and aesthetic experience of music, characterized in terms of the absolute expressionism of Leonard Meyer. Although expectation is only one of the mechanisms by which music may induce emotions in a listener, it is the one most closely related to musical structure and therefore provides the most natural basis for investigating influences of enculturated implicit knowledge acquired through exposure to a musical style. In spite of well-established theoretical links between expectation and emotional experience, there has been a dearth of concrete empirical evidence which may be related to the lack of a formal model for characterizing musical expectations. Aesthetic experience was characterized as satisfaction arising from attention to the form and content of a piece of music while research in empirical aesthetics has emphasized the role of structural complexity, positing an inverted-U shaped relationship between complexity and aesthetic satisfaction. However, once again, empirical research has been hampered by the lack of a sophisticated model of perceived musical complexity.

Addressing these issues, empirical research using IDyOM has demonstrated a relationship between expectations based on probabilistic prediction and levels of valence and arousal experienced during musical listening while also establishing that information content and entropy accurately simulate complexity perception. Meanwhile, pleasure experienced during musical listening does appear to show the hypothesized inverted-U shaped relationship with complexity (modelled as mean information content) and this relationship depends on the uncertainty (entropy) of the prediction.

In sum, the results presented in this chapter suggest that statistical learning and probabilistic prediction (as simulated by IDyOM) can account for perceived musical complexity as well as aspects of the emotional and aesthetic experience of music. The next chapter examines whether other psychological processes at play while listening to music can be accounted for in terms of probabilistic prediction based on statistical learning, with a focus on two general-purpose psychological processes (memory and similarity perception) and two psychological mechanisms for organizing mental representations of auditory stimuli (grouping structure and stream segregation).

6 Beyond Expectation

In Chapters 3 and 4, it was shown that expectations for pitch, timing, and harmony can be simulated by a process of statistical learning and probabilistic prediction as implemented within the IDyOM model from Chapter 2. In Chapter 5 this approach was extended to provide a theoretically motivated and empirically substantiated account of expectation-related effects on emotional and aesthetic experience of music. The present chapter addresses the question of whether other cognitive processes at play during musical listening can be simulated in terms of the same underlying psychological processes of statistical learning and probabilistic prediction. In other words, can these same psychological mechanisms provide the basis of a more general theory of music cognition? As an initial foray into addressing this question, I will consider three general areas of cognition that are not typically linked with expectation in the psychological literature and consider whether there might exist a common underlying account in terms of statistical learning and probabilistic prediction. The first domain is auditory memory where it is important to ask whether the knowledge acquired by IDyOM through statistical learning provides a plausible account of how listeners encode, store, and access representations of musical sequences in memory. The second domain relates to the psychological processes involved in dynamically creating structured cognitive representations during listening to music: segmenting music into sequential groups or chunks (e.g., motifs and phrases) and segregating simultaneous musical events into parallel perceptual streams (e.g., segregating a melody from its accompaniment). The third domain is the perception of similarity between pieces of music as well as between the component parts of an individual piece.

6.1 Auditory memory

Learning is not possible without memory and therefore any theory of learning—statistical learning in the present case—necessarily implies a theory of memory, either explicitly or implicitly. Furthermore, there can be no generation of expectations based on past experience without memory. The role of memory is especially important for auditory perception because of its ephemeral nature—no sooner than they arrive, sounds disappear entirely in a way that visual objects tend not to. While listening to a piece of music, the preceding musical sequence providing the context for prediction must first be stored in memory and then compared with stored memory representations of previously experienced music (over timescales ranging from the past few seconds up to decades) in order to estimate the probability of different possible continuations. This is true both for IDyOM and for listeners, which raises the question of whether IDyOM's learned representations of musical structure can tell us anything meaningful about human memory for music.

Taken at face value, IDyOM posits two memory stores: a short-term store that survives for the length of a composition and a long-term store with capacity for a large number of compositions, with progressive transfer of material event-by-event from the short-term store to the long-term store (for the LTM+ and BOTH+ configurations). In Chapter 3, it was argued that the long-term model plausibly represents schematic knowledge held in long-term memory while the short-term model represents dynamic knowledge stored in short-term memory (see Table 3.1). Even if the timescales over which information is encoded and retained might not be wholly accurate, the presence of a long-term and short-term store with transfer of material from one to the other seems compatible with classical theories of memory, which posit a number of different stores, operating at different timescales and at different stages of auditory processing, each with different underlying neural substrates (Atkinson & Shiffrin, 1971; Cowan, 2008; Huron, 2006; Snyder, 2016).

At the lowest level, *echoic memory* consists of a sensory image covering unprocessed, detailed, and immediate auditory input within a time window of up to a few hundred milliseconds (Cowan, 1984, 1988). Sensory information is then passed into *auditory short-term memory* in a form that integrates different features (pitch, timbre, loudness) and chunks elements together into sequential groups, covering time windows of up to several tens of seconds. In order for IDyOM (and the listeners it simulates) to generate expectations, the current auditory context held in short-term memory must be matched with a context (the *schema*) from the preceding experience of the listener held in *long-term memory*. These contexts are retrieved from long-term memory into short-term memory for comparison with the current context. This comparison with information in long-term memory yields so-called *schematic expectations* while corresponding information held in short-term or intermediate-term memory can be used to generate so-called *dynamic expectations* (see Chapter 3, Table 3.1).

Finally, information about the current context held in short-term memory is dynamically and continuously passed into long-term memory.

This picture of memory reflects knowledge acquired through learning the regularities of a sensory environment, consistent with IDyOM's conception as a model of statistical learning. In general, the fundamental knowledge acquired through statistical learning incorporates some representation of distributional associations or co-occurrences between salient subcomponents of the sensory environment, which should allow more efficient or effective processing of future sensory input within the same environment. In the present case, this involves learning a model of the frequency with which auditory events appear following a wide range of variable-order contexts, so as to improve prediction performance in auditory contexts encountered in the future. The memory mechanisms underlying this learning are examined further in § 6.1.1.

There is a distinction to be made between this conception of the memory mechanisms underlying statistical learning and the psychological processes and representations involved in committing specific stimuli to *episodic memory* for later retrieval, which has received much greater attention in the literature on memory (e.g., Cowan, 2008; Crowder, 1993). Recall from Chapter 3 that the distinction between veridical and non-veridical memory does not align exactly with Crowder's (1993) distinction between *episodic auditory memory* and *auditory knowledge*. In Crowder's terms, echoic, dynamic, schematic, and veridical memory, as conceived here, all correspond to auditory knowledge with episodic memory reserved for memories of auditory experiences that are tied to a concretely specific spatial and temporal context, as discussed in § 3.1. However, Huron (2006, pp. 221–224) provides a detailed illustration of how a succession of episodic memories formed through repeated experiences of a particular piece of music in different contexts could be combined into a veridical memory that retains its richly vivid and specific character but loses its extra-musical, episodic associations. Although veridical memory is less germane to the present investigation, there are nonetheless interesting relationships between veridical and schematic memory that are considered further in § 6.1.2.

6.1.1 Modelling memory constraints

While IDyOM's implications for musical memory seem plausible at a general level, it also has several limitations as a model of auditory memory. Notably, there is no simulation of the limited temporal or informational capacities of echoic and short-term memory; indeed there is no distinction at all between echoic memory and short-term memory. Furthermore, the suffix trees making up the short-term and long-term models retain perfect memory of every sequence that they encounter (indeed every suffix of every sequence they encounter). This does not seem plausible, since information in short-term and long-term memory should fade over time, due to either decay or interference (Hardt et al., 2013; Lewandowsky et al., 2009).

These limitations have been addressed in a model called PPM-Decay, developed and tested by Peter Harrison, which operates very similarly to IDyOM, using the same prediction by partial match (PPM) algorithm for prediction from a learned statistical model, except that the stored frequency counts decay over time (cf. Chapter 2). Whereas in IDyOM's PPM implementation, each observation of a symbol in a given context is recorded as a single count (i.e., the frequency count is incremented by one), with no distinction made between observations on the basis of recency, PPM-Decay implements a customizable decay kernel controlling how the frequency count from a given observation decays over time. Harrison et al. (2020) showed, in a first experiment, that the addition of an exponential decay kernel improved prediction performance for artificially generated sequences where the underlying generative model switched periodically between different first-order Markov models. A second experiment simulated the harmonic expectations of an individual listening to a single piece of music from one of three styles (pop, jazz, chorales) every day for 100 days. Given that musical styles possess structure in common across individual pieces, the kernel decayed not to zero as in the previous experiment but to a non-zero asymptote. This rationale is similar to the motivation for including long- and short-term models in traditional IDyOM. The exponential decay produces a focus on recent short-term experience while the asymptote retains in memory some traces of long-term historical experiences. The results showed that for pop music, prediction performance was improved by exponential decay but not further improved by a non-zero asymptote, suggesting a predominance of local regularities over long-term schematic structure. For jazz and chorales, however, exponential decay alone did not improve prediction performance but only when combined with a non-zero asymptote, suggesting benefit in exploiting both short- and long-term regularities.

These results are interesting because they suggest that memory constraints can actually improve prediction performance in non-stationary environments whose statistical structure changes over time. However, more intuitively, constrained memory can also impair performance and PPM-Decay has successfully simulated effects of memory decay on listeners' performance in behavioural experiments. These experiments have used a regularity-detection paradigm developed by Nicolas Barascud and Maria Chait (Barascud et al., 2016). In this paradigm, listeners are presented with a series of rapid tone sequences consisting of tones drawn from a pool of 20 frequencies with equal logarithmic intervals between 222 and 2,000 Hz presented in random order at a rate of 20 Hz (corresponding to a tone duration of 50 ms). Sequences were constructed from a randomly selected subset of this pool of tones. While some of these sequences were sequentially random (termed RAN) others were regular (REG) in that a sequence of *r* tones was repeated consecutively a number of times (*r* being the *cycle length*). The critical experimental stimuli (RANREG) consist of an initial RAN portion and a subsequent REG portion. The task for participants on each trial is to detect the transition from RAN to REG as quickly and accurately as possible (upon the second appearance of the regular cycle).

Barascud et al. (2016) hypothesized that regularity detection relies on monitoring the predictability of the signal based on an internal predictive model acquired dynamically through statistical learning of the stimulus sequences. To simulate this process, IDyOM was configured to predict pitch using a long-term model with ongoing learning (LTM+; see Chapter 2) and then presented with the full stimulus set for one participant in the study. In this case, IDyOM acts as an ideal observer, with perfect memory and attention, establishing an upper bound on regularity detection. The IDyOM simulations showed that the RANREG transitions are characterized by a sharp drop in information content, with detection taking place by approximately the fourth tone of the second cycle of the regular sequence. In a magnetoencephalography (MEG) experiment using behavioural and neural measures of regularity detection (reaction time and MEG sustained amplitude response), listeners showed similar detection times for cycle lengths of five and ten. However, detection times for greater cycle lengths of 15 or 20 tones were longer than those predicted by IDyOM, raising the question of memory capacity limitations, since the repeating sequence must be held in short-term memory for comparison with stored representations of the first occurrence of the same sequence.

Harrison et al. (2020, Experiment 3) further investigated this possibility by orthogonally varying the tone duration (25, 50, and 75 ms) and cycle length (10 and 20 tones), producing regular patterns that ranged from 250 to 1,500 ms. The results showed significantly faster detection of 10-tone than 20-tone regular patterns, even when pattern duration was held constant by comparing the 10-tone, 50 ms condition with the 20-tone, 25 ms condition. These two conditions yield regular patterns of exactly the same duration but consisting of different numbers of tones, suggesting an informational rather than temporal capacity limitation on echoic memory. By comparison, effects of tone duration were almost non-existent, limited to stimuli with a 20-tone cycle, where the number of tones required for detection was very slightly higher for 75 ms than 50 ms tone durations. The PPM-Decay model introduced above failed to distinguish between the 10-tone and 20-tone cycle lengths, outperforming participants for the 20-tone cycle length with short tone duration. This motivated the addition of a sensory buffer consisting of a sliding window of 15 tones for which there is no weight decay, followed by exponential decay to an asymptote as before. The addition of the buffer accurately simulated the behavioural data, showing longer regularity-detection times for 20-tone than 10-tone cycles. The model also incorporated noise when counts were retrieved from memory to simulate inaccuracies in human memory.

PPM-Decay was further investigated in a series of experiments designed by Roberta Bianco, which introduced a long-term memory element to the regularity-detection paradigm (Bianco et al., 2020). Unbeknownst to the participants, three of the regular patterns (called RANREGr, with a 20-tone cycle length) reoccurred across five blocks of the experimental session (approximately 40 minutes). Detection times for RANREGr, but not RANREG, decreased across the session suggesting long-term memory for the regular patterns making up the RANREGr stimuli. Furthermore, the memory effect was almost completely obliterated for stimuli in which the REGr patterns appeared in non-adjacent positions within a random tone sequence. Finally, the memory effect was weakened but not obliterated by interrupting blocks of stimuli that did not contain the RANREGr stimuli. These results were simulated with a PPM-Decay model whose decay kernel incorporated an additional phase of slow exponential decay to zero following the initial high-fidelity buffer and short-term exponential decay to a non-zero asymptote. Whereas a PPM model without memory constraints learned the RANREGr patterns perfectly, showing much faster detection times than listeners, regardless of the adjacency of the regular pattern or the presence of intervening blocks, PPM-Decay closely simulated listeners' trajectory of improved detection times for RANREGr patterns, also showing impaired performance for non-adjacent regular patterns and in the presence of intervening blocks of trials.

In these experiments, the RANREGr memory effect was largely implicit because the per-participant reaction time (RT) advantage for reoccurring RANREGr stimuli did not correlate with performance on a separate task requiring explicit classification of the stimuli according to familiarity. The effect was obliterated when the regular pattern was reversed demonstrating the ordered sequential nature of the underlying memory representation. However, the RT advantage for RANREGr stimuli persisted even when the pattern was rotated, so that it started and ended on different elements, suggesting that the pattern is not stored in memory as a fixed, unitary chunk. Finally, the memory effect was robust to transposition of the stimulus to higher or lower frequencies, suggesting storage in terms of relative rather than absolute pitch representations (see Chapter 3). All of these findings are consistent with IDyOM's model of auditory memory in which multiple variable-order subsequences of the auditory input are stored, while allowing for abstract representations (of pitch in this case).

The PPM-Decay model also has some limitations as a complete model of auditory memory. For example, while the RANREGr memory effect in humans persisted (rather surprisingly) for up to six months without further exposure to the stimuli (Bianco et al., 2023), the decay kernel that simulated the RANREG and RANREGr effects on the scale of a single experimental session did not capture these very-long-term effects. It remains to be seen whether such long-term storage reflects a distinct psychological process of memory consolidation or whether it can be captured using a more sophisticated kernel. More generally, the parameters defining the kernel need to be examined in the context of a much wider range of experimental conditions, so as to pin them down more precisely and understand how they vary across participants and experimental tasks. Finally, it is not immediately clear how the model could replicate sensory memory for non-symbolic acoustic structure, in a way that is captured by the sensory model of Leman (2000), for example, as shown in Chapter 4. One possibility would be to discretise continuous acoustic features for input to symbolic modelling with IDyOM or PPM-Decay (see Harrison & Pearce, 2018). Another possibility is to characterize the sensory buffer in terms of information capacity, rather than number of events, which could be achieved in different ways depending on the nature of the input.

The portrait of auditory memory painted by this research has interesting theoretical implications. According to this emerging picture, auditory sequences are encoded into subsequences or *n*-grams of multiple orders and stored along with a representation of the frequency with which they have appeared, which decays over time. The profile of decay, captured in PPM-Decay's kernel, appears to be non-linear consisting of three phases as shown in Figure 6.1: first, a high-fidelity window with limited informational and temporal capacity, corresponding to sensory memory; second, a period of fast exponential decay corresponding to short-term memory over a period of seconds; and third, a period of slow exponential decay corresponding to long-term memory over periods of tens of minutes.

It is interesting to consider how these different phases of the decay kernel might be implemented as cognitive systems in the brain. The traditional approach would be to associate the different phases of the decay kernel with different dedicated memory stores corresponding to distinct neural substrates with different information-processing characteristics (Cowan,

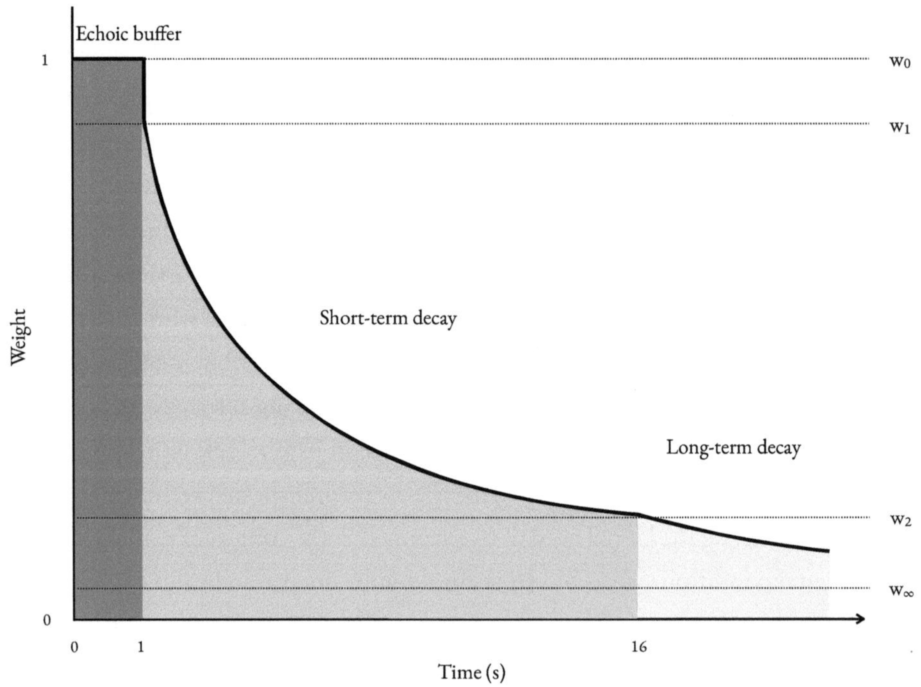

Figure 6.1: A schematic representation of the decay kernel implemented in PPM-Decay illustrating how the weight associated with *n*-gram frequency counts could decay over time. The weight can vary between 0 and 1 corresponding to having no memory or full memory of the *n*-gram respectively. The kernel is divided into three phases: an initial echoic buffer; a short-term phase with exponential decay; and a long-term phase. The kernel can be customized in terms of the duration of the buffer and short-term phase as well as the weights associated with each of the three phases (w_0, w_1, w_2, and w_∞). The buffer can also optionally be limited in terms of information capacity by the number of events it can hold. In the example shown, the buffer has a duration of 1 s followed by an immediate but small decline in weight; the short-term memory phase shows exponential decay over 15 s to a relatively low weight; and finally, the long-term phase exhibits a slower process of exponential decay to a long-term asymptote.

2008; Snyder, 2016). Under this view, we might expect to find sets of model parameters (including the decay kernel) defining the properties of sensory, short-, and long-term memory stores that generalize across a wide range of stimuli and tasks. An alternative to this *library metaphor* of memory storage is a *proceduralist* approach in which "memory is depicted as *the persistence that is a by-product of some original mental activity*, which activity, in turn, comprised the original learning episode" (Crowder, 1993, p. 115). Under this view, memories are stored in the underlying neural substrate responsible for learning, which may show different characteristics depending on the modality of the stimulus and nature of the task.

As a result, different kinds of perception and cognition would result in different kinds of memories, with different characteristics.

Consistent with this conception is evidence that statistical learning reflects a set of domain-general computational principles with modality- or domain-specific implementations subject to different constraints related to the particular psychological and neural characteristics of sensory processing in each modality. For example, statistical learning shows qualitative differences between domains, limited transfer of learning across domains, and little correlation in individual statistical learning performance between domains (Frost et al., 2015; Siegelman et al., 2018). There are also echoes of the proceduralist position in the temporal-ratio model of memory proposed by Brown and Chater (2007), which emphasizes similarities between the psychological mechanisms underlying memory retrieval over different timescales, rather than categorical differences between distinct long- and short-term memory stores corresponding to those timescales. For present purposes, it is important to note that the proceduralist perspective allows for constructing different kernels for PPM-Decay corresponding to the different stimuli and task demands involved in different experimental scenarios. In other words, the account of auditory memory proposed here is potentially compatible with both the library and proceduralist conceptions of memory and it is worth noting that the two conceptions are not mutually exclusive but may coexist in various forms.

Finally, while the PPM-Decay model focuses on decay of the frequency counts over time, it is worth noting that IDyOM and PPM-Decay inherently reflect interference effects, which are also thought to be an important part of memory degradation (see Hardt et al., 2013; Lewandowsky et al., 2009). Consider, for example, a fifth-order deterministic prediction in which a context of five events has been followed by just a single event in the experience of the model. In this case IDyOM can make a very precise low-entropy prediction for the continuation. If the same context reoccurs followed by a different event, two continuations will be possible and the prediction will be less precise (i.e., higher in entropy). Continuing the thought experiment, imagine that the context keeps on reoccurring with different continuations until each possible continuation has appeared equally often. In this case, each possible continuation is equiprobable and IDyOM's prediction has maximum entropy. In other words, the memory of the original highly predictable continuation has been degraded through interference. The general point is that interference arises through the way in which statistical learning is implemented in these models. Furthermore, in PPM-Decay there should be an interaction between interference and decay, such that contexts which have decayed more substantially would result in lower proactive interference, consistent with empirical observations in other areas of memory research (Altmann & Gray, 2002; Altmann & Schunn, 2012). Combined modelling and experimentation on the ways in which interference and decay combine to determine human memory degradation provides a worthy topic for future investigation.

My purpose here has been to make explicit the memory mechanisms encapsulated within IDyOM and relate them to theoretical and empirical observations about human auditory memory. While I believe this provides a compelling, if embryonic, picture of how knowledge acquired through statistical learning of musical structure might be represented in human cognition, it is also clear that much remains to be understood about the memory mechanisms underlying statistical learning and probabilistic prediction in the context of music perception.

6.1.2 Veridical musical memory

The kinds of auditory memory considered in § 6.1.1 relate to very short fragments of recent auditory context in the case of sensory knowledge, distributional representations of statistical structure in the current auditory episode (with decay) in the case of dynamic knowledge, and representations of statistical regularities in long-term musical experience in the case of schematic knowledge. None of these forms of knowledge contain distinct veridical representations of complete, individual pieces of music. Clearly, IDyOM cannot simulate truly episodic memories for music since it has no way of representing the non-musical elements that make up the episodes in question. However, it may have some ability to simulate non-episodic veridical memory for music because entire pieces of music are represented within IDyOM's suffix tree (see § 2.2.3). These representations coexist, and may be overlaid, with shorter suffixes of the same piece as well as suffix representations of other pieces of music. Therefore, there is no clear distinction in IDyOM between veridical and schematic long-term memory for music. If a piece of music has not been heard before, matching paths in the suffix tree from other pieces of music will be activated and schematic expectations will be generated on the basis of matching contexts that are shorter than the actual context in short-term memory. If the piece has been heard before then expectations will be based on an exact trace of the piece stored in the suffix tree (assuming unbounded order and no decay), such that predictions towards the end of the piece will be based specifically on the context of the entire piece and will, therefore, tend towards determinism (though the smoothing mechanism ensures that IDyOM also entertains the schematic possibility of other continuations based on shorter contexts). The difference between schematic and veridical expectations then arises through the degree of deterministic prediction arising from differences in the length of matching context, with longer contexts tending to allow for more deterministic predictions.[1]

Does this vision of veridical memory for music have any explanatory value? In contrast to the experimental designs discussed thus far for assessing schematic, dynamic, and sensory knowledge, veridical memories for music are generally assessed experimentally using recognition tasks in which music heard earlier in an experimental session is subsequently presented again (usually after something like 10–30 min) for recognition judgements along with foils not presented earlier (Halpern & Bartlett, 2010; Snyder, 2016). This task probes veridical memory for particular pieces of music but it goes further in requiring episodic memory of whether or not the piece occurred earlier in the experimental session. While performance is often poor for novel music first heard during the experimental session, it is typically better for music that was known to a participant prior to the experimental session. For example, Bartlett et al. (1995) found higher hit rates (correct recognition) and lower false alarm rates (incorrect recognition) for well-known than novel music each presented in separate blocks. Interestingly, however, when both well-known and novel music were presented in randomized order within the same block, the false alarm rates were no different. This was explained in terms of different recognition thresholds for a measure of generalized familiarity, combining experience during and prior to the experiment. When the entire block contained well-known music, listeners could use a stringent threshold for recognizing music presented earlier in the experiment. However, when the well-known music was mixed up with novel music, it was not possible to consistently apply such a high recognition threshold and the prior familiarity of the well-known tunes was mistakenly attributed to them having appeared earlier in the experimental session.

This points to a veridical memory system supporting a generalized sense of overall familiarity, which has been contrasted with genuine episodic memory for music in an elegant experiment run by McAuley et al. (2004). The experiment consisted of two exposure phases followed by a test phase. In the first exposure phase, participants listened to a set of 24 melodies, consisting of equal numbers of well-known and novel music (the latter composed especially to match the well-known stimuli) with each melody occurring once or three times. The familiar songs were generally nursery rhymes (e.g., Daisy Bell), Christmas carols (e.g., Silent Night), or film music (e.g., Singing in the Rain). The following day, participants returned to the lab to repeat the exposure session with a different set of 24 well-known and

[1]See Schubert and Pearce (2016) for a related conception of veridical musical memory.

novel melodies. Ten minutes later, they were presented with two memory tests, one for half of the stimuli asking for recency judgements (today vs yesterday) and one for the other half of the stimuli asking for frequency judgements (once vs thrice).

The results showed that while well-known melodies afforded accurate judgements of recency and frequency, novel melodies afforded accurate judgements of frequency only. Judgements of recency for novel melodies showed an interaction between recency and frequency such that, for example, there were more incorrect "Today" responses for yesterday/thrice melodies (false alarm rate: .59) than for both yesterday/once melodies (false alarm rate: .38) and even for correct "Today" judgements for today/once melodies (hit rate: .46). Again, this supports the existence of a generalized memory effect with frequency appearing to have a stronger effect than recency. Contrasting effects were found for recency judgements of previously familiar melodies which, for example, showed no difference whatsoever in false alarm rate for yesterday/once and yesterday/thrice melodies, suggesting a robust episodic mechanism for well-known melodies. Furthermore, recency performance for familiar melodies (the difference between hit rate and false alarm rate) correlated ($r = .53$) with the probability of correctly recalling the title or lyrics of the melody, further suggesting an episodic memory mechanism involving nameability (see Halpern & Bartlett, 2010, for further discussion).

IDyOM is clearly not designed to simulate episodic memory effects for well-known music associated with knowledge of song titles or lyrics. However, the generalized familiarity effects on veridical memory observed in these experiments fit quite well with the conception of veridical memory outlined above based on IDyOM. For example, the PPM-Decay model might be expected to show exactly these kinds of effects for novel melodies, mirroring the empirically observed confusability of familiarity based on frequency and recency. An experiment run by Kat Agres brings this picture into sharper focus (Agres et al., 2018). The experiment used artificially constructed tone sequences which varied systematically in information-theoretic complexity. These stimuli were presented to 23 participants across three blocks each comprising an exposure phase containing all 24 sequences, followed by a recognition test consisting of eight of the previously presented sequences together with eight foils generated in the same way but not previously presented. To simulate listeners' memory performance, an IDyOM LTM+ model with no prior training was exposed to the stimulus set, learning the structure of the artificial style dynamically throughout the course of an experimental session (just as we hypothesized that listeners would).

Memory performance in this experiment was moderately poor with hit rates and correct rejection rates consistently ranging between .63 and .70 across the three sessions. Interestingly, however, the pattern of errors changed markedly across the course of the experiment. In the first session, listeners' memory performance for familiar tone sequences—measured by *d′* scores—did not correlate significantly with the average information content of the stimuli. Over time, however, listeners learned the structure of the artificial musical style to the extent that by the third session information content accounted for 85% of the variance in memory performance, such that hit rates were higher for predictable (low information content) than unpredictable (high information content) stimuli. In other words, unpredictable (or stylistically incongruent) tone sequences tended to be incorrectly judged as being unfamiliar even though they had in fact been presented in the exposure phase. These results clearly illustrate the effect of generalized familiarity on veridical memory, in this case reflecting increasing predictability resulting from incremental statistical learning of an artificial musical style resulting in long-term schematic knowledge of the style (see also § 7.2.3 for evidence that schematic knowledge cannot easily be overridden by veridical knowledge arising from recent experience with schematically unexpected stimuli).

There are limitations to this conception of veridical memory. For example, the work of Jay Dowling, James Bartlett, and others has shown that while pitch contour is more important than pitch interval for immediate recognition of melodic stimuli, this pattern reverses when the time interval between the target and comparison is extended to approximately 30 s (Dewitt & Robert, 1986; Dowling, 1991; Dowling & Bartlett, 1981). In other words, the representation in memory of a musical sequence seems to develop over time, somewhat paradoxically, from an abstract representation of contour into a more detailed representation of specific pitch intervals. Furthermore, these effects are dependent on the tonality of the sequence, such that improvement of recognition based on differences in pitch interval was greater for strongly tonal than weakly tonal or atonal melodies. Although, the IDyOM framework allows for multiple representations of music, it is not immediately clear how this effect can be simulated. One possibility is that listeners maintain multiple representations simultaneously in memory corresponding to different auditory features, such as pitch contour, pitch interval, and scale degree, with different models corresponding to each representation (or possibly combinations of representations) showing different characteristic decay parameters (see also § 3.5). Another possibility, not necessarily mutually exclusive from the first, is that there exists a pipeline of information processing during encoding of the auditory signal such that contour information is encoded and stored first, followed by a more detailed representation of pitch interval, which perhaps takes longer to encode. The investigation of these possibilities presents interesting challenges for future research.

6.2 Grouping structure: Segmentation and segregation

There is a debate in the literature on statistical learning about whether the knowledge acquired is stored in terms of probabilistic regularities of sequential structure (Aslin et al., 1998; Saffran et al., 1996) or rather in terms of frequently occurring chunks of sequential information (Perruchet, 2019; Thiessen, 2017). In many experimental contexts, the two accounts make indistinguishable predictions but there is some specific experimental evidence for both probabilistic representations (e.g., Endress & Mehler, 2009; Franco & Destrebecqz, 2012) and chunk-based representations (e.g., Giroux & Rey, 2009; Perruchet & Poulin-Charronnat, 2012). Therefore, it seems quite possible that these perspectives are not mutually exclusive and both kinds of representation may be computed in parallel to differing degrees depending on individual differences and task demands (see also McCauley & Christiansen, 2019; Siegelman et al., 2019). Furthermore, it is possible that both chunk-based and probabilistic representations could derive from an underlying cognitive representation of sequential regularities similar (at some level of cognitive description) to that learned by IDyOM, consisting of a network (specifically, a tree) of symbol sequences constructed from the sequential input with associated frequency information. This representation can be used to compute conditional probabilities for expectation generation (amply illustrated in Chapters 3 and 4) but equally it can be used to identify frequently occurring subsequences within the input domain. Consider for example the suffix tree depicted in Figure 2.4. The branches labelled BRA indicate that this subsequence of the input has not only occurred twice but has appeared deterministically as a "chunk" in that B has only ever been followed by R, and BR only ever by A. In other words, the suffix tree encodes representations that can be used simultaneously to derive probabilistic representations *and* chunk-like representations, with the common ground being that chunks consist of chains of high-frequency, low-entropy (or, at the extreme, deterministic) transitions delineated by lower-frequency, higher-entropy symbol transitions.

Chunking involves the grouping of elements in a sequence together or, from a complementary perspective, identifying boundaries between the sequential groups thus formed. Such grouping plays a significant role in short-term memory, allowing the encoding of longer sequences of symbols by recoding them into subsequences (Cowan, 2000; Miller, 1956).

Figure 6.2: Sequential segmentation of the opening melody from the first movement of Mozart's Symphony No. 40 in G minor. After Lerdahl and Jackendoff (1983, Figure 3.1).

Grouping plays an important role in perception and production of natural language, both written and spoken, and is equally important in music, where compositions are frequently composed of motifs, phrases, sections, movements, and other formal elements. Examples include the musical phrases making up many folk songs, the verse, chorus, and bridge characteristic of much popular music and the component sections of sonata form. The goal in § 6.2.1 is to investigate whether probabilistic prediction based on statistical learning plausibly plays a role in the dynamic formation of mental representations of sequential grouping structure during musical listening. Another way in which musical events are grouped perceptually is into simultaneous, parallel streams (Bregman, 1990) and the question of whether probabilistic prediction based on statistical learning might underlie perceptual stream segregation is addressed in § 6.2.2.

6.2.1 Sequential segmentation

Just as listeners segment continuous speech into words and phrases, so they also segment music into sequential groups of motifs, phrases, and sections (see Figure 6.2). The most influential account of how listeners achieve this process of musical segmentation is that of Lerdahl and Jackendoff (1983), who included grouping as one of two modules of basic processing of musical structure (the other focusing on metre) in their generative theory of tonal music (see Chapter 1). According to Lerdahl and Jackendoff, listeners achieve the process of segmenting music through the application of a number of grouping well-formedness rules (GWFRs), placing constraints on possible grouping structures, and grouping preference rules (GPRs), giving preference to some grouping structures over others. The GWFRs ensure that the grouping structure of a piece of music exhaustively partitions the piece into contiguous groups and subgroups (although overlaps of one note between contiguous groups are allowed in certain prescribed circumstances).

The most prominent GPRs are inspired by Gestalt principles of proximity and similarity. They predict that grouping boundaries will be perceived between two melodic events whose temporal proximity is less than that of the immediately neighbouring events due to a slur, a rest (GPR 2a) or a relatively long inter-onset interval (IOI; GPR 2b) or when the transition between two events involves a greater change in register (GPR 3a), dynamics (GPR 3b), articulation (GPR 3c), or duration (GPR 3d) than the immediately neighbouring transitions. GPR 4 states that where the effects of GPRs 2 and 3 coincide and are relatively more pronounced, a larger level grouping boundary may be perceived. Empirical research has provided evidence that listeners do tend to perceive boundaries at points predicted by the GPRs, though primarily only those related to temporal proximity (i.e., GPR 2) (Deliège, 1987; Frankland & Cohen, 2004; Peretz, 1989). GPR5 prefers subdivision of groups into subgroups of equal length while GPR 6 prefers grouping boundaries that accord with musical parallelism (e.g., appearing at the same point in the bar or at similar points with respect to a repeated motif).

Two prominent computational models of musical segmentation are, like the GPRs, inspired by Gestalt principles of proximity and similarity. Cambouropoulos (2001) introduced the *Local Boundary Detection Model* (LBDM) which combines a proximity rule (reflecting the size of an interval) with a change rule (reflecting the distance between successive intervals) applied to pitch interval, IOI, and rests which are weighted and combined into a boundary strength for each note in a melody. The LBDM predicts that boundary strength increases with the size of the pitch interval, IOI, or rest and also with the extent to which these change from the preceding pitch interval, IOI, or rest. Temperley (2001) introduced a related rule-based model called *Grouper* comprising three preference rules: 1) prefer to locate grouping boundaries after long IOIs or rests (see GPR2); 2) prefer groups of about 10 notes; 3) prefer to begin successive groups at parallel points in the metrical hierarchy (e.g., on the same beat of the bar; see GPR6).

It is evident that many of the rules defined in these models are rather specific to musical melody and may not characterize all musical styles equally well (e.g., sound-based music; Olsen et al., 2016). An approach which offers the possibility of transcending these limitations arises from the cognitive-scientific literature on statistical learning. In a seminal paper introducing the simple recurrent neural network (SRN), Jeff Elman illustrated the relationship between prediction and segmentation:

> Thus, one can ask whether the notion "word" (or something which maps on to this concept) could emerge as a consequence of learning the sequential

structure of letter sequences which form words and sentences (but in which word boundaries are not marked).

(Elman, 1990, p. 192)

The SRN is a three-layer feed-forward connectionist network which includes a context layer recording the state of the hidden layer on the previous step which then feeds into the current hidden layer. When configured to predict the next letter in a sequence of letters making up an artificial text, the SRN yields high error at the beginning of words and declining error within words as the sequence becomes increasingly predictable based on the statistics of letter co-occurrences in the network's input.

One can interpret this relationship between prediction and segmentation in terms of probabilistic models. As discussed in Chapter 7, the literature on statistical learning has provided evidence that infants and adults are capable of dynamically learning transition probabilities in artificial continuous phoneme and tone sequences and using these for recognition of segmented chunks with high internal transition probability delimited by symbol transitions with low probability (Frost et al., 2019; Saffran & Kirkham, 2018). Inspired by early statistical learning results (Saffran et al., 1996), Brent (1999) implemented a model of word discovery in phonemic transcripts of infant-directed speech. The model places a word boundary between phonemes whenever the digram transition probability at x_i is lower than at both x_{i-1} and x_{i+1}. Brent also examined a related model that replaces transition probabilities with *pointwise mutual information*:

$$I(x_i, x_{i-1}) = \log_2 \frac{p(x_{i-1}^{i})}{p(x_i)p(x_{i-1})}, \tag{6.1}$$

which is symmetrical with respect to the order of the two events and measures how much the occurrence of one event reduces uncertainty about the co-occurrence of the other event. This latter model outperformed the former in predicting word boundaries in a corpus of phonemically encoded infant-directed speech.[2]

This research on language provides general motivation for applying the same principles to modelling perception of boundaries in music. Pearce et al. (2010a, 2010b) developed an online segmentation model based on statistical learning and probabilistic prediction using IDyOM. The model predicts a boundary before an event whose information content is greater than that of both the immediately preceding and following events, and is also

[2] It should be noted that both models were outperformed by a more sophisticated explicitly generative model of entire phoneme sequences (Brent, 1999).

Model	1,705 Folk Songs			15 Pop Songs		
	Precision	Recall	*F*1	Precision	Recall	*F*1
Grouper	.71	.62	.66	.86	.82	.83
LBDM	.70	.60	.63	.79	.81	.78
IDyOM	.76	.50	.58	.57	.73	.64
GPR2a	.99	.45	.58	.70	.54	.58
GPR2b	.47	.42	.39	.47	.45	.43
GPR3a	.29	.46	.35	.26	.43	.30
GPR3d	.66	.22	.31	.17	.11	.11
PMI	.16	.32	.21	.24	.49	.31
TP	.17	.19	.17	.25	.45	.31
Always	.13	1.0	.22	.13	1.0	.23
Never	0	0	0	0	0	0

Table 6.1: Summary of results presented by Pearce et al. (2010a, 2010b) in order of *F*1 scores. The segmentation models are Grouper (Temperley, 2001), Local Boundary Detection Model (LBDM) (Cambouropoulos, 2001), the Grouping Preference Rules (GPRs) of GTTM (Lerdahl & Jackendoff, 1983), simple statistical models based on transition probabilities (TP) and pointwise mutual information (PMI) (Brent, 1999; Saffran et al., 1999) and two baseline models which respectively predict boundaries for every note (Always) and for no notes (Never). *F*1 is the unweighted harmonic mean of precision and recall.

high relative to the local context (more than *k* standard deviations greater than the mean information content computed in a linearly weighted window by event from the beginning of the current piece, where *k* is a free parameter of the model). IDyOM was configured with both short-term model (STM) and long-term model (LTM) predicting chromatic pitch, IOI, and rests and trained on the corpus shown in Table 3.2. The model was evaluated by comparison with phrase boundaries perceived by 25 participants in 15 pop and folk songs (Pearce et al., 2010b) and 1,705 German folk songs comprising 78,995 notes of which about 12% fell on phrase boundaries annotated by a musicologist (Pearce et al., 2010a).

In both studies, the model was compared with several other boundary detection models including the GPRs, LBDM, and Grouper as shown in Table 6.1. In general, performance was better than simple statistical models based on transition probabilities and the GPRs but worse than the rule-based models Grouper and the LBDM. However, it is worth noting that these models, unlike IDyOM, were designed and optimized specifically for phrase boundary detection in Western melody. As such, their component rules do not always apply naturally

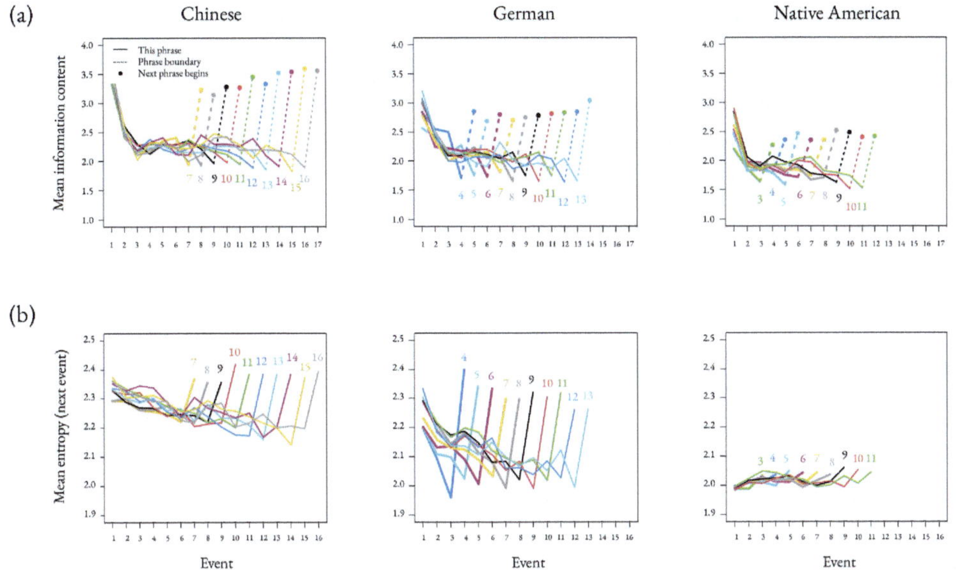

Figure 6.3: Mean information content and entropy with respect to phrase position for corpora of Chinese (N = 2,230), German (5,332), and Native North American (1,638) folk songs. (a) Average information content for each phrase position in phrases of differing length plus the first note of the subsequent phrase. (b) Average entropy in predicting the next note for each phrase position in phrases of differing length. The German and Chinese corpora are from the Essen folk song collection (Schaffrath, 1995) available at https://kern.ccarh.org; the Densmore Collection of Native American Folksongs (Shanahan & Shanahan, 2014) is available at https://github.com/shanahdt/densmore. Phrases from two to 20 notes were considered subject to a minimum of 500 phrases per corpus for each phrase length. IDyOM was configured to predict pitch with both the short-term model and the long-term model, the latter trained on each dataset with 10-fold cross validation, using the source viewpoint `interval⊗scale-degree` (see Chapter 3). Plots produced by Niels Chr. Hansen and used with permission.

to non-melodic aspects of music, in the way that information-theoretic boundary detection could be applied to harmonic movement for example (see Chapter 4). That said, there is a convincing effect of temporal proximity, illustrated clearly by GPR2a but also present in Grouper and the LBDM. The strong performance of Grouper suggests that metrical alignment and constraints on phrase length also play a role.

Niels Chr. Hansen extended these findings in a corpus analysis of more than 9,000 traditional German, Chinese, and Native North American songs (comprising more than 55,000 annotated phrases), focusing particularly on the entropy and information content of pitch predictions. Across a wide range of phrase lengths (from three to 16), phrases from these cultures consistently show a pattern of decreasing average information content and entropy throughout the phrase, followed by an increase in entropy while prospectively predicting—

across a phrase boundary—the initial note of a new phrase, which is also characterized retrospectively by higher information content (see Figure 6.3). In a series of logistic regression analyses, phrase boundaries were significantly predicted by prospective entropy change and retrospective information content change (where change indicates the difference between a note and the mean of the preceding notes).

The significance of pitch information content and entropy in perception of boundaries between melodic segments has been reinforced by the results of an experimental study conducted by Niels Chr. Hansen and Haley Kragness (Hansen et al., 2021). In a first experiment, participants listened to 56 extracts taken from the chorale melodies harmonized by J. S. Bach. Listening was self-paced with a press of the space bar on a keyboard sounding the next note in the melody, allowing for analysis of so-called *dwell times*—the amount of time spent processing (or dwelling on) each element in a self-paced sequence of stimuli. Increased dwell times have been observed at boundaries between segments of visually presented action sequences such as cleaning a room or making breakfast (Hard et al., 2011). In the present case, the stimuli consisted of a complete musical phrase (as notated in the score using fermatas) plus the first note of the subsequent phrase (see Figure 6.4). There were four categories of stimulus varying in terms of the target note which was either the first note of the subsequent phrase (the final note of the stimulus) or the final note of the phrase (the penultimate note of the stimulus) and the entropy of IDyOM's prediction for the note following the target note (high vs low entropy). Crucially all stimuli were matched for the information content of the target note. As shown in Figure 6.4, the results showed longer dwell times for high-entropy target notes regardless of whether they had originally appeared at the beginning or end of a phrase.

Entropy is computed after the target note appears, reflecting prospective predictive uncertainty for the next note based on the preceding context. Because information content and phrase position of the target note were controlled, these results provide concrete evidence for a prospective anticipatory effect of predictive uncertainty. This result is quite striking in the context of the majority of experimental investigations of musical expectation which rely on retrospective responses to musical events (see Chapters 3 and 4) and, therefore, fail to demonstrate *anticipatory* prediction. A subsequent analysis across all notes showed that dwell times increased with both greater information content and greater entropy. A second experiment presented the same stimuli with a fixed note duration (400 ms), truncated if necessary so that the target note was always the final note, to a distinct group of 31 participants, who rated how complete each stimulus sounded. The results corroborated those of the first

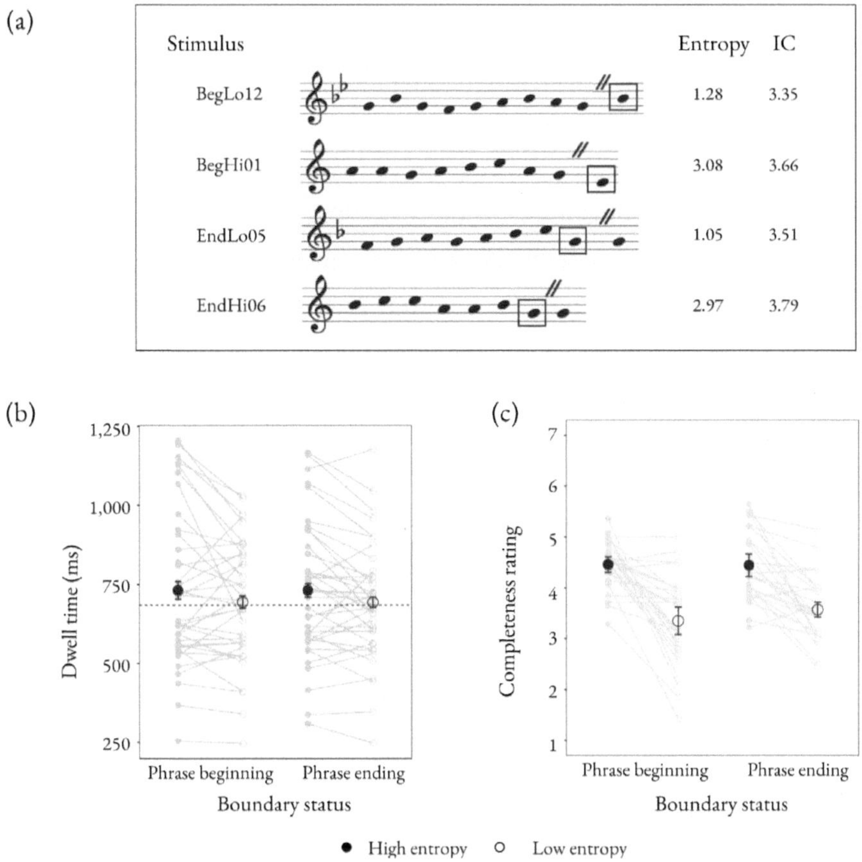

Figure 6.4: Influence of entropy on auditory boundary perception. (a) Stimulus categories from the experiments reported by Hansen et al. (2021). Stimuli were extracts from chorale melodies consisting of a complete phrase plus the first note of the subsequent phrase. Phrase boundaries are indicated by a double slash. The target note was either the first note (Beg) or last note (End) of a phrase while IDyOM's prediction for the next note was either high (Hi) or low (Lo) in entropy. Four example stimuli are shown, one for each category, illustrating the differences in entropy and also the fact that information content did not vary between stimulus categories. (b) Dwell times were greater for high- than low-entropy target notes regardless of the phrase position of the note in the original composition (Hansen et al., 2021, Experiment 1). (c) High-entropy target notes elicited stronger completeness ratings regardless of the phrase position of the note in the original composition (Hansen et al., 2021, Experiment 2). Figures created by Niels Chr. Hansen and reproduced with permission of the authors. Hansen, N.C., Kragness, H., Vuust, P., Trainor, L., & Pearce, M. T. (2021). Predictive uncertainty underlies auditory boundary perception. *Psychological Science*, 32, 1416–1425. Copyright ©2021 The Authors. DOI: 10.1177/0956797621997349

experiment, showing greater perception of completeness for high-entropy notes (i.e., those for which the next note was uncertain) regardless of phrase location (beginning or end).

These experimental results confirm that listeners are sensitive to the information-theoretic structure of music. If memory is constrained by an informational capacity based not only on the number of events but also on their predictability (i.e., their information content), then these dwell times can be interpreted as an effort to smooth out the rate of information processed over time, avoiding information peaks which might overload perceptual processing (Temperley, 2019). For example, in spoken language unpredictable words tend to be articulated more slowly (Aylett & Turk, 2004). This is consistent with findings reviewed in Chapter 5 demonstrating slowing down of tempo during expressive performance in anticipation of high information content events (Gingras et al., 2016). It is possible that phrase-final lengthening of musical notes during expressive performance (Honing, 2003) fulfils a similar role in smoothing out the information rate.

Overall, these results suggest a predictive process during perception where uncertainty and surprisal generally decrease with increasing context length. However, dramatic increases in uncertainty or surprisal flag that the preceding context has not been beneficial for prediction, indicating an informational boundary and the start of a new context for prediction. If segmentation is required by the experimental task or otherwise beneficial to perceptual processing, chunks may be formed at these informational boundaries in the musical sequence.

6.2.2 Stream segregation and musical voicing

Thus far, music has been conceived as a temporal sequence of discrete events, be they individual tones in a melody or individual vertical sonorities in a harmonic sequence. However, many pieces of music—indeed, more generally, many environmental auditory scenes—consist of multiple sequential components unfolding simultaneously in time. Therefore, in addition to grouping auditory events into sequential segments or chunks as discussed above, listeners also segregate a polyphonic musical scene into separate parallel perceptual streams, which may appear, disappear or merge with other streams dynamically throughout listening. This is a specific instance of the psychological phenomenon of auditory scene analysis (Bregman, 1990; Darwin, 1997), which allows the auditory system to make sense of complex auditory scenes by attempting to identify and separate auditory components corresponding to different sound-producing sources in the environment. Examples are shown in Figure 6.5.

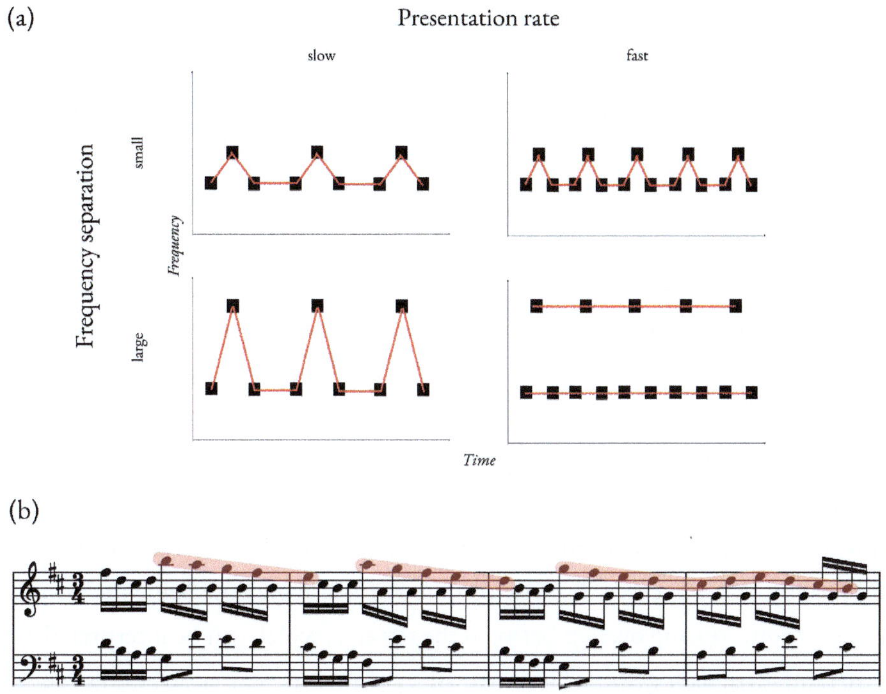

Figure 6.5: (a) An illustration of auditory streams perceived when listening to an ABA pattern formed of high and low tones (van Noorden, 1975). When the presentation rate is slow or the frequency separation between high and low tones is small, a single stream with a characteristic galloping rhythm (ba-da-dum ba-da-dum . . .) is heard. When the presentation rate is sufficiently fast and the frequency separation sufficiently large, two distinct streams are heard, one containing isochronous high tones and the other isochronous low tones. (b) An illustration of stream segregation in Prelude 15 from J. S. Bach's Well-Tempered Clavier (Book II, bars 9–12). Adapted from Pearce (2023) with permission.

Perceptual segregation of an auditory scene into parallel streams has traditionally been conceived as a psychological process of discrimination based on properties such as frequency separation, timbral similarity, and spatial location—once again inspired by Gestalt psychology (see Bregman & Campbell, 1971; Miller & Heise, 1950). More recent computational models of auditory stream segregation have emphasized prediction, assigning incoming auditory events to the stream whose predicted extension provides the closest match (Elhilali & Shamma, 2008; Mill et al., 2013). However, such models have been evaluated on simple artificial stimuli such as the ABA pattern (van Noorden, 1975), which alternates notes with pitches A and B of duration *d* followed by a silence of the same duration (i.e., ABA*d*ABA*d*...), with a galloping rhythm being perceived at slow tempi and two isochronous A and B streams at higher tempi (see Figure 6.5). It is not clear if these models can generalize beyond these examples to the much more complex examples of stream segregation that occur in polyphonic music, where top-down schematic influences of prior musical learning may also have an influence.

Sarah Sauvé (2018) developed an extension of IDyOM to polyphonic stream segregation based on the idea that events are assigned to the stream for which they make the most predictable extension given both long-term schematic models and short-term dynamic models learned while processing each stream as it is extracted. IDyOM was evaluated with a range of viewpoints but the best performing was a viewpoint linking pitch interval and scale degree, which is reported here for consistency with the results from Chapters 3 and 4. For the long-term model, IDyOM was trained on all four voices in the training set to ensure that learning was entirely unsupervised without relying on any labelling of voice in the training data. The slice-based representation introduced in Chapter 4 was used and four initial streams were assumed. At each slice, the model appends to each inferred stream the pitch with the lowest information content for that stream. This ensures that every stream is represented by a note in every slice but also allows for a single note being included in more than one stream, creating overlap between the streams, which can be interpreted as two or more streams merging together. Three heuristic constraints were added to the model: 1) for the first five slices, notes are assigned to streams by pitch proximity; 2) if a pitch recurs in successive slices, it is assigned to the same streams in the second slice as those assigned in the first slice; 3) if more than one note has the same information content for a given stream (or the same pitch proximity in the first five slices), all interpretations are maintained and the one with the lowest degree of overlap between the streams is chosen after the entire piece has been processed.

The system was evaluated on a voice separation task with 371 chorales harmonized by J. S. Bach (1875), which were divided randomly into a training set of 315, a validation set of 35, and a test set of 19 (after removing two chorales containing rests or with more than four parts). The test set was not used in developing the model to provide a robust measure of generalization performance. The voice separation task provides a proxy for perceptual stream segregation. It is important to note that musical voices (often corresponding to different instruments or performers) do not necessarily map directly on to perceptual streams since events from more than one voice may fuse into a single perceptual stream and, conversely, events in a single voice may segregate into different perceptual streams. However, David Huron (2001, 2016) used principles of auditory stream analysis to derive the traditional rules of voice-leading in Western musical practice with a focus on music written for up to four parts.[3] These rules place constraints on how the individual voices of a composition move from note to note, in terms of principles of auditory scene analysis, arguing that effective voice-leading in polyphonic music requires clear stream segregation between voices and integration within voices to create perceptually independent musical lines. For example, rules stating that pitches common to successive chords should be retained in the same voice (the common tone rule) and that large pitch intervals should be avoided (the avoid leaps rule) are derived from the empirically observed effects of pitch proximity on the coherence of an auditory stream. This provides some justification for the use of voice separation as a well-defined starting point for investigating stream segregation in music.

Performance of the stream segregation model based on IDyOM was assessed in terms of degree of overlap between the extracted voices (the proportion of notes assigned to just one voice for each slice) and the degree of overlap with the voicing indicated in the score (soprano, alto, tenor, and bass). These measures reflect respectively the degree of voice separation and the extent to which it matches the score. The results indicated relatively good performance on the validation set with an average voice overlap of 23% and an average of 72.8% of notes assigned to the correct voice. Performance on the test set was less convincing, but still better than chance, with 43.7% of slices containing overlap between the voices and an average of 51.1% of notes assigned to the correct voice. These initial results, though preliminary, provide some encouragement for a computational theory of stream segregation in perception of complex musical stimuli based on probabilistic prediction resulting from both long- and short-term statistical learning.

6.2.3 Outlook

The results presented in this section suggest that musical phrase structure is associated with characteristic changes in information content and entropy, which decline throughout a phrase and then increase across phrase boundaries, while listeners appear to be sensitive to these informational trajectories. Furthermore, a model based on these principles shows above chance performance in identifying notated and perceived phrase boundaries. However, the model does not perform as well as rule-based models that, for example, associate phrase boundaries with large temporal intervals between events and require phrase boundaries to be metrically aligned. If the information dynamics of predictive processing does influence

[3] Focusing on the music of J. S. Bach, Huron (2001) argued that pseudo-polyphony (or compound melody) in two- and three-part music gives way to more homophonic textures in music with four or more parts, to avoid exceeding a listener's limited perceptual capacity to represent multiple concurrent streams.

perception of grouping boundaries in music, it seems likely that this would be just one factor of potentially many, including temporal proximity, metrical alignment, and memory constraints. Evidence for a role of prediction in stream segregation of polyphonic music is even more preliminary. While the model performs above chance on the voice separation task, its performance leaves much to be desired and, again, it seems likely that Gestalt effects would continue to play a role. Furthermore, perceptual evaluation of the model's predictions is sorely needed since listeners' perception of musical streams might deviate significantly from voicing in the musical score.

6.3 Perceptual similarity

6.3.1 Modelling similarity

Accurate (albeit constrained) memory is essential for recognition of previously experienced musical material which, in turn, provides a basis for both statistical learning of regularities within this material and prediction of new musical material during listening. However, accurate memory has a potential disadvantage in that the more detailed the memory representation of a sensory event, the less likely that representation is to match incoming sensory information, which is rarely repeated precisely, even when the environmental source remains unchanged. Roger Shepard (1986) highlighted the fundamental importance of generalization for successful perception in enabling an individual to draw on past experience given the extremely small chance of any object or situation ever recurring in precisely the same perceptual configuration or context. We have already seen examples of generalization in IDyOM. When it encounters an event that it has not seen before in a given context (e.g., abra→d), the PPM algorithm reviewed in Chapter 2 generalizes by using a shorter suffix of that context in which the event has appeared in its prior experience (e.g., a→d). Furthermore, a guiding principle behind multiple viewpoint representations is to improve prediction performance by constructing equivalence classes containing sets of compositions that have different absolute pitch and timing but are recognized as being fundamentally similar due to having, for example, the same pitch interval structure (i.e., transpositional invariance) or the same pattern of IOI ratios (i.e., invariance to diminution and augmentation of absolute duration). These more abstract representations allow IDyOM to generalize its learning across stimuli that fall within the same equivalence class according to the viewpoint in question.

These examples of generalization relate directly to matching musical contexts for statistical learning and probabilistic prediction, forming—in machine learning parlance—part of the *inductive bias* of IDyOM. Shepard, on the other hand, was primarily interested in generalization on a different scale, using a similarity metric to place different perceptual experiences into the same category, allowing for generalization of responses to all members of the category. He proposed a *universal law of generalization* according to which the probability of categorizing two phenomena together is a negative exponential function of the distance between them in a Euclidean psychological space. However, Chater and Vitányi (2001) criticized this formulation on the grounds that a Euclidean space does not naturally accommodate phenomena, such as a narrative, that have internal structure and cannot be represented as a set of features or numerical values along orthogonal dimensions. Music is just such a phenomenon. The structural similarity between two pieces of music can certainly be influenced by overlap in global properties such as tonality and metre but is more likely to be related to the extent to which the two pieces share structural patterns—motivic structure in other words—at a range of different scales even if this doesn't involve exact repetition but rather variation on a theme. In the general case, it is hard to see how pieces of music, which can vary widely in length and contain a broad range of different kinds of internal structure relating to melody, harmony, voice leading and rhythm amongst other parameters, can be represented elegantly in a Euclidean space for the purposes of similarity estimation.

To address this concern (in the general case, rather than specifically with respect to music), Chater and Vitányi (2001) introduced a more general similarity metric for simulating psychological generalization, the *generalized universal law of generalization*. Given two sequences a and b, the conditional Kolmogorov complexity $K(a|b)$ is the length in bits of the shortest binary program that can generate a as its only output from b, while $K(a)$ is the special case when b is the empty sequence. The *information distance* between a and b can be defined as the shortest binary program that computes a given b and also computes b given a.

Since the Kolmogorov complexity is non-computable, a compression algorithm such as Lempel-Ziv or PPM is typically used to provide an approximate estimate by computing the length of compressed encodings of each sequence given a model trained on the other sequence. Using compression to approximate Kolmogorov complexity, similarity can be computed using a version of the information distance known as normalized compression distance or NCD (Li et al., 2004). The NCD between a and b is given by:

$$D_{NCD}(a, b) = \frac{\max(C(a|b), C(b|a))}{\max(C(a), C(b))} \tag{6.2}$$

where $C(a)$ and $C(b)$ are the length of compressed encodings of a and b respectively, $C(a|b)$ is the length of a compressed encoding of a given a model trained on b and $C(b|a)$ is the length of a compressed encoding of b given a model trained on a. NCD satisfies the properties of a metric:

$$D_{NCD}(a, b) = 0 \iff a = b \text{ (the identity axiom);}$$
$$D_{NCD}(a, b) + D_{NCD}(b, c) \geq D(a, c) \text{ (the triangle inequality);}$$
$$D_{NCD}(a, b) = D(b, a) \text{ (the symmetry axiom).}$$

IDyOM can be used to implement compression-based similarity measures as follows. Recall from Chapter 2 that IDyOM estimates the conditional probability $\mathrm{p}(\mathrm{x} \mid \overline{\mathrm{x}})$ of events in a sequence x_1^k of length k (see Equation 2.1). The information content of the event given the model, $\mathrm{h}(\mathrm{x} \mid \overline{\mathrm{x}})$, is the negative log, to the base 2, of the probability (see Equation 2.2) and represents a lower bound on the number of bits required to encode a compressed representation of x_1^k (Bell et al., 1990). Here we denote the information content resulting from the application of a model m to each event in a sequence x_1^k as $\mathrm{h}_m(\mathrm{x}_i \mid \overline{\mathrm{x}}), \forall i \in \{1...k\}$. Assuming that the IDyOM model m is initially empty (like the short-term model), $C(a)$ in Equation 6.2 can be estimated by summing the information content of each event in x_1^k:

$$C(a) = \sum_{i=1}^{k} \mathrm{h}_m(\mathrm{x}_i \mid \overline{\mathrm{x}}).$$

$C(a|b)$, the compression distance between a and another sequence b, is obtained using a model m_b with prior training on b, yielding an unnormalized, asymmetric compression distance:

$$\begin{aligned} D_1(a|b) &= C(a|b) \\ &= \sum_{i=1}^{k} \mathrm{h}_{m_b}(\mathrm{x}_i \mid \overline{\mathrm{x}}). \end{aligned} \quad (6.3)$$

Since the two sequences being compared can be of different lengths, NCD normalizes the compression distance between two sequences x and y with respect to the largest of their individual compressed lengths (see Equation 6.2). It is also possible to normalize directly with respect to length. Li et al. (2004) consider this possibility but note that it raises the question of whether to normalize with respect to the length of a or b (or the sum of the two) and also that the resulting measure does not satisfy the triangle inequality. The first question may be addressed by dividing the sum expressed in Equation 6.3 by k, yielding the average per-event compression distance:

$$D_2(a|b) = \frac{1}{k}\sum_{i=1}^{k} \mathrm{h}_{m_b}(\mathrm{x}_i \mid \overline{\mathrm{x}}). \tag{6.4}$$

This is equivalent to the cross-entropy measures used to evaluate the prediction performance of IDyOM in Chapter 2 (see Equation 2.11). A symmetric version of this distance follows naturally:

$$D_3(a|b) = max(D_2(a|b), D_2(b|a)). \tag{6.5}$$

This has efficiency advantages over $D_{NCD}(a|b)$ since $C(a)$ and $C(b)$ need not be computed. Furthermore, the failure to satisfy the triangle inequality need not necessarily concern us here, given that the goal is to model psychological similarity which may also violate the triangle inequality (e.g., Tversky & Gati, 1982).

6.3.2 Perception of musical similarity

Do these compression-based similarity measures accurately simulate human perception of similarity? Pearce and Müllensiefen (2017) compared the similarity measures D_1, D_2, D_3, and D_{NCD} with human judgements of the similarity between excerpts from Western pop songs (7–10 bars; 15–20 s) and versions with various kinds of alteration: 1. rhythm errors; 2. pitch errors not affecting pitch contour; 3. pitch errors changing contour; 4. errors in phrase order; and 5. modulation errors. In Experiment 1, 82 participants judged similarity for 84 stimuli consisting of 14 reference melodies each paired with six altered versions; in Experiment 2, 16 participants judged similarity for 24 stimuli consisting of two reference melodies each with 12 altered versions; and in Experiment 3, 10 participants judged similarity for 32 stimuli consisting of four reference melodies each paired with eight altered melodies. In Experiments 2 and 3, the altered versions were always transposed versions of the reference melody while in Experiment 1, the altered versions were not transposed.

The results indicated very high correlations between the compression-based similarity measures and the mean perceptual similarity ratings, with D_3 providing the strongest correlations overall (see Figure 6.6). The viewpoints yielding the strongest correlation varied between the experiments: `pitch` in Experiment 1 where there was no transposition ($r = .89$); `pitch`, `interval`, and `onset` in Experiment 2 ($r = .94$); and `pitch`, `interval`, `onset`, and `ioi-contour` in Experiment 3 ($r = .87$). To validate these best-fitting models, Pearce and Müllensiefen (2017) applied them unchanged to a very different set of data. Every year

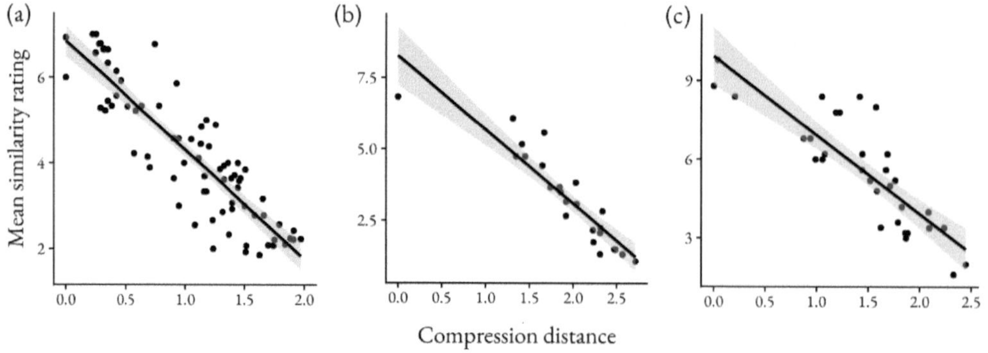

Figure 6.6: Correlations between compression distance (D_3) and mean perceptual similarity ratings in the three experiments of Pearce and Müllensiefen (2017).

the Music Information Retrieval (MIR) community organizes the Music Information Retrieval Evaluation eXchange (MIREX; Downie, 2008; Downie et al., 2010): a set of tests for comparing MIR algorithms in a number of different areas, including melodic similarity. In 2005, the challenge was to develop a computer program that ranks the similarity of 558 melodic incipits (the *candidates*) to each of 11 incipits (the *queries*) from the RISM A/II bibliographic database of post-1600 music manuscripts.[4] Algorithms were evaluated by comparison to human performance in which 35 musical experts ranked about 50 of the candidates by similarity to each query. Using the designer of the task's preferred performance measure (Typke et al., 2005), IDyOM's compression distance D_3 with viewpoints optimized on Experiment 2 (`pitch`, `interval`, `onset`) outperformed all other algorithms even though these in some cases had prior optimization on a comparable dataset (which is no longer available).

Overall, these findings demonstrate strong support for D_3 as a model of human musical similarity perception. The results suggest that perceptual similarity for music is best captured by a measure that is both normalized, reflecting the average per-event compression distance, and symmetric, reflecting the largest distance between pairs of stimuli presented in either of the two possible orders. The results also suggest that listeners are able to use different representations, depending on the nature of the stimulus comparisons presented to them. This relates neatly to the idea that listeners implicitly track multiple features of auditory stimuli and use those features that allow them to most effectively perform a given task with given stimuli (see § 3.5).

[4] Répertoire International des Sources Musicales (RISM). Serie A/II, manuscrits musicaux après 1600. https://www.rism.info/.

6.3.3 Perception of thematic structure in music

One area of music perception in which similarity plays a major role is perception of thematic structure; while listening to a piece of music, repetitions, variations, and developments of themes are often perceived and appreciated. However, since they are generally not exact repetitions of the original theme, identification of these approximate thematic reoccurrences depends on perception of similarity rather than exact matching. Perception of these elements of thematic structure allows relationships between the different parts of a composition to be apprehended, thereby lending a piece of music a sense of coherence or unity. Using IDyOM as a platform, Edward Hall developed a computational model of the perception of large-scale thematic structure in music (Hall & Pearce, 2021) which implements the idea that thematic coherence emerges through approximate repetition of material.

Several quantitative measures of thematic structure can be derived from the model including *internal predictability* (the mean STM information content of all notes in a piece) and *stylistic predictability* (the mean LTM information content given training on a stylistically representative corpus). Hall and Pearce (2025) reported a pair of experiments designed to test whether these measures account for human perception of thematic structure, both of which used a melodic corpus of European art music spanning more than 300 years from the 18th to 20th centuries—from Handel to Vaughan-Williams—specially constructed for the purpose (see Figure 6.7).[5] In the first experiment, 40 participants listened to 100 pairs of melodies consisting of a theme and a repetition (according to the model), rating the extent to which the two excerpts sounded as if they came from the same piece. The results showed that ratings were significantly predicted by the compression distance between the two excerpts with a smaller effect of the difference in stylistic predictability between them.

While the first experiment examined perception of thematic structure on a very local timescale (essentially an immediate comparison of a theme as originally presented and subsequently restated), the second experiment examined perception of thematic structure across longer timescales. Forty participants listened to 40 two-minute-long melodies, which were selected so as to distinguish the model-derived measures of thematic structure described above. During the second half of each melody, participants were visually cued four times to report whether they had heard the music sounding at the time of the cue at any point

[5]The MCCC is available at http://osf.io/dg7ms/.

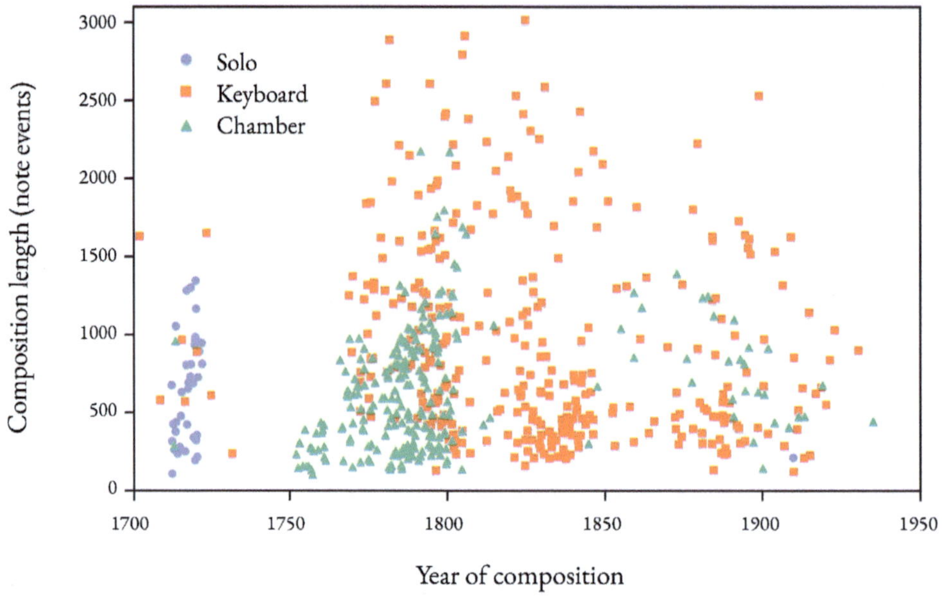

Figure 6.7: The Monophonic Corpus of Complete Compositions (MCCC) showing the distribution by composition year, number of note events and instrumentation type. Created by Edward Hall and reproduced from Hall and Pearce (2021, Figure 1) under the terms of the Creative Commons attribution licence (CC-BY 4.0).

previously in the piece. After listening to the whole piece, participants rated their perception of the unity of the piece.[6]

The results showed significant effects of internal predictability for both pitch and rhythm on both experimental measures of thematic structure: recognition memory and coherence. Musical passages that were more unpredictable for IDyOM's short-term model—making predictions in the context of learning from the preceding portion of the piece—were less likely to be identified as having been heard before, and pieces with higher overall information content for IDyOM's short-term model were rated as less unified. In other words, passages having lower compression distance (i.e., greater similarity) with the material appearing earlier in the piece were more likely to be rated as having occurred before, leading to perception of a more unified composition.

[6]Following Tan et al. (2006), unity was defined for participants as follows: "For a unified piece, even though there may be many different ideas in it, the music still sounds like one well-integrated, whole, single composition. A piece that is not unified, on the other hand, is one that sounds like unconnected fragments of music that do not seem to belong together, so that they do not hold together as one well-integrated, whole, single piece of music."

These findings were confirmed by two subsequent experiments using even longer pieces taken from the same corpus (Hall, 2023). In the first experiment, participants listened to 20 two-minute and 20 four-minute extracts containing equal numbers of monophonic and polyphonic pieces; after each piece, they rated three possible melodic continuations each lasting approximately 10 seconds for how well they would continue the context. The results showed that higher ratings were given for continuations with lower information content from an IDyOM model trained on the context. In a subsequent experiment, the same 40 stimuli were divided into segments (7–15 per piece) and a separate group of participants arranged them into the order they felt made the most coherent piece of music. The internal unpredictability of the resulting compositions (from IDyOM's short-term model) was significantly lower than that expected by chance (where chance levels were estimated by Monte Carlo simulation over all possible permutations of orders). Together these experiments reveal the role of perceptual similarity—simulated by IDyOM's compression distance—on perception of thematic structure in music over both short and long timescales within a composition.

6.4 Conclusion

The goal in this chapter was to consider whether the mechanisms of statistical learning and probabilistic prediction, as implemented in IDyOM, are capable of simulating a broader range of cognitive processes involved in music perception: memory, grouping structure, and similarity. The overall finding that the same psychological processes of statistical learning and probabilistic prediction may be capable of simulating all these apparently distinct psychological phenomena lends support both to the model itself and to the hypothesis that statistical learning and predictive processing play a fundamental underpinning role in music perception.

7 Musical Enculturation

The empirical work considered thus far has focused on adults from cultures in which Western music is prevalent. In most cases, the analyses presented involve training the IDyOM model from Chapter 2 on a large corpus of Western music and examining the extent to which its responses to musical stimuli capture the responses of these listeners to the same stimuli. The results corroborate the model-based simulations across a wide range of musical stimuli and psychological domains. This evidence is consistent with an underlying process of probabilistic prediction based on statistical learning, according to which listeners have acquired an internal probabilistic model reflecting the statistical structure of the music of their culture, which informs their perception and experience of music. However, since the evidence is fundamentally correlational, it does not provide causal corroboration of the statistical learning hypothesis from Chapter 1. In other words, we have thus far focused on listeners who are already strongly *enculturated* in a musical style rather than examining the process of *enculturation* itself.

According to the present account, enculturation is a fundamental psychological process underpinning musical experience, playing an important role in both perception of structure and experience of pleasure when listening to music. To situate this claim more clearly, it is worth considering a counter-hypothesis that adult listeners (with similar psychological profiles in terms of personality, intelligence, and other individual differences) would perceive and experience music in exactly the same way regardless of differences in their lifetime experience of music (and other auditory stimuli such as speech and environmental sounds). The evidence to be covered in the present chapter suggests that this is not the case. This does not, however, imply that there is no role for universal psychological constraints, impervious to experience, in the process by which an individual becomes enculturated in a musical style. In contrast, it is very likely that a whole host of universal factors constrain learning including perceptual factors (e.g., constraints on the range and resolution of pitch and time perception) as well as constraints on learning and memory (e.g., representational or capacity limitations). These factors are considered further in Chapter 9.

The present chapter is concerned with direct assessment of the statistical learning hypothesis by considering whether perception and experience of music vary systematically as a function of differences in musical experience. There are three main ways such differences can be investigated experimentally. First, and perhaps most obviously, through cross-cultural comparisons, it is possible to test whether individuals from two (or more) different musical cultures show characteristic differences in perception that can be related systematically to differences in the music of those cultures. Second, through examination of developmental trajectories across the lifespan of listeners living within a given musical culture, it is possible to test whether listeners show changes in perception that are systematically related to their accumulating experience of the music of that culture. The third approach combines elements of the first two in a controlled, self-contained experimental paradigm known as artificial grammar learning. During an exposure phase, listeners are presented with sequences of sounds created according to an artificially constructed grammar designed to generate structures that are different from those present in music. This provides a kind of *tabula rasa* for investigating the process of statistical learning, which can be examined precisely and in detail as it unfolds during exposure. During a subsequent test phase, listeners can be tested as to whether they have learned the structure of the grammar, which can include comparisons between listeners exposed to different grammars, who should show characteristically different responses systematically related to their differential exposure.

We will consider each of these three approaches in turn, starting with artificial grammar learning before turning to cross-cultural comparisons, and ending with developmental trajectories. This chapter is concerned exclusively with the implicit effects of passive, incidental exposure to music whereas Chapter 8 considers the effects of explicit musical training.

7.1 Artificial grammar learning

According to the statistical learning hypothesis, enculturation occurs through implicit statistical learning during incidental exposure to music resulting in cognitive representations of the structural and statistical regularities making up the syntax of the musical style. Research using artificial grammar learning has demonstrated the basic psychological mechanisms of implicit statistical learning underlying this hypothesized process of enculturation. The goal of artificial grammar learning research is to investigate the mechanisms by which individuals implicitly learn the structure of an artificial domain through experience.[1] In this approach a set of artificial auditory sequences is generated, conforming to some set of structural principles that are designed to differ from those of actual music. Participants in the experiment are exposed to these sequences and subsequently tested for any learning that may have taken place during exposure. Rather than relying on an individual's previous experience outside the lab, which is inherently uncontrollable, the power of this approach lies in its ability to demonstrate causal effects of exposure, through interventions that are entirely self-contained within the Petri dish of the experimental session.

There is a large and long-standing literature on artificial grammar learning, dating back to Erwin A. Esper in the 1920s (see Christiansen, 2019), through the landmark studies of Miller (1958), Reber (1967), and Saffran et al. (1996) to the present day (for reviews, see Frost et al., 2019; Saffran & Kirkham, 2018). Here, I will focus on three influential studies that are most relevant to the question of musical enculturation specifically.

Tillmann and Poulin-Charronnat (2010) constructed a finite-state grammar with six states whose structure defined the probability of transitioning from one state to another. Each connected pair of states was associated with a specific tone (from the non-diatonic set: a, a#, c, d, f#) which was added to the tone sequence generated on transitioning from one to the other (all transitions between states were unidirectional). The structure of the grammar was designed to allow for a range of initial and final notes. Participants were divided into two groups: an experimental group who listened to 37 grammatical melodies during an exposure phase prior to the test phase and a control group who moved straight to the test phase without prior exposure to melodies generated by the grammar. During the test phase, participants were presented with the 37 grammatical melodies randomly shuffled with 37 ungrammatical melodies created by changing one tone in each of the grammatical melodies. Each of these melodies was presented in two versions: the original and a version with one of the tones mistuned by 52 cents. While listening to each melody, a visual cue indicated the arrival of the target note and participants responded as to whether or not this target note sounded mistuned. Importantly, the participants were unaware of the presence of grammatical and ungrammatical melodies generated according to a grammar. The results showed that the experimental group responded more quickly and accurately for

[1]For simplicity, I will use the term artificial grammar learning to refer to research examining learning of artificially constructed stimuli, even when the grammar is not specified explicitly. The fields of research using artificial grammar learning methods are traditionally referred to as *implicit learning* and *statistical learning* or, more recently, *implicit statistical learning* as a unifying umbrella term (Christiansen, 2019). In this chapter, it makes sense to emphasize the artificial nature of the stimuli which differentiates artificial grammar learning from cross-cultural or developmental approaches, while implicit statistical learning is common to all three.

the grammatical melodies than the ungrammatical melodies, while the control group showed no such difference. Notably, the performance of the experimental group, but not the control group, was related to digram and trigram frequencies in the exposure set of grammatical melodies.

Rohrmeier et al. (2011) constructed a more complex finite-state grammar consisting of 13 states that emitted pairs of tones, designed to ensure that the resulting melodies were predominantly composed of small intervals but otherwise contained few prototypical Western tonal patterns. During an exposure phase, participants in an experimental group listened to 17 of the 33 melodies generated by the grammar. Participants in a separate control group received no exposure to the melodies from the grammar. In a subsequent testing phase, rather than using an implicit measure of learning like Tillmann and Poulin-Charronnat (2010), participants gave forced choice familiarity judgements for all 33 melodies, randomly shuffled with 33 ungrammatical melodies. The results showed that the experimental group distinguished the grammatical melodies from the ungrammatical melodies more accurately than the control group. Unexpectedly, however, the control group also performed significantly above chance, with their performance increasing from initial chance levels throughout the testing session as they learned from the test melodies.

Loui et al. (2010) went a step further in developing artificial musical styles that would be unfamiliar to their participants. As well as creating two finite-state grammars, the terminals of the grammars were also selected from a scale that differed substantially from the familiar division of an octave (a doubling of frequency) into 12 (roughly) equal logarithmic steps. Instead, the stimuli were created using the Bohlen-Pierce scale, which divides a tritave (a tripling of frequency) into 13 logarithmically equal steps. In an exposure phase, participants listened to 400 eight-tone melodies generated by one of the two grammars. A subsequent test phase was designed to assess effects of exposure on both expectation, like Tillmann and Poulin-Charronnat (2010), and familiarity, like Rohrmeier et al. (2011). Rather than using an implicit measure of expectation, like Tillmann and Poulin-Charronnat (2010), explicit goodness-of-fit ratings for the final probe tone were taken before and after the exposure phase. The post-exposure ratings showed stronger correlations with the distribution of tones in the exposure set than the pre-exposure ratings. Interestingly, the pre-exposure probe-tone ratings were themselves significantly correlated with tone frequencies in the exposure set of stimuli but this correlation disappeared after accounting for the frequency of tones in the stimulus context, suggesting that participants learned even from the eight tones making up the probe-tone context.

Loui et al. also asked participants to listen to pairs of stimuli and choose the most familiar. In one condition, designed to test recognition memory, the pair was made up of one melody from the exposure set and another generated by the alternate grammar. In another condition, designed to test generalization, the first of the pair was a grammatical melody that the participant had not previously heard. In both cases, participants selected grammatical melodies at a rate significantly above chance, indicating successful recognition memory but also successful generalization of learning to novel grammatical melodies.

It is striking that these studies found evidence for statistical learning within individual stimuli (Loui et al., 2010) as well as longer term dynamic learning across stimuli presented within the experimental session (Rohrmeier et al., 2011). These map neatly onto the statistical learning shown by IDyOM's short-term and long-term models respectively. Furthermore, the fact that learning occurred within stimuli and across the test session even for participants without prior exposure to the grammar speaks to the dynamic, implicit, and obligatory nature of the learning. Subsequent research using implicit measures of statistical learning (i.e., reaction times or recall errors) points to logarithmic trajectories of knowledge acquisition across an experimental session (Krishnan et al., 2022; Siegelman et al., 2018). It is noteworthy that neither Loui et al. (2010) nor Rohrmeier et al. (2011) found any difference in implicit statistical learning between musicians and non-musicians, suggesting that this ability is orthogonal to any knowledge or expertise acquired through explicit musical training (see Chapter 8 for further discussion).

The results reported by Tillmann and Poulin-Charronnat (2010), Rohrmeier et al. (2011), and Loui et al. (2010) demonstrate the human capacity for statistical learning of tone sequences, just as hypothesized and implemented in IDyOM (see Chapters 1 and 2). Given this, we would expect IDyOM to be able to predict listeners' responses in such artificial grammar learning experiments. This was demonstrated in an experiment run by Ioanna Zioga, using the stimuli generated by the artificial grammar of Rohrmeier et al. (Zioga et al., 2020). When self-transitions are removed (i.e., transitions from a state back to itself), this grammar generates 18 melodies that vary in length from eight to 22 notes. Across four sessions each taking place on a separate day, 40 participants were extensively exposed to 12 of these melodies through a combination of passive listening and active recall. The fourth and final session included a generalization task, like that of Loui et al. (2010), in which participants were presented with stimuli derived from the remaining six grammatical melodies that they had not experienced so far. IDyOM was used to select notes whose probability was in the highest and lowest 30% across all notes in all 18 melodies: 16 of the resulting high-probability notes and eight of the low-probability notes were from the six generalization melodies. These melodies were truncated such that the selected note became the final note of the stimulus and were presented to participants along with versions whose final note had been replaced with a note that was ungrammatical according to the grammar. This yielded melodies with three kinds of final note: high probability, low probability (but still grammatical), and ungrammatical. Participants listened to these stimuli and rated how surprising they found the final note. The results showed that low-probability final notes were more often judged as surprising than high-probability notes, while ungrammatical notes were more often judged as surprising than grammatical notes (both high and low probability).

A major advantage of the artificial grammar learning approach is that the structure of the artificial domain can be carefully designed to test hypotheses about the learnability of different structures while the rate, duration, and conditions under which the artificial stimuli are presented can be precisely controlled to test hypotheses about the nature of the exposure required for learning to take place. There are two main disadvantages of this method, as a way of understanding musical enculturation. First, the structure of the grammar (and the stimuli it generates) must necessarily differ from the structure of real-world musical styles. Second, the conditions under which learning takes place are far more structured and concentrated than is typically the case with musical exposure during enculturation. As a result, while artificial grammar learning provides evidence that listeners are capable of statistical learning, it has low ecological validity and does not demonstrate that musical syntax is actually acquired via implicit statistical learning during incidental exposure. For this we must turn to cross-cultural and developmental research.

7.2 Cross-cultural comparisons

Artificial grammar learning studies necessarily lack ecological validity, since the stimuli must not resemble existing musical stimuli, and are limited to relatively simple structural regularities that can be learned over the short timescales possible within an experimental study. A more direct, albeit logistically more challenging, way of investigating musical enculturation is to compare the perception of listeners from different musical cultures. The most robust results arise from *fully comparative* cross-cultural studies which include both participants and music from at least two distinct musical cultures (Patel & Demorest, 2013), ideally accompanied by music from a third culture equally unfamiliar to the first two cultures acting as a control condition (Demorest et al., 2008). Evidence for enculturation results from differences in some aspect of music perception between the cultures that can be predicted on the basis of stylistic differences between the music of those cultures.

7.2.1 Cross-cultural differences in music perception

Inspired by the writings of Leonard Meyer, John R. Platt proposed a thought experiment:

> If the playing of 'Three Blind Mice' were interrupted at one point or another, a Western 6-year-old child with a good ear and some musical experience could probably pick out many of the next fulfilment-notes on a piano, even if he had not heard the tune before (if such a combination of experience and musical naïveté is possible in our society). But a comparable Senegalese child or a Chinese child, from a different culture with a different musical scale, probably could not do this nearly as well, except perhaps for the identification of the universal keynote at the end. On the other hand, these other children would have expectations for their own music that we would be unable to guess.
>
> (Platt, 1961, p. 425)

Two studies published in 1984 did almost exactly as Platt had proposed two decades earlier, using the probe-tone method to compare the pitch expectations of, respectively, Indian and US participants listening to North Indian music (Castellano et al., 1984), and Balinese and US participants listening to Balinese *gamelan* music (Kessler et al., 1984). The comparison is not as easy at it might first appear due to the need to distinguish schematic knowledge, acquired through prior cultural experience, from dynamic knowledge, acquired during the experiment through experience of the stimuli. Recall the results of Loui et al. (2010), discussed in § 7.1, showing that listeners develop expectations based on rapid learning of the distribution of tones *within* a stimulus. Oram and Cuddy (1995) investigated this phenomenon directly in a probe-tone experiment using artificially generated tone sequences in which the relative frequency of each possible tone was carefully controlled. Listeners gave higher goodness-of-fit ratings for tones that appeared more frequently in the preceding context than tones that appeared less frequently. This presents a difficulty for cross-cultural research because listeners who are unfamiliar with a musical culture may nevertheless learn the distributional properties of stimuli presented in an experiment, even when they are taken from an unfamiliar musical culture. This learning will, in turn, influence their expectations. To the extent that these distributional properties mirror the overall distributional properties of the style, it may be difficult to distinguish such dynamic knowledge acquired through exposure to the experimental stimuli from schematic knowledge resulting from long-term enculturation (see § 3.1 for the distinction between dynamic and schematic knowledge).

Effects of schematic and dynamic knowledge were carefully distinguished by Castellano et al. (1984), who compared probe-tone responses of 16 participants in the US, half of whom had extensive experience of North Indian classical music while the other half had no experience, while both groups had extensive experience of Western music. The stimuli were created to reflect North Indian classical music in which a rāg is a collection of melodic features, including a scale (or thāt), used as a basis for improvisatory performance. The stimulus contexts were themes from 10 different rāgs (reflecting eight different thāts) and the probes were the 12 chromatic tones within an octave. The resulting probe-tone profiles for the two groups were quite similar. However, the ratings of the Indian listeners showed significant influences of both thāt membership—probe tones that were members of the thāt received higher ratings than those that were not—and the duration with which the probe tone appeared in the stimulus context, while those of the US listeners showed significant effects only of tone duration. Thus, while both groups showed dynamic statistical learning from the stimuli, only the Indian group showed an additional schematic influence of inferred tonality (see § 4.3.2) reflecting their enculturation in North Indian classical music.

Kessler et al. (1984) went a step or two further in conducting a fully comparative cross-cultural probe-tone study including Balinese and US participants listening in their native contexts to both Western and Balinese stimuli. They studied two groups of listeners in Bali: 26 students in a music conservatory called Kokar and 27 residents of a remote village with very little, if any, exposure to Western music. The story of how this village was identified makes for one of the more colourful methods sections in the literature—there cannot be many that combine such disparately exotic words as motorbike, dysentery and volcano:

> In order to find a village with these rare properties, one of us (C. H.) and an interpreter (her husband Putra), after having collected the data at Kokar, set out on motorbike toward the remote northeast part of the island of Bali. At each stop, local villagers were asked "What is the most isolated village in this area? Does it have a gamelan?" Then, when the indicated village was reached, the same question was repeated and the quest resumed. After various mishaps—flat tires, narrowly missed collisions, mechanical break-downs, strained knees, motorbike spills, skin gouges, thorn punctures, infections, dysentery and fever—this search finally ended at a tiny thatch-roofed village (which we refer to as "Keker"), reachable only by footpath, on the slopes of the volcano Gunung

> Agung (whose 1963 eruption had destroyed and forced abandonment of the village for 3 years).
>
> (Kessler et al., 1984, p. 137)

The stimulus contexts consisted of Western diatonic melodies and Balinese melodies in two different Balinese tuning systems and scales: *pelog* and *slendro*. The probe tones were the complete set of tones within one octave of the corresponding scale: the 12 chromatic tones for diatonic contexts, seven and five tones respectively for *pelog* and *slendro* contexts. A hierarchy of tonal stability in the Balinese music was modelled by distinguishing scale tones from non-scale tones and the *gong tone*, which is emphasized at important structural locations in the music such as phrase endings, from the remaining scale tones.[2] The analysis was complicated by enormous diversity of response patterns amongst the Keker villagers, some of whose responses reflected simply the size of the pitch interval between the probe tone and final tone of the context (i.e., pitch proximity). However, the responses for the *slendro* contexts showed evidence of enculturation: the ratings of participants from the USA correlated with the frequency with which the tone appeared in the context but not with tonal hierarchy while those of all Balinese listeners correlated positively with tonal hierarchy but not (or negatively) with frequency of occurrence.[3]

Krumhansl et al. (2000) conducted an experiment using traditional spiritual songs (*yoiks*) of the indigenous Scandinavian people (Sami) presented in a probe-tone study to European listeners (with no *yoik* experience), Finnish undergraduates in music education from the University of Jyvaäskylä (with a little *yoik* experience) and Sami yoikers (with extensive *yoik* experience). Although there were significant correlations between the expectations expressed by the three groups of listeners, there were also significant differences. Expectations of Finnish students were more strongly correlated than those of the Sami listeners with a measure of schematic Western tonal expectations (the probe-tone responses of US adults in single-interval contexts collected by Krumhansl, 1995b). The responses of all listeners were correlated with statistical regularities in a corpus of 18 *yoiks* from which the eight stimulus contexts used in the probe-tone experiment were selected, presumably based on dynamic statistical learning of the stimuli in the case of the non-Sami participants. However, the Sami

[2] In the *pelog* tuning system, only five out of the seven scale degrees are typically used in an individual piece whereas in *slendro* all five tones are usually used, so only the gong tone was distinguished.

[3] For the diatonic and *pelog* contexts, the responses of the US participants, Kokar students and a subgroup of the Keker villagers showed comparable correlations with frequency of occurrence and tonal hierarchy but not pitch proximity, while the responses of a second Keker subgroup correlated with pitch height, and those of a third subgroup showed no significant correlations with any predictor.

were much better able to predict the actual next tone than the other groups. This effect was greater for *yoiks* that were familiar to the Sami, suggesting an effect of veridical knowledge (see § 3.1). Nonetheless, it remained even for unfamiliar *yoiks*, demonstrating an effect of stylistic enculturation.

While the probe-tone method provides a very detailed lens for examining pitch expectations, in the context of cross-cultural differences it suffers from the challenge of robustly differentiating effects of long-term statistical learning through enculturation from short-term statistical learning within the experiment. One way of circumventing this problem is to use a recognition memory paradigm in which participants are presented with unfamiliar music from both familiar and unfamiliar cultures and subsequently asked to recognize target excerpts from both cultures along with comparable foils that did not appear. This approach produces focused insight into perceptual processing of fragments of music rather than individual notes, making it harder to discern which musical features contribute to any resulting differences in recognition memory between cultures. However, the approach has two significant advantages: first, stimuli from both familiar and unfamiliar cultures are equally represented within the experiment, so any differences in recognition memory between cultures can be attributed robustly to effects of long-term enculturation, since general memory ability should not vary between cultures; and second, complex, ecological stimuli can be used (Patel & Demorest, 2013).

In a fully comparative cross-cultural study, Demorest et al. (2008) found that Turkish and US adults' recognition memory for novel Western, Turkish, and Chinese music was better for music of the native culture. While the US participants showed better memory performance for Western music than both Chinese and Turkish music, the Turkish participants showed better performance for the Turkish music than the Western music but also better performance for the Western music than the Chinese music. Morrison et al. (2008) replicated the enculturation effect for US adults and children (10–11 years old) using the same stimuli (recorded ensemble pieces) as well as simpler stimuli (solo instrumental melodies). Going further, Demorest et al. (2016) found that the enculturation effect for US adults was equally strong for the original full-ensemble recordings, mechanically performed transcribed melodies and isochronous versions of the transcribed melodies with no rhythmic structure. These results suggest that the differential effects of enculturation on memory performance between Turkish and US listeners depend primarily on sequential pitch structure.

The Turkish participants in the experiment of Demorest et al. (2008) had exposure to both Turkish and Western tonal musical cultures. Their secondary enculturation in Western music is reflected in the fact that they showed significantly better recognition memory performance for Western music than for Chinese music, albeit weaker than for Turkish music.[4] This result suggests that listeners can become enculturated in more than one musical style in much the same way as multilingual individuals become native speakers of more than one language. Wong et al. (2009) examined this phenomenon of *bimusicalism* directly, hypothesizing that equal degrees of musical exposure to two musical cultures would result in equal sensitivity to both cultures. This *bimusical hypothesis* was contrasted with the *early exposure* and *current exposure* hypotheses, which predict an advantage for cultures experienced, respectively, early in life, or most recently and frequently in the current environment. In two separate experiments, they compared recognition memory performance and tension judgements for unfamiliar Indian and Western music in three groups of listeners: monomusical adults born and raised in the US with no experience of Indian music; monomusical Indians recruited from rural villages in North-Eastern India; and bimusical Indians who were equally familiar with Western and Indian musical cultures. The bimusical Indians were US residents born either in India or the US and had early exposure to Indian music but currently listened to equal amounts of Indian and Western music, the latter being dominant in the ambient musical environment. While the monomusical participants showed better recognition memory and lower experience of tension for their native cultures, the performance of the bimusical Indians was no different for Indian and Western music, corroborating the bimusical hypothesis.

These results provide evidence for enculturation effects in bimusicalism but they do not shed much light on the underlying knowledge representations acquired during enculturation in multiple musical styles, leaving many questions unresolved. Is the knowledge acquired during enculturation stored separately for the two cultures or in a single representation? If the former, are there interactions between the two representations? Is the knowledge acquired by a bimusical individual for one of their musical cultures represented in the same way as the knowledge acquired for that culture by a monomusical individual? David Huron (2006, Chapter 11) suggested that listeners form different schematic knowledge representations for the different musical styles that they experience (see Chapter 1). When listening to a piece of music, an individual will use any cues available to identify the style as quickly as possible. These cues might include surface characteristics such as the timbral palette along with the non-musical context to activate rapidly the corresponding schematic knowledge stored in

[4] See also Matsunaga et al. (2018) in which Japanese listeners responded to Western melodies very similarly to US listeners, presumably due to cultural exposure to Western music in Japan.

long-term memory. If the music disconfirms the listener's expectations derived from the selected schema, this might prompt switching to a new schema corresponding to a different musical style or, if no appropriate existing schema exists, the creation of a new one. Huron proposed that these different stylistic schemas are separated by cognitive barriers or *firewalls* that allow a bimusical person to avoid confusion resulting from leakage from one musical culture into perception of another. The strength of these barriers could depend on several factors related to the extent and context of experience with the two musical cultures involved.

Within the modelling framework proposed in this book, the scenario outlined by Huron could be simulated with multiple long-term IDyOM models, one for each of the styles with which a listener is familiar, combined with mechanisms for dynamically selecting and switching between stylistic models. If these models were associated with timbral signatures of the corresponding style, for example, then they could be activated by recognition of timbral features. Another possibility would be to develop a mechanism that switched models when information content, given the current stylistic model, remained high for a sustained period of time (see § 5.5.3).

7.2.2 IDyOM simulations

While the empirical evidence from cross-cultural studies demonstrates enculturation effects, it does not provide detailed understanding of the psychological mechanisms underlying these effects. This is where computational modelling comes to the fore. According to the statistical learning hypothesis, musical structure is learned through exposure resulting in observable differences between enculturated listeners from different musical cultures. Leonard Meyer (1957, p. 420) suggested that the separation between two cultures would have an impact on communication of meaning: "There appears to be positive correlation between cultural distance, whether historical or anthropological, and cultural noise. That is, the more distant a culture is from our present set of habit responses, the greater the amount of cultural noise involved in communication." Extending and generalizing this idea, Demorest and Morrison (2016, p. 189) proposed the *cultural distance hypothesis*: "the degree to which the musics of any two cultures differ in the statistical patterns of pitch and rhythm will predict how well a person from one of the cultures can process the music of the other." IDyOM provides an effective way of testing this hypothesis.

To this end, IDyOM models were trained on corpora reflecting different musical cultures, simulating listeners from those cultures. Listeners with enculturation in Western music were simulated by training a model on a corpus of German folk songs (the *German model*) and

Chinese listeners by training a model on a corpus of Chinese folk songs (the *Chinese model*). Each model was used to make both within-culture and between-culture predictions. For the within-culture predictions (i.e., the German model processing German folk songs or the Chinese model processing Chinese folk songs), IDyOM was used to estimate the information content of every event in every composition in the corpus (using 10-fold cross-validation to create training and test sets from the same corpus). For between-culture predictions, IDyOM was first trained on the within-culture corpus (e.g., the German folk song corpus for the German model) and then used to estimate the information content of every note in every composition in the other corpus representing the comparison culture (e.g., the Chinese corpus for the German model). IDyOM was configured to use only its long-term model trained on the appropriate corpus; the short-term model was not used. In all cases, information content was averaged across notes yielding a mean information content value representing the unpredictability of each composition for each model. The results are shown in Figure 7.1 and Table 7.1.

For the comparison between cultures (German vs Chinese), the data is plotted in Figure 7.1 for each composition in the two corresponding corpora: information content for one model is plotted on the abscissa while information content for the second model is plotted on the ordinate. The line of equality ($x = y$) indicates equivalence between the two models: compositions lying on this line are equally predictable for each model and do not distinguish the two cultures; in other words, they should be equally familiar and predictable to listeners enculturated in each musical style. Positions near the origin represent compositions that are predictable within both cultures while positions far from the origin represent compositions that are unpredictable within both cultures. Positions further away from the line of equality represent compositions that are predictable for the simulated model of one culture but unpredictable for the simulated model of the other culture. Distance from the line of equality, therefore, provides a quantitative measure of cultural distance (as conceived by Demorest & Morrison, 2016) based on probabilistic modelling of musical enculturation. Figure 7.1(a) illustrates how cultural distance is computed for a comparison between IDyOM models trained on German and Chinese corpora and, by rotating the data points through 45°, Figure 7.1(b) shows the same data with cultural distance on the ordinate and culture-neutral complexity on the abscissa. In this example, IDyOM correctly classifies 98.52% of the folk songs by culture (Chinese vs German). Moreover, classification accuracy and cultural distance are greater for IDyOM models configured to predict both pitch and time than models configured to predict pitch or time in isolation (see Table 7.1), suggesting

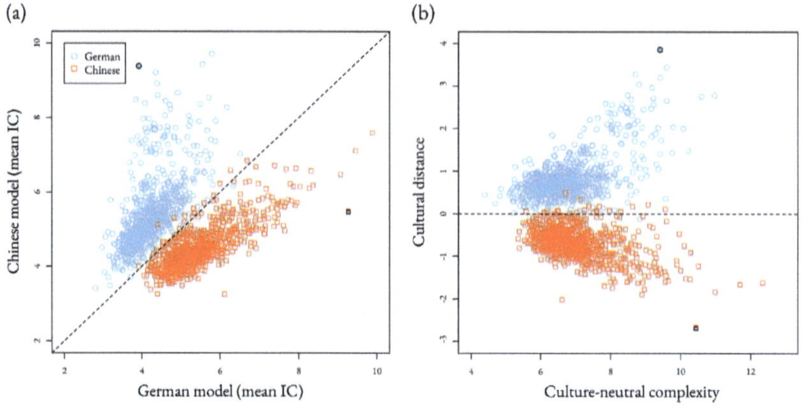

German melody (deut1445) with greatest cultural distance from the Chinese corpus

Chinese melody (han0118) with greatest cultural distance from the German corpus

Figure 7.1: Simulating cultural distance between German and Chinese listeners. (a) The information content (IC) of the German model plotted against that of the Chinese model with the line of equality shown. (b) A 45° rotation of (a) such that the ordinate represents cultural distance and the abscissa culture-neutral complexity. For each style, the composition with most extreme cultural distance is highlighted and corresponding musical scores are shown for these two melodies. The German corpus consists of 769 German folk songs from the Essen folk song collection (Schaffrath, 1995, datasets fink and erk). The Chinese corpus consists of 858 Chinese folk songs from the Essen folk song collection (datasets han and natmin). In a prior step, duplicate compositions were removed from the full datasets using a conservative procedure that considers two compositions duplicates if they share the same opening four melodic pitch intervals regardless of rhythm. IDyOM is configured to predict pitch with a viewpoint linking pitch interval with scale degree (`interval⊗scale-degree`) and onset with the `ioi-ratio` viewpoint using the long-term model only trained on the German and Chinese corpora respectively for the German and Chinese models.

	German example (deut1445)			Chinese example (han0418)			Overall	
	IC	Chinese IC	Cultural distance	German IC	Chinese IC	Cultural distance	Cultural distance	Accuracy
Pitch	2.44	6.53	2.89	4.77	2.36	1.70	0.62	97.91
Onset	1.49	2.86	0.97	4.51	3.11	0.99	0.15	84.27
Pitch + Onset	3.93	9.39	3.86	9.27	5.48	2.69	0.77	98.52

Table 7.1: IDyOM simulations of cultural distance between the Chinese and German corpora (see Figure 7.1). Results are shown for IDyOM models configured to predict pitch only (using a viewpoint linking pitch interval with scale degree, `interval⊗scale-degree`), onset only (using the viewpoint `ioi-ratio`), and both pitch and onset. Overall classification accuracy for the entire corpus and mean cultural distance (across all pieces in the corpus; that is, all points plotted in Figure 7.1) are shown as well as results for the German and Chinese pieces with greatest cultural distance (see Figure 7.1) including the information content (IC) for the German and Chinese models (trained on the German and Chinese corpora respectively) and cultural distance.

both that a combination of temporal and pitch regularities distinguish the styles, and also that IDyOM is capable of learning such distinctive regularities in pitch and timing.

This approach provides a formal, computational model of enculturation, which guides the proposition of hypotheses about cultural familiarity and processing fluency. For example, referring to the examples shown in Figure 7.1, stimuli with strongly positive cultural distance should prove culturally familiar and easy to process for German listeners but culturally unfamiliar and difficult to process for Chinese listeners and vice versa for stimuli with strongly negative cultural distance. Stimuli close to the line of equality should prove culturally ambiguous and difficult to classify.

The predictions of the cultural distance model have been tested in an experiment run by Mathias Klarlund and colleagues (Klarlund et al., 2023). Chinese and German participants listened to Chinese and German folk melodies (performed on a harp, as an instrument present in both musical cultures) selected according to IDyOM's cultural distance measure to have pitch structure that was culturally ambiguous (low cultural distance) or unambiguous (high cultural distance) as shown in Figure 7.2(a). The participants judged whether each stimulus sounded Chinese or Western and gave a confidence rating on a five-point scale. The results, shown in Figure 7.2(b), confirmed an in-culture bias such that accuracy was higher for music from the participants' own culture but also showed that performance was significantly better for unambiguous than ambiguous melodies. This provides concrete evidence that the quantitative measures of cultural distance generated by IDyOM simulate differences

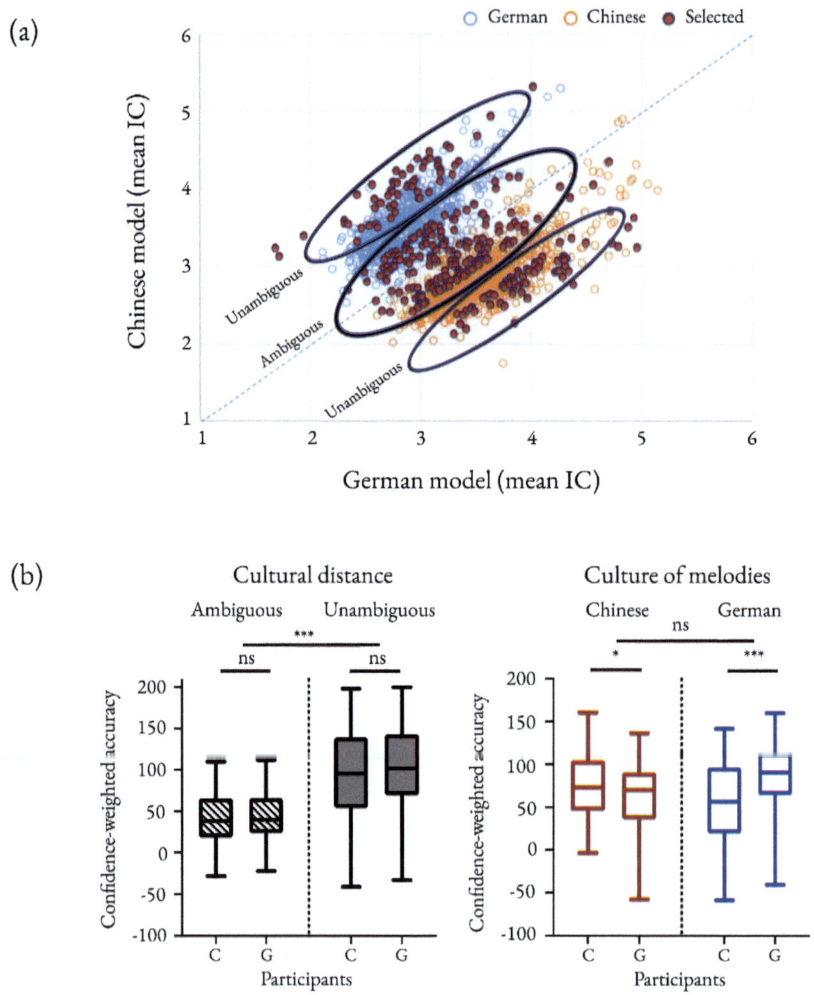

Figure 7.2: The stimuli and results of Klarlund et al. (2023). (a) Selection of Chinese and German folk song melodies according to cultural distance, represented as perpendicular distance from the blue dashed line of equality. The abscissa represents average pitch information content (IC) according to an IDyOM model trained on a corpus of German folk songs (the German model) while the ordinate represents pitch information content according to an IDyOM model trained on a corpus of Chinese folk songs (the Chinese model). From a total of 381 melodies, 80 melodies (40 Chinese and 40 German) were selected based on low and high cultural distance (ambiguous or unambiguous) while ensuring equal variance in culture-neutral melodic predictability (indicated by the line of equality with melodies further to the top right being more unpredictable). Blue- and orange-outlined circles depict German and Chinese melodies respectively. Red-filled circles represent melodies chosen for the experiment. (b) Differences in confidence-weighted accuracy ($N_{correct}$ x $Confidence_{correct}$ - $N_{incorrect}$ x $Confidence_{incorrect}$) between Chinese and German participants as a function of cultural distance (left) and stimulus culture (right). Adapted from Klarlund et al. (2023, Figures 1 and 2) created by Marina Kliuchko and Mathias Klarlund, licensed under CC-BY 4.0.

(a)

	Corpus	
Model	German	Chinese
German	1.34	1.72
Chinese	1.70	1.49

(b)

	Corpus	
Model	German	Chinese
German	73%	72%
Chinese	47%	75%

Table 7.2: Simulations of cultural distance between Chinese and German corpora using metrical inference models configured to predict note onset only (van der Weij et al., 2017). (a) Cross-entropy (i.e., mean information content) of the Chinese and German test corpora for the Chinese and German models. (b) Accuracy in predicting notated time signatures in the Chinese and German test corpora for the Chinese and German models.

in perception of music not just between cultures but also within a culture, depending on cultural distance.

These simulations of cultural distance should also apply beyond pitch to the perception of temporal structure in music. Bastiaan van der Weij developed empirical simulations of the effects of enculturation on metrical inference, using the computational model of metrical inference described in Chapter 4 (van der Weij et al., 2017). A model trained on 1,136 German folk songs (the *German model*) was compared with a model trained on 1,136 Chinese folk songs (the *Chinese model*), with all stimuli taken from the Essen folk song collection (Schaffrath, 1995). When tested on 200 unseen folk songs from each culture, the Chinese model showed greater cross-entropy (i.e., mean information content or prediction error) for the German corpus than for the Chinese corpus, while the German model showed greater cross-entropy for the Chinese corpus than the German corpus—see Table 7.2(a). Furthermore, the Chinese model also exhibited better metre-induction performance (see § 4.3.1) for the Chinese corpus than the German corpus, while the German model performed better on the German corpus than the Chinese corpus—see Table 7.2(b). These findings were replicated by van der Weij (2020) in a comparison between German folk songs and Turkish *makam* music using the comparison between German and Dutch folk songs as a baseline with no cultural distance (see Table 7.3). These simulations of rhythmic cultural distance await empirical experiments to test their predictions for perception.

7.2.3 Effects of veridical repetition

During enculturation, individuals' experience of a musical culture increases in two ways: first, they experience more pieces of music within the set of musical styles that make up the musical culture; second, they experience particular pieces of music repeatedly. Repeated

(a)

	Corpus	
Model	German	Dutch
German	1.38	1.46
Dutch	1.43	1.38

(b)

	Corpus	
Model	European	Turkish
European	1.41	1.88
Turkish	1.60	1.49

Table 7.3: Simulations of cultural distance between German and Dutch corpora (a), and European (combined German and Dutch) and Turkish corpora (b) using metrical inference models configured to predict note onset only (van der Weij, 2020). (a) Mean cross-entropy for the German and Dutch models predicting the German and Dutch test corpora, illustrating low cultural distance. (b) Mean cross-entropy for the European and Turkish models predicting the European and Turkish test corpora, illustrating high cultural distance.

listening to individual pieces of music is likely to have an impact on enculturation through statistical learning in that popular pieces of music will be heard more often and will therefore contribute more to a listener's expanding internal model of stylistic syntax than pieces of music that are less popular, and therefore heard less frequently. Thus repeated listening to individual pieces of music will have an impact on schematic knowledge of music acquired during enculturation, representing generalized structural and statistical regularities making up the syntax of the style.

However, repetition will also have an effect on veridical knowledge, representing specific memory traces for familiar pieces of music (see §§ 3.1 and 6.1.2). Deployment of veridical knowledge can be characterized in terms of the subjective experience of hearing a piece of music and knowing instantly the title and composer, being able to recall previous times that you listened to the piece, and being able to recall forthcoming sections of the music. I suspect, however, that veridical knowledge is not necessarily always explicit. Consider, for example, the subjective experience of hearing a piece of music accompanied by the feeling of having heard it before without being able to say anything more about it. If the piece is in a familiar style, it might be very hard to ascertain whether this experience reflects implicit veridical knowledge or implicit schematic knowledge (see § 6.1.2). It is certainly the case that listeners can be lured into thinking that an unfamiliar piece of music in a familiar style has actually been heard before (e.g., Bartlett et al., 1995). However, repeated listening to specific pieces of music will more firmly establish specific veridical memories for those pieces, whether this knowledge is implicit or explicit.

Two studies have compared effects of veridical knowledge, elicited through repeated presentation of stimuli within an experimental session, with effects of schematic knowledge. Justus and Bharucha (2001) presented participants with pairs of chords and asked them to make intonation judgements for the second chord, some of which contained mistunings. The results demonstrated facilitated processing (i.e., faster reaction times) for chords that were harmonically related to the preceding chord, even when unrelated chord pairs had been presented more frequently within the experimental session. Tillmann and Bigand (2010) noted that the very short stimuli in this experiment might advantage schematic over veridical knowledge and set out to design an experiment that would favour the formation of veridical memory traces. First, they used longer musical materials comprising sequences of eight chords (in which the final chord was either schematically related or unrelated and was the target of a timbre discrimination task). Second, they exposed one group of participants to the 12 stimuli with schematically unrelated endings prior to the testing session. This prior exposure to the unrelated test stimuli failed to extinguish the effect of schematic relatedness on processing speed, although it did produce greater facilitation (faster reaction times) for the less related endings than observed in a control group with prior exposure to the 12 stimuli with related endings.

This evidence suggests that schematic knowledge cannot easily be overridden by veridical knowledge deriving from recent experience with the stimuli. This contrasts with the very vivid experience of veridical knowledge for music that many people experience in real life. We can recognize familiar pieces of music within a few hundred milliseconds (Krumhansl, 2010; Schellenberg et al., 1999) and are often able to recall very complete and detailed representations of the music. However, these real-life cases differ from the research scenarios reviewed above in two important ways. First, real music is a much richer stimulus than the eight-chord sequences used by Tillmann and Bigand (2010), providing a host of detailed timbral, dynamic, melodic, contrapuntal, and linguistic cues that serve to distinguish one piece of music from others. Second, we might listen to our favourite pieces of music hundreds or perhaps even thousands of times over periods of years compared with the matter of minutes that it would take to listen to 24 eight-chord sequences in the experiments of Tillmann and Bigand (2010). For these reasons, it may prove difficult to replicate genuine instances of veridical knowledge overriding schematic knowledge in laboratory experiments.

7.2.4 Emotional and aesthetic experience

The evidence discussed above suggests that musical enculturation influences music perception such that individuals developing in different musical cultures perceive music differently, in ways that are related systematically to implicit statistical learning of the music of the culture. The results of Chapter 5 would predict corresponding differences in emotional and aesthetic experience of music. In fact, cross-cultural research on emotion recognition and induction has tended to emphasize commonalities between cultures (Balkwill & Thompson, 1999; Balkwill et al., 2004; Fritz et al., 2009). In part, this reflects the evidence but it does not tell the whole story because the results almost always show substantial differences between cultures too. For example, in their study of the culturally isolated Mafa tribe from Northern Cameroon, Fritz et al. (2009) emphasize that the Mafa participants (without access to radios) were able to recognize at above chance levels the intended emotional expression (happy, sad, scared) of specially composed pieces of Western piano music (Vieillard et al., 2008) and showed greater dislike for spectrally and temporally manipulated versions of both Mafa music and Western music, as did German and Canadian participants.[5] However, it is also the case that the Mafa recognized the intended emotion expressed by Western music significantly less accurately than European participants. Furthermore, their valence ratings discriminated significantly less than European listeners between original and manipulated versions of the Western music and even, to a lesser extent, their own music.

As another example, Egermann et al. (2015) examined the emotional experience of music in an isolated population of Mebenzélé Pygmies living in the Congolese rainforest. Using both Pygmy and Western music (the latter comprising excerpts from mostly 19th- and 20th-century orchestral music plus three film scores), they compared emotional experience between Pygmies with no experience of Western music and Canadian participants with no experience of Congolese Pygmy music. The results showed some correspondence between the arousal ratings of the two cultural groups, in that Canadian arousal was a significant predictor of Pygmy arousal for the Western music only. Other than this, however, the two groups showed very different subjective emotional experience. Each group rated their own music as more arousing than the other group and Pygmy arousal was a *negative* predictor of Canadian arousal for Pygmy music. Furthermore, there was no correspondence whatsoever in the valence ratings of the two groups for either Western or Pygmy music.

These two studies are notable for their investigation of remote cultures with very little experience of Western music. However, they also illustrate the considerable challenges associated with conducting empirical psychological research in such cultures. In both Mebenzélé and Mafa cultures, music functions in the context of participatory rituals and ceremonies. Mafa music is always accompanied by dance and, in fact, there is no distinct word for music in the Mafa language, while the comments of the participants indicate the social context in

[5]The Western music comprised excerpts from joyful dance music in styles ranging from Baroque to Tango and Jazz created by Koelsch et al. (2006).

which they conceive of music: "I know this, this is from the people of the Gouzda village, I really don't like how they play the flutes" (Fritz et al., 2009, supplementary materials). Perhaps as a result of this circumscribed social function, the music of these cultures does not show the broad range of emotional expressivity found in Western musical culture (Fritz, 2013). For example, the Mebenzélé participants studied by Egermann et al. (2015) rated all their music as having positive valence and arousal. Finally, while the experimenters in both these studies went to great lengths to adapt their experimental procedures to make them meaningful for their non-Western participants, a lack of familiarity with experimental testing, both conceptually and practically, can make it difficult to gather meaningful subjective responses. For example, Egermann et al. (2015, p. 8) report that the Pygmies "always expressed appreciation for the Western music ('It's good music!') but when asked what it expresses, they answered 'I don't know, it's your music, you should know.'" These observations also make it clear that musical enculturation does not materialize in isolation from other elements of culture, highlighting the importance of understanding the culture as a whole when conducting this kind of research.

A final difficulty with this research is that analyses of commonalities and differences in emotional experience between cultures have tended to focus on stimulus properties such as mode, tempo, and timbre rather than psychological mechanisms such as expectation (see Chapter 5). It seems quite possible that although listeners from different cultures respond emotionally to music in different ways, their responses might reflect similar psychological mechanisms of learning that result in diverging experience through exposure to (and interaction with) different kinds of music across their lifespan.

A focus on underlying mechanisms becomes especially important when considering aesthetic experience of music, where it is paramount that the experience is linked psychologically to perception of the form and content of the music (see Chapter 5). This is well illustrated by the case of pleasure induced by consonant compared with dissonant sounds for which it has become apparent that multiple components of consonance exist (see Harrison & Pearce, 2020, for a review): roughness, reflecting spectral components close in frequency that interfere during peripheral auditory processing (producing beating or masking); harmonicity, reflecting spectral components of a sound that form a harmonic series (i.e., multiples of a fundamental frequency); and cultural familiarity, reflecting frequency of exposure. Cross-cultural studies have shown that psychological processing of these aspects of consonance dissociate.

McDermott et al. (2016) investigated preferences for consonance in the Tsimané, a native Amazonian culture with little exposure to Western culture. The Tsimané who participated in the experiment lived in a village with no electricity, tap water, or road access. Their liking for a series of simple musical stimuli was compared with town- and city-dwelling Bolivians and US adults. Surprisingly, the Tsimané showed no preference for consonant chords over dissonant chords (whether synthesized or sung), nor for harmonic over inharmonic synthesized complex tones. The US adults, as expected, strongly preferred consonant chords and harmonic tones while the town- and city-dwelling Bolivians fell somewhere in between. In contrast to their apparent lack of sensitivity to harmonicity, the Tsimané showed clear dislike of sensory roughness that was similar to US adults. Specifically, they showed greater dislike for tone pairs with very close frequencies when they were presented diotically (maximizing spectral interference in the peripheral auditory system and, therefore, the experience of roughness) than when they were presented dichotically (thereby reducing spectral interference and roughness).

These findings have been corroborated in subsequent experimental comparisons of remote villagers from Northern Pakistan with British non-musicians (Lahdelma & Athanasopoulos, 2021), and villagers from North-Eastern Papua New Guinea with musicians and non-musicians from Australia (Milne et al., 2023). This consistency suggests a universal negative effect of sensory roughness and a cross-culturally variable effect of harmonicity. An additional effect of cultural familiarity is apparent in, for example, cross-cultural differences in preference for major and minor triads which have similar roughness and harmonicity. British non-musicians prefer major to minor triads, reflecting their greater prevalence in Western music, whereas villagers from Northern Pakistan show the opposite pattern, reflecting the greater prevalence of minor relative to major chords in their musical culture (Lahdelma & Athanasopoulos, 2021). Furthermore, in a within-culture comparison of five different sets of experimental data on Western listeners' perception of consonance in musical chords, Harrison and Pearce (2020) found significant effects of cultural familiarity operationalized as frequency of occurrence in the Billboard corpus of popular music (see §§ 5.4.3 and 4.2.3), that were independent of roughness and harmonicity.

Looking to extend these results to more complex components of aesthetic experience, a central question for future research is whether differences in aesthetic experience of music between cultures can be systematically related to differences in expectation based on implicit syntactic knowledge acquired during enculturation.

7.3 Enculturation across the lifespan

Cross-cultural studies provide a snapshot of the differences between cultures at one particular point of enculturation, usually in adulthood representing a stable point of cultural maturity. However, this approach does not provide a direct window into the process of musical enculturation that results in such differences between cultures. For this we need to examine changes in perception of music across the lifespan and investigate whether and how these changes can be related to changes in musical exposure, most obviously the increasing volume and variety of music experienced. In principle, observed changes in music perception that can be systematically related to changes in musical experience over time would provide evidence for the statistical learning hypothesis. The direct way of assessing such changes is to track groups of people over time in a longitudinal design but most research follows a cross-sectional approach, comparing listeners of different ages (and therefore presumably different degrees of musical exposure and enculturation) at a single point in time. Generally speaking, research has fallen into one of two categories: first, cross-sectional examination of developmental trajectories in infants and children, including comparisons with adults; second, cross-sectional comparisons of older and younger adults. I will consider each of these in turn.

7.3.1 Developmental trajectories in infants and children

Humans appear to be born with a basic initial toolkit of auditory processing abilities. Newborn babies with intact hearing prefer music made up of consonant than dissonant harmonic intervals (reflecting both roughness and harmonicity) even when they are born to deaf parents (Masataka, 2006) and show larger electrophysiological neural responses (the mismatch negativity amplitude; see § 3.4) to omitted notes on the downbeat than other locations in simple drum rhythms (Winkler et al., 2009). However, the abilities of infants (even newborns) require careful interpretation in light of demonstrated effects of *in-utero* auditory experience. For example, Hepper (1991) compared an experimental group of newborn babies whose mothers had watched an Australian soap opera called *Neighbours* daily during pregnancy with a control group of babies born to mothers who never watched *Neighbours*. Having not heard the *Neighbours* theme tune since birth, the experimental group aged 2–4 days showed reduced heart rate and movement when played the *Neighbours* theme tune, whereas the control group showed no change. Furthermore, no effect was found when the *Neighbours* theme tune was reversed or when newborns were played the theme tune from a different soap opera (*Coronation Street*) that their mothers had never watched during pregnancy.

Other aspects of music perception appear to emerge within the first months of life. Infants as young as 2 months of age can discriminate simple rhythmic patterns (Chang & Trehub, 1977) while 6-month-old infants can recognize a transposed version of a melody (Plantinga & Trainor, 2005). However, there remain many aspects of music perception that emerge over longer timescales. For example, 8-month-old infants are still equally able to detect changes to a melody regardless of the tonal stability of the replacement tone whereas North American adults are better at detecting out-of-key than within-key replacements (Trainor & Trehub, 1992). Furthermore, while North American adults can better identify mistunings in melodies based on a Western scale than melodies based on a Balinese scale, 6-month-olds perform equally well in either case (Lynch et al., 1990). Therefore, infants appear to show equally accurate perception of musical stimuli regardless of tonality. However, by the age of 4 or 5 years, North American children, like their parents, more accurately detect changes in tonal than atonal melodies (Trehub et al., 1986) and out-of-key changes than within-key changes (Corrigall & Trainor, 2010). By 7 years of age, North American children are better at detecting changed within-key notes in a Western melody that violate implied harmony while 5-year-olds do not show such sensitivity to implied harmony (Trainor & Trehub, 1994).

These effects of age on perception of tonality have an impact on pitch expectations. Probe-tone studies with European, North American, and Japanese participants suggest that differentiation of scale degrees based on tonal stability starts at the age of 6 to 8 years, with increasingly differentiated representations of tonal hierarchy emerging with increasing age up to 13 years (Cuddy & Badertscher, 1987; Krumhansl & Keil, 1982; Lamont & Cross, 1994; Matsunaga et al., 2020; Speer & Meeks, 1985). However, implicit knowledge of tonality has been demonstrated at even younger ages. Harmonic priming experiments have shown faster reaction times to chords that are tonally stable in the context of the preceding sequence in 8-year-olds (Schellenberg et al., 2005) and even 4- and 5-year-olds (Marin, 2009). Furthermore, in an EEG study using sequences of five chords, Koelsch et al. (2003) found that 5-year-olds showed differential neural responses (an increased early right anterior negativity or ERAN; see § 3.4) to chords that grossly violated principles of Western tonal harmony but not to more subtle violations.

Given evidence of statistical learning in infancy, it is perhaps surprising that sensitivity to tonality emerges so late in development. In considering this question, Krumhansl and

Cuddy (2010, p. 60) suggested that "although the infant brain has developed to the extent of extracting simple regularities in sound patterns, it has not yet developed those memory resources that along with musical experience allow the extraction of hierarchical regularities among tones". A complementary possibility is that the cognitive mechanisms or sufficient data (or both) are not yet in place to support robust probabilistic inference of tonality, as presented in § 4.3.2, until several years of life have passed.

Age-related differences have also been observed in aspects of pitch expectation beyond tonality. Schellenberg et al. (2002) found that Canadian 8- and 11-year-olds showed evidence of more complex melodic pitch expectations than 5-year-olds, in a task that involved singing continuations to a two-tone context. In particular, while the older group exhibited influences of both pitch proximity and pitch reversal (see Chapter 3), the younger group showed an influence only of pitch proximity. Furthermore, in a separate probe-tone experiment with contexts taken from Canadian folk songs, pitch reversal was found to be underdeveloped in 8- and 11-year-olds compared with adults. Pitch proximity simply reflects the size of the interval formed with the preceding note in a melody, larger intervals being less expected than smaller intervals—that is, a first-order prediction. Pitch reversal is more complex in that it requires the listener to relate the size and direction of the pitch interval to the preceding pitch interval—that is, a second-order prediction. Therefore, while the younger children showed expectations based on a context of the previous tone only, the older children and adults appeared to be increasingly able to form expectations based on contexts of two notes.[6]

It is certainly plausible that these results arise from older listeners having greater knowledge of musical structure acquired through implicit statistical learning, resulting in a more accurate stylistic model of musical structure. In other words, the older listeners may be better enculturated as a result of having greater musical exposure. One can imagine a process whereby children first acquire an internal representation of the very frequent occurrence of small intervals in music, then subsequently learn the less frequent and more complex higher-order regularity that when large intervals do occur, they tend to be followed by smaller intervals in the opposite pitch direction.

However, there are other plausible explanations (Schellenberg et al., 2002). For example, the lack of evidence for pitch reversal in the younger children may reflect typical intonation patterns in child-directed speech which contains exaggerated intervals, more often followed by movement in the same pitch direction than speech directed at older children and adults.

[6]Note that in the present account, pitch proximity and pitch reversal reflect patterns in music that are internalized through statistical learning rather than Gestalt rules of perception (see Chapter 3).

This mechanism still relies on implicit statistical learning during enculturation but involves transfer of implicitly acquired knowledge of sequential pitch regularities from language to music. A suggestive example is given by Stobart and Cross (2000) who reported an analysis of the music of the indigenous people of Northern Potosí in Bolivia. Rather than viewing relatively long events as having metrical salience (as native English speakers typically do when listening to music), the Potosíans take the first event of a piece as initiating the metrical framework, which may be related to the fact that the only fixed position prosodic stress in their language occurs on the first syllable of a word.

Empirical evidence for such transfer comes from a pair of studies (Iversen et al., 2008; Yoshida et al., 2010) examining whether listeners hear a sequence of alternating long and short tones as short-long (an *iamb*) or long-short (a *trochee*). While adult native English speakers from the USA very clearly hear an iambic pattern, Japanese adults tend to perceive a trochaic pattern, although a significant minority perceive the iambic pattern (Iversen et al., 2008). It was suggested that this result may reflect differences in stress patterns related to word order in English and Japanese. In English, short function words tend to precede their corresponding content word (e.g, "the book", "to write", "my desk") whereas in Japanese such function words typically follow their corresponding content word. A subsequent experiment found that 5- and 6-month-old infants from Canada and Japan showed no preference for iambs or trochees while 7- and 8-month-old infants showed adult patterns of preference in their respective cultures (Yoshida et al., 2010). The transition coincides with the age at which infants are able to distinguish culture-specific differences in word order (Gervain et al., 2008). These results suggest that native language may influence rhythm perception in music just as it does for perception of speech rhythm (Smit & Rathcke, 2024).

The other alternative explanation considered by Schellenberg et al. (2002) is that the developmental trajectory they observe reflects domain-general cognitive development and maturation rather than (or, probably more likely, alongside) enculturation. More complex expectations in older children may result from more precise perceptual differentiation of environmental stimuli or increases in memory capacity. Cognitive maturation significantly complicates the use of developmental research to examine enculturation and its effects might vary between different aspects of music perception. For example, while the developmental trajectory in perception of tonality seems unlikely to reflect transfer from language exposure, if tonality perception relies on an ability to represent hierarchies that itself emerges developmentally, it might emerge later than other aspects of pitch perception (e.g., transpositional invariance) that do not depend on hierarchical representation, regardless of experience.[7] How can we discern whether age-related developmental trajectories are the result of increasing enculturation or increasing cognitive maturation?

One powerful way of teasing apart the two explanations is to combine developmental and cross-cultural methods to relate developmental trajectories specifically to differences between the musical cultures involved rather than cognitive abilities, which should develop comparably between cultures. There is relatively little work in this vein but an excellent example is provided by a series of experiments conducted by Erin Hannon and her colleagues, in which listeners detected changes to rhythmic patterns that either preserve or violate metrical structure. These experiments rely on the fact that non-isochronous metres (e.g., 5/4 or 9/8) are common in the folk, popular, and classical music of the Eastern Mediterranean (e.g., the Balkans and Turkey) but relatively rare in Western music (especially popular music) which is dominated by isochronous metres (e.g., 4/4 or 6/8).[8] Correspondingly, North American adults showed better detection for rhythms with isochronous metres than non-isochronous metres while adults from Bulgaria and Macedonia, who also had exposure to Western music, showed no such difference (Hannon & Trehub, 2005b). In subsequent experiments (Hannon et al., 2012), equivalent processing of isochronous and non-isochronous metres by Turkish adults was evident only for non-isochronous metres that actually appear in Turkish *makam* music (e.g., 7/8) and not for those that do not appear (e.g., 15/8). This suggests a specific effect of enculturation.

These experiments were complemented by corresponding experiments with North American infants, who were hypothesized to be equally amenable to isochronous and non-isochronous metres, given their lack of enculturation relative to the North American adults. In fact, while North American 6-month-olds showed no such difference in processing of isochronous and non-isochronous metres, 12-month-old North American infants performed like North American adults, showing better discrimination for isochronous than non-isochronous metres (Hannon & Trehub, 2005a). It is hard to ascribe these effects to cognitive maturation rather than enculturation, especially since the performance dif-

[7] In a different area of cognition, evidence suggests that children show developmental trajectories in the ability to create, store, and process hierarchically organized spatial representations, not reaching adult levels of performance until the age of 10 (Sandberg, 1999).

[8] Isochronous metres are those in which a metrical cycle is subdivided into equally spaced groups of two or three and multiples thereof (e.g., 4/4 subdivides 4 into 1 + 1 + 1 + 1; 6/8 subdivides 6 into 3 + 3); non-isochronous metres are those in which the subdivisions are not evenly spaced (e.g., 5/4 could be subdivided into 3 + 2 as in the well-known jazz standard "Take Five" written by Paul Desmond from the Dave Brubeck Quartet; or the Ottoman *aksak*, which subdivides 9 into 2 + 2 + 2 + 3).

ference between isochronous and non-isochronous stimuli shown by the 12-month-olds was eliminated by two weeks of listening to recordings of Balkan folk music containing non-isochronous metres. Notably, this was not the case for the North American adults, who remained at chance level for the rhythms with non-isochronous metre. Overall, this research provides convincing evidence for an enculturation effect in metre perception within the second semester of life. The age range within which the developing brain retains sufficient plasticity to enable enculturation remains unclear but it seems to have ended by early adulthood for non-isochronous metres specifically.

7.3.2 IDyOM simulation

To what extent can developmental trajectories observed in infants and children be simulated by IDyOM? Some preliminary evidence is provided by a study run by Nina Politimou working in the laboratory of Fabia Franco; the experiment used an implicit paradigm to examine pitch expectations in 40 5- and 6-year-old children, listening to five-note phrases taken from the Essen folk song collection (Schaffrath, 1995), rendered on a synthesized piano (Politimou et al., 2021). For half of the stimuli, the timbre of the final note was changed to a synthesized trumpet sound and the children's task was to identify the instrument that played the final note. Unbeknownst to them, half of the stimuli had a final note whose pitch had been changed while the remaining stimuli retained the original pitch for the final note. For the changed notes, IDyOM was used to sample a new pitch from the distribution predicting the final note of the stimulus, such that the replacement had lower conditional probability than the original pitch. The sampling was constrained such that equal numbers of modified melodies had final notes that formed large (six or more semitones) and small intervals with the penultimate note of the melody. IDyOM was configured to use the long-term model trained on the corpora shown in Table 3.2 using the `interval⊗scale-degree` viewpoint.

The stimuli were presented to the children on an iPad in the form of a game, using an app designed by Pedro Douglass-Kirk, working with Lauren Stewart, and illustrated by his wife Chloe, who by happy circumstance is an illustrator and author of children's books. In the game, children were introduced to a drawing of a bear called Mr Giles, who is a conductor but, unfortunately, is rather hard of hearing. They were asked to help Mr Giles recognize the sound of the instrument at the end of each song by tapping on a picture of a piano or trumpet. The results showed that the children responded significantly faster and more accurately to high-probability than low-probability notes. Furthermore, there was evidence that performance on the task was related to the children's musical environment at home. Individual differences in task performance for small intervals, measured as the mean difference in accuracy between high- and low-probability stimuli, correlated significantly with a measure of the richness of musical experiences at home (in particular, parent initiation of musical behaviour and breadth of musical exposure).

These results suggest that 5- and 6-year-olds possess the implicit knowledge to distinguish stylistically probable from stylistically improbable phrase endings in a way that cannot be put down to pitch proximity alone (cf. Schellenberg et al., 2002). Furthermore, this knowledge seems to be related to the breadth of their musical experience, suggesting an effect of musical enculturation. These results contribute to a growing body of research showing that children do show increasingly complex cognitive representations of music with increasing age, related in part to their increasing cultural exposure to music (see § 7.3.1).

While these results are indicative, the trajectories for many aspects of music perception are unknown or incomplete and the cognitive representations underlying them are poorly understood. Most importantly, as noted above, there is very little research comparing developmental trajectories across cultures. Therefore, there is a clear need for further research to establish more complete developmental trajectories for various aspects of music perception thought to be sensitive to enculturation. This includes expectations based on statistical learning, acquired tonal and metrical models, and other factors such as grouping structure. Such research will prove particularly fruitful if it includes comparisons between children developing within different musical cultures, demonstrating diverging trajectories as a result of diverging musical and linguistic experience. This presents a significant research challenge and a coordinated approach will be necessary in order to relate the results of different experiments into coherent and comparable culture-specific developmental trajectories. Aspects of developmental trajectories that remain constant across cultures may be attributed to maturation of general-purpose cognitive or perceptual abilities or culture-invariant aspects of music. Teasing these factors apart remains a significant challenge, which can be partly addressed through the concurrent testing of, for example, increasing working memory and perceptual acuity. Finally, the question of how musical enculturation is influenced by linguistic enculturation must form part of an integrated understanding of developmental trajectories of musical enculturation. Computational modelling provides an important way of simulating not only behaviour but also underlying cognitive representations and processes, generating testable hypotheses for the developmental trajectories expected given increasing exposure to a musical culture.

7.3.3 Music perception in older adults

The rate at which enculturation continues through the lifespan remains an open question. It seems possible that perception of music would continue to develop during adulthood, albeit at a slower rate, as a result of increasing experience of familiar styles and exposure to new styles, along with changes in general aspects of cognition such as memory. The truth is that understanding of such lifespan trajectories is itself at an early stage of development. The results of Hannon and Trehub (2005a) suggest that 12-month-old US infants are far more capable of acquiring unfamiliar isochronous metrical representations than adults over a two-week period of exposure (see § 7.3.1). However, Wong et al. (2009) showed that bimusical Indians living in America exhibit equal enculturation effects for Indian music, experienced to a greater extent during childhood, and Western music, experienced to a greater extent during adulthood (see § 7.2.1). Furthermore, artificial grammar learning experiments (Loui et al., 2010; Rohrmeier et al., 2011; Tillmann & Poulin-Charronnat, 2010) and cross-cultural studies of expectation (Castellano et al., 1984; Kessler et al., 1984; Krumhansl et al., 2000), reviewed in §§ 7.1 and 7.2.1 respectively, all speak to the remarkable capacity of adults to engage statistical learning mechanisms when confronted with unfamiliar musical styles. If this is the case, we would expect enculturation to continue throughout the lifespan, with increasing exposure to music continuing to develop and refine an individual's internal schematic models of stylistic musical structure.

A useful starting point for investigating these questions is to examine differences in perception between healthy older and younger adults. There is relatively little research on the music perception of healthy older adults and, again, all of the evidence that does exist comes from experiments with cross-sectional comparisons of older and younger adults (see Halpern, 2020, for a review). This causes more difficulty here than in developmental research because of the greater cultural differences between, for example, an 80-year-old born in the 1940s and a 20-year-old born in the 2000s. Music itself and the way in which it is recorded and experienced have changed radically in that time, let alone broader aspects of culture which may have shaped perception and experience of music in different ways. In addition, research on older adults must address a further problem that parallels the question of cognitive maturation in empirical studies of enculturation in infants and children. Older adults may show differences in music perception related to decline of general cognitive capacities such as working memory that counteract any effects of increased exposure to music. Such effects may work against any potential benefit of greater enculturation. A final challenge is that an older adult will already have greater enculturation than a child, so the relative impact of the same amount of additional enculturation is likely to be smaller for familiar musical styles.

In spite of these difficulties, there is a small literature on changes in music perception in healthy older adults, bringing useful evidence to bear on the question of whether increased musical experience throughout adulthood has any impact. Halpern et al. (1996) hypothesized that implicit knowledge of tonal hierarchies would be better established in older than younger US adults as a result of their greater experience of Western tonal music. This hypothesis was tested in a probe-tone study using a strongly tonal context, a major triad, followed by a probe tone, one of the 12 chromatic scale degrees. The results suggested that older adults (aged 60–80) possess a very similar representation of the tonal hierarchy to that of younger participants (aged 15–21) once effects of pitch height were accounted for.

A subsequent study with older and younger non-musicians used more musical contexts to elicit expectations which were simulated with IDyOM (Halpern et al., 2017). The stimuli were specially composed eight-note melodies whose final note was either stylistically predictable (high information content) or unpredictable (low information content). Both high and low information content endings were diatonic scale degrees in the key of the context and the size and direction of the final interval was counterbalanced across the two stimulus categories. Listeners rated the expectedness of the final note of each melody and expectations were simulated with two IDyOM long-term models trained on the corpora shown in Table 3.2 using either `interval` or `scale-degree` viewpoints. The results showed that listeners found the high information content endings more expected than the low information content endings, as hypothesized, but no overall difference between older and younger listeners. However, the pitch expectations of the older adults correlated more strongly with the `scale-degree` IDyOM model, while the expectations of the younger adults correlated more strongly with the `interval` IDyOM model. This suggests that the expectations of the older group were more influenced by general tonal representations acquired through long-term exposure, while the expectations of the younger participants were more influenced by long-term statistical learning of concrete melodic patterns.

These results are indicative of continuing changes in music perception as a result of changes in enculturation across the lifespan. However, much more work needs to be done to understand the details of such trajectories, along with the factors that contribute to them.

7.4 Conclusion

Until the present chapter, the focus has been on perception and experience of music in adults with enculturation in Western music. While this has produced evidence consistent with underlying mechanisms of probabilistic prediction based on statistical learning, the present chapter focused on causal evidence for statistical learning as a psychological mechanism underlying musical enculturation. Three kinds of experimental approach were considered. First, artificial grammar learning studies demonstrate clearly that adults and infants are capable of learning the statistical structure of unfamiliar melodic styles and that such learning proceeds in an apparently obligatory way.

Second, cross-cultural research has revealed differences in music perception between individuals from different cultures that are systematically related to stylistic differences in the music of those cultures. Such enculturation effects influence pitch expectations and recognition memory. Importantly, they appear over and above the ability of individuals to rapidly acquire the statistical properties of unfamiliar musical stimuli in unfamiliar styles during an experimental session. Individuals can be enculturated in more than one musical culture with evidence suggesting that secondary enculturation can occur in adulthood as well as childhood. There is strong evidence for differences in emotional and aesthetic experience between different musical cultures, consistent with the statistical learning hypothesis, but this research has not specifically related these differences to expectations arising from acquired structural knowledge of musical styles during enculturation. The evidence to date suggests that schematic knowledge takes priority over veridical knowledge but this may reflect the difficulty of experimentally replicating real-world levels of veridical familiarity with particular pieces of music. Experimentally teasing apart the effects of these different kinds of knowledge remains an important question for future research.

Third, research has shed light on the acquisition of various aspects of music perception across the lifespan. Of particular relevance to the present enquiry is evidence that culturally sensitive perception of tonality is not apparent before the age of about 4 and then emerges progressively over several years; and also evidence that pitch expectations show evidence of being sensitive to increasingly long contexts with increasing age during childhood. In rhythm perception, evidence suggests that 12-month-old infants already show signs of enculturation in the metrical structure of the music of their surrounding culture, while also exhibiting greater plasticity than adults when presented with music exhibiting unfamiliar metrical structures. Although relatively little research has examined music perception in older adults, there is some evidence that enculturation continues through adulthood, with representations of statistical knowledge showing differences between older and younger adults.

Finally, research suggests that IDyOM can predict the results of artificial grammar learning studies, capture cross-cultural effects of stylistic musical exposure, and account for the experience dependent pitch expectations of young children as well as differences between older and younger adults. However, there is a pressing need for research that integrates cross-cultural and developmental experiments with computational modelling to provide detailed understanding of the lifespan trajectories of change in music perception and experience that occur with increasing exposure to musical styles throughout enculturation.

8 Musical Training and Expertise

Chapter 7 focused on empirical approaches for testing the statistical learning hypothesis by looking for changes in music perception that can be attributed directly to the effects of exposure. The experimental approaches considered—artificial grammar learning, cross-cultural comparisons of ordinary listeners, and analysis of developmental trajectories in music perception across the lifespan—generally involve implicit learning of musical structure through listening to music. This listening might be incidental, in which case it is not specifically selected by the listener and may not be explicitly attended to. On the other hand, an individual might choose to listen to particular pieces of music at particular times, in which case they are more likely to attend to it more closely. In either case, however, the vast majority of individuals with little or no musical training will be unaware of the structural knowledge of stylistic musical syntax that they are internalizing or bringing to bear on their listening experience, in the same way that most people can understand a natural language fluently with little explicit awareness of the syntactic structure of the language.[1] However, there is another widespread form of culturally embedded musical experience—formal training and the acquisition of expertise on a musical instrument—which might be expected to influence stylistic knowledge and, therefore, music perception. This chapter examines the role that formal musical training and expertise play in enculturation through statistical learning and how this role relates to the effects of musical exposure simulated by IDyOM in previous chapters.

The central questions of interest in this chapter are whether musical training and expertise have effects on the empirical indicators of enculturation discussed in previous chapters over and above effects of exposure and, if so, whether these additional effects are captured by the statistical learning mechanisms simulated within IDyOM. Sections 8.2 and 8.3 focus on whether expertise influences two measures of enculturation—expectation (cf. Chapters 3 and

[1]While this particular asymmetry is shared between these two cultural domains, musical cultures often exhibit a further asymmetry, not usually shared with language, such that a majority of individuals have much better perceptual than production capacities, being able to appreciate music that they cannot produce.

4) and aesthetic experience (cf. Chapter 5) respectively—while § 8.4 considers whether any of these effects could be explained by other psychological factors distinguishing musicians and non-musicians, such as enhanced auditory perception or statistical learning ability. Before coming to these questions, it is important to understand the nature of musical training itself, the expert knowledge so acquired, and how effects of musicianship can be investigated.

8.1 What is musical expertise?

In cognitive science, expertise is typically characterized by the possession of a large body of specialized, non-transferable domain-limited knowledge and skills, at least partly available to conscious reflection, acquired through a process that goes beyond mere experience of the domain, often involving years of focused, deliberate study and rehearsal (Feltovich et al., 2006; Lehmann & Gruber, 2006). In the context of musical enculturation, the most striking difference between musicians and non-musicians is the manner in which musical knowledge is acquired, with musicians often dedicating a huge amount of time to explicit theoretical and practical training, rehearsal, performance experience, and assessment. Indeed, because musical training is easier to quantify than expertise, much of the empirical literature measures musical expertise in terms of number of years of formal training.

However, it is important to distinguish musical expertise from musical training, since one can exist without the other. Musical training, like any form of training, does not always lead to expertise and it is possible—in the case of certain music critics, for example—to have a high level of formal musical knowledge with very little instrumental expertise. Conversely, expertise can be attained without explicit formal training (Sloboda, 1991) and amount of deliberate practice, rather than formal training, is generally the better predictor of expertise (Sloboda et al., 1996). In fact, the process that leads to musical expertise is not a single uniform entity but varies enormously not only in degree but also qualitatively along several different dimensions from individual to individual and culture to culture. How much sense does it make to put a self-taught professional blues guitarist in the same category as a highly trained professional opera singer? In a technical sense, such musicians may have equal proficiency over their instruments in spite of achieving that mastery in very different ways. How might we compare musicians from more geographically widespread cultures whose pathway to expertise could vary in even more fundamental ways (e.g., in terms of use of improvisation, written notation, non-lexical vocables)?

Therefore, rather than focusing on the process of knowledge acquisition, it may be more meaningful to concentrate on how musicians and non-musicians differ in terms of the nature of their musical knowledge. In the context of the present enquiry, we might expect musicians to have more extensive memory representations for music, allowing for more detailed and accurate models of stylistic syntax and therefore more accurate predictive processing. Furthermore, it is reasonable to hypothesize that the knowledge acquired during musical training differs from that acquired via implicit statistical learning while listening to music (either passively or attentively). Referring back to Table 3.1, we might expect musicians to have greater explicit knowledge across the board: schematic knowledge of the regularities making up the syntactic structure of a musical style; veridical knowledge of particular pieces of music (acquired, for example, through rehearsal and performance of those pieces); greater ability to acquire explicit knowledge of the dynamically unfolding thematic structure of an unfamiliar piece of music; and possibly even greater conscious access to the contents of echoic memory while listening to music.

Of these potential effects of musical training, schematic knowledge is particularly interesting from the perspective of enculturation through statistical learning. As noted above, expertise is associated with consciously accessible domain knowledge which, for a musician, usually extends to the stylistic regularities making up the syntax of the musical styles with which they are familiar. However, it is also highly likely that musicians listen to more music throughout their lives than non-musicians, through individual practice, group rehearsal, critical listening to music, concert attendance, and other activities. Furthermore, it seems eminently possible that some of this additional knowledge would be acquired and represented implicitly. If that is the case, musicians may possess both greater explicit and also implicit schematic musical knowledge than non-musicians as a result of the activities undertaken in order to acquire expertise.

Can such effects of musical training be captured within the statistical learning hypothesis, as implemented by IDyOM? Up to this point, IDyOM has been used largely to simulate the perception and experience of non-musicians, with an assumption that the schematic and dynamic knowledge possessed by these listeners is mostly implicit. However, there is nothing in the theoretical position, nor within IDyOM, that requires or depends upon this knowledge being implicit. Therefore, it is quite possible that the greater implicit or explicit schematic knowledge acquired by musicians would result in more accurate IDyOM simulations for musicians than for non-musicians.

Another widely acknowledged difficulty with studying musical expertise arises from the fact that the vast majority of studies compare the performance of musicians and non-musicians on the task of interest. This has the advantage of being procedurally straightforward but the underlying assumption that any performance advantage for musicians indicates a causal effect of musical training does not necessarily hold (see, e.g., Swaminathan & Schellenberg, 2019). In fact, causality may run in the opposite direction: a performance advantage on some task may indicate a pre-existing ability that proved advantageous for developing musical expertise, making it more likely that individuals would persist with musical training to become what we would call *musicians*. In other words, the evidence thus acquired is correlational rather than causal, which makes it difficult to know whether differences between musicians and non-musicians reflect a result or a cause of musical expertise achieved through formal training. The extent of the problem depends on the specificity of the task with respect to the musical styles in which the musicians in question possess expertise. For tasks relating to general auditory perception—including sensory or dynamic auditory memory—it is almost impossible to distinguish cause from consequence. However, for tasks more specifically related to the structure of the musical style in which expertise has been obtained—including tests of schematic knowledge—it is less straightforward to argue persuasively for an enhanced pre-existing ability that would have facilitated performance. In other words, it is difficult to convince oneself that musicians would have had innately greater implicit schematic knowledge of a particular musical style, prior to onset of training, than non-musicians. However, given differences in statistical learning ability between individuals (Frost et al., 2015), it is perfectly possible that some children could acquire greater schematic knowledge through the same degree of exposure, prior to onset of musical training.

I will return to this issue of cause and consequence in § 8.4 after considering the evidence for effects of musical training and expertise on enculturation, over and above the musical exposure that is common to both musicians and non-musicians. A final issue with the literature on effects of musical training is that different studies have used different operational definitions of musical training to distinguish musicians and non-musicians. Even when formal musical training is used as a criterion, different levels and kinds of training have been considered, and different numbers of years of training have been used as thresholds for characterizing non-musicians and musicians. This is important when drawing integrated conclusions from the literature, since it qualifies the degree to which direct comparisons can be drawn between studies: at the extremes, one study's musician might be another study's non-musician and vice versa. In recognition of this issue, I have taken care below to report the criteria used to distinguish musicians and non-musicians for each study reviewed.

8.2 Expertise-related effects on musical expectation

The extent to which musical training influences perception and experience of music has been the subject of differing opinion with some authors focusing on evidence that musicians and non-musicians show little difference in performance across a range of psychological tasks (Bigand & Poulin-Charronnat, 2006) while other researchers have focused on enhanced abilities in musicians compared with non-musicians, which have been especially apparent in neuroimaging experiments (Hannon & Trainor, 2007). This is not the place for a comprehensive review of the large and somewhat complex literature on effects of musical training on musical and non-musical abilities. However, the empirical literature considered in the preceding chapters has, in some cases, included comparisons between individuals with different degrees of musical training, providing evidence on whether effects of musical training exist over and above effects of exposure during enculturation.

8.2.1 Differences in musical expectation between musicians and non-musicians

Early studies using the probe-tone paradigm to investigate pitch expectations (reviewed in Chapter 3) tended to report no differences between musicians and non-musicians. For example, Cuddy and Lunney (1995) compared the responses of participants with university-level music education, with an average of 13 years of training, with non-musicians, who had an average of two years of musical training. The responses of the two groups were highly correlated ($r = .88$) and there were no significant differences in expectation between the groups. This study used two-note contexts, leaving open the possibility that using longer contexts would provide a greater challenge capable of exposing differences between musicians and non-musicians. However, Schellenberg (1997) used longer contexts taken from actual music and also reported strong correlations between the pitch expectations of participants with different degrees of musical training. In a first experiment, using extracts from British folk songs, pitch expectations of listeners with moderate training (defined as having at least six years of studying or playing music, with regular playing or studying in the past two years) correlated strongly ($r = .91$) with those of listeners with little training (no playing/studying in the past two years and less than six years overall). In a second experiment, using extracts from non-tonal *Lieder* by Anton Webern, pitch expectations of musicians, who averaged 12 years of music lessons, correlated strongly ($r = .85$) with those of non-musicians, with an average of three years of music lessons, and no lessons or regular playing of music in the past two years.

These results suggest that non-musicians show robust effects of enculturation. However, while the correlations between musicians and non-musicians are strong, they leave scope for subtle effects of musical training, and the results of other probe-tone experiments have yielded differences in musical expectation between musicians and non-musicians. For example, Oram and Cuddy (1995) found that, compared with participants without any music training, the pitch expectations of musically trained individuals (defined as having attained a Grade 9 Royal Conservatory certificate) were more affected by the frequency of occurrence of the probe tone in the context and internalized tonal schema. This is consistent with the probe-tone results of Halpern et al. (1996), who reported that musicians (defined as having at least eight years of private lessons) showed a more differentiated tonal hierarchy than non-musicians (with no more than two years of lessons). Correspondingly, in a temporal probe-tone task, Palmer and Krumhansl (1990) found that musicians (with at least five years of formal instrumental training) showed evidence of deeper metrical hierarchies affecting their temporal expectations than non-musicians (with less than two years of training).

Overall, these results suggest that while there is generally considerable overlap between the expectations of musicians and non-musicians, presumably reflecting common implicit processes of enculturation, there are also differences, which might possibly reflect additional knowledge acquired by musicians during training. We can now turn to the question of whether this additional knowledge is explicit or implicit.

8.2.2 Explicit vs implicit knowledge

It seems that differences in expectation do exist between musicians and non-musicians, especially related to schematic knowledge of musical structure. But to what extent is this additional knowledge an explicit, consciously accessible result of formal music education or an implicit result of increased range and depth of musical exposure? Several prominent studies using the priming paradigm to test for longer reaction times to harmonically unexpected continuations of a chord sequence (see Chapter 4) have found no difference in harmonic expectations between musicians and non-musicians (Bigand & Poulin-Charronnat, 2006). In their study of harmonic priming effects, for example, Bigand et al. (2003) found that while musicians (musicology students with an average of 12 years of musical training) gave faster and more accurate responses overall than non-musicians (with no musical training or instrumental ability whatsoever), differences between expected and less expected continuations of a seven-chord context were actually greater for non-musicians than musicians. Furthermore, responses of both musicians and non-musicians were dominated by cognitive rather than sensory effects, though at the fastest tempi (75 ms per chord) non-musicians were more influenced by sensory priming than musicians (see § 4.2 for further discussion of this experiment).

Bigand et al. (2003) attributed the fact that musicians were faster and more accurate overall than non-musicians to the task of making consonant/dissonant judgements being familiar to musicians from their formal musical training. In accordance with this conjecture, research using implicit tasks that are equally familiar for musicians and non-musicians has found no main effects of musical training. Specifically, Bigand et al. (2001) found that musicians (Conservatory graduate students) and non-musicians (with no formal musical training or instrumental experience) showed no difference in speed or accuracy of discriminating sung phonemes paired with the final chord of an eight-chord sequence. However, both groups showed faster responses when the chord sequence ended on the more expected tonic rather than the less expected subdominant. The implicit nature of the effect is emphasized by the rather surprising fact reported in the paper that none of the musicians noticed whether the chord sequences ended on a tonic or subdominant.

In a follow up study, Poulin-Charronnat et al. (2005) paired the same chord sequences with sung eight-syllable sentences and asked musicians and non-musicians (defined as in the previous study) to make a lexical decision as quickly as possible as to whether the final sung syllable was a word or non-word. The final words were either semantically related or unrelated to the preceding context of the sentence. Both groups of participants showed, first, a semantic priming effect, with faster responses for semantically related words, and second, a musical priming effect, with responses faster for tonic than subdominant endings. For both groups, there was also a significant interaction between semantic and musical priming such that semantic priming was stronger for chord sequences ending on the tonic than the dominant. The only difference between the groups was that the non-musicians showed a stronger semantic priming effect; there was no difference in musical priming between musicians and non-musicians.

These results suggest that when participants are not required to explicitly verbalize their expectations, non-musicians appear to possess equivalent implicit knowledge of harmonic structure as musicians, at least for the level of musical complexity represented by the stimuli used in these experiments. Therefore, it may be that differences between musicians and non-musicians when making explicit expectedness ratings (as reviewed in § 8.2.1) result from musicians having greater explicit knowledge of musical structure. This is consistent with evidence from the study reported by Demorest et al. (2008) which assessed Turkish and US participants' recognition memory for music in culturally familiar and unfamiliar styles (see Chapter 7). Although both groups showed better recognition memory for culturally familiar music, no differences were observed between musicians (who had completed or were completing a music degree and had more than six years of formal training) and non-musicians (with less than four years of any kind of active participation in music).

Concrete evidence for this hypothesis was provided by a pair of experiments run by David Sears, which elicited both implicit and explicit measures of expectation from musicians and non-musicians listening to cadences extracted from Mozart's keyboard sonatas (Sears et al., 2019, reviewed in detail in Chapter 4). The musicians had musical training to second-year undergraduate level or beyond while the non-musicians had less than one year of musical training. Recall from Chapter 4 that the stimuli were selected to represent different cadence types, reflecting progressively lower degrees of expectedness: perfect and imperfect authentic cadences, half cadences, deceptive cadences, and evaded cadences. The first experiment presented participants with complete stimuli including the final cadence, asking for goodness-of-fit judgements for the final chord. Participants were also presented with truncated stimuli in which the final chord was omitted, and asked to provide ratings of expectation strength, expectation specificity, and phrase completion (how strongly they felt that the ending of the passage was imminent). The second experiment used only the complete non-truncated stimuli, both in their original form and with the final chord mistuned, measuring reaction times on a mistuning detection task for the final chord.

The results of the first experiment showed no differences between musicians and non-musicians for explicit goodness-of-fit, expectation strength or phrase completion ratings. However, there was a difference between the groups for expectation specificity, with musicians showing more specific expectations and greater differentiation amongst the cadence types. In the second experiment, both groups exhibited significant effects of cadence type, with authentic cadences (perfect and imperfect) generally eliciting faster responses than cadential deviations (deceptive and evaded cadences) and musicians showing no greater effect than non-musicians. Therefore, musicians showed more specific explicit expectations in spite of performing similarly to non-musicians on an implicit measure of expectedness. The comparison is robust because the experiments used exactly the same stimuli for the implicit and explicit task. This is consistent with musicians being able to draw on greater explicit knowledge of musical structure in the explicit but not the implicit task.

Although there was no evidence of musicians showing an advantage on implicit tasks in Sears et al. (2019), several studies have reported greater neural sensitivity to unexpected musical events in musicians than non-musicians (Besson & Faïta, 1995; Di Liberto et al., 2020; Jentschke & Koelsch, 2009; Kim et al., 2011; Koelsch et al., 2002, 2007; Park et al., 2018; Regnault et al., 2001). For example, Koelsch et al. (2002) reported that musicians (with Conservatory education in musicology or performance, and at least 10 years of instrumental training) show a larger early right anterior negativity (ERAN; see § 3.4) to syntactically unexpected chords than non-musicians (with no instrumental training or extra-curricula musical education), suggesting that they find such chords more unexpected than non-musicians. Since these studies do not explicitly ask participants to report their expectations, they can be viewed as implicit measures, especially when the design incorporates a secondary task such as detecting occasional changes from staccato to legato (Kim et al., 2011) or from piano to another instrument (Koelsch et al., 2002). However, differences in peak amplitude of the ERAN between musicians and non-musicians are usually small and not always statistically significant (e.g., Koelsch & Jentschke, 2008; Koelsch & Sammler, 2008; Miranda & Ullman, 2007; Steinbeis et al., 2006). Furthermore, it is difficult to be sure whether these differences in neural markers of expectation between musicians and non-musicians reflect implicit or explicit knowledge, or a combination of the two.

Evidence that IDyOM captures effects of greater musical knowledge on musicians' expectations comes from a study by Cheung et al. (2023). In this experiment, which was reviewed in detail in Chapter 4, age-matched musicians and non-musicians gave continuous ratings of perceived surprise for each chord in a collection of 30 chord sequences taken from commercially successful pop songs appearing in the US Billboard Hot 100 chart between 1958 and 1991. Musicians had extra-curricular music lessons at or before the age of 10, at least seven years of continuous musical training, and had actively practised their instruments once a week on average for three years prior to taking part. Non-musicians had no musical training before the age of 7, or in the previous seven years, and less than 10 years of continuous training in total. Musicians and non-musicians were respectively at the 86th and 15th percentile of the musical training subscale of the Goldsmiths Musical Sophistication Index (Gold-MSI), based on a sample of 147,633 British individuals (Müllensiefen et al., 2014). Musicians showed significantly stronger linear association with information content derived from IDyOM than non-musicians, whereas there was no difference for the periodicity pitch model (Leman, 2000), a low-level model of sensory pitch memory (see Chapter 4).

In summary, these results suggest that musicians do in some cases show expectations that differ from non-musicians and that these differences are due to explicitly available knowledge (since the differences appear with explicit but not implicit behavioural tasks) and possibly also implicit knowledge (since differences have been observed in some EEG and MEG experiments, which may reflect implicit knowledge). Furthermore, when differences exist, musicians appear to be more accurately simulated by IDyOM than non-musicians, suggesting that IDyOM captures the effects on expectation of musicians' additional stylistic knowledge of musical structure (whether explicit or implicit). The next section formulates a more specific hypothesis for these effects and assesses the extent to which it is supported by the empirical literature.

8.2.3 Expertise effects on predictive precision

It is reasonable to hypothesize that formal musical training would lead expert musicians to possess more detailed and accurate cognitive models of the syntactic structure of the musical styles for which they possess expert knowledge. If this is the case, we would expect such an optimized model to generate more certain (i.e., lower entropy) predictions than the higher-entropy predictions generated by the less detailed and accurate models of non-musicians. As a result, low-probability events would be expected to generate larger prediction errors and be experienced as more unexpected by musicians than non-musicians. We would also expect the lower-entropy predictions generated by musicians to be correct more often than not, producing very high expectedness for high-probability events. Referring back to Figure 2.3, it is as if musicians are able to use their musical knowledge (whether implicit or explicit) to make highly certain predictions at the 5% entropy quantile, such that events are either highly expected, when a highly certain prediction for a specific continuation turns out to be correct, or highly unexpected, when the prediction turns out to be incorrect. Non-musicians, on the other hand, would generate more uncertain predictions such that all possible continuations would be similarly, and moderately, unexpected (taking the 95% entropy quantile as an extreme for illustration).

This hypothesis was proposed by Niels Chr. Hansen (Hansen, 2016; Hansen & Pearce, 2014) and I will refer to it here as the *expert prediction hypothesis*. This hypothesized scenario makes two key predictions: first, that musicians should generate more certain predictions than non-musicians; and second, musicians should show more extreme degrees of expectedness (high and low) for musical events than non-musicians.[2] The empirical data reviewed above already suggest some specific support for the hypothesis, beyond the general observation of differences in expectation between musicians and non-musicians. For example, probe-tone experiments have revealed flatter, less differentiated expectation profiles for non-musicians than musicians across levels of tonal stability (Halpern et al., 1996) and metrical stability (Palmer & Krumhansl, 1990). Furthermore, Sears et al. (2019) reported that musicians exhibited more specific expectations than non-musicians when predicting cadential endings in Mozart keyboard sonatas. These results are consistent with musicians being able to generate lower-entropy predictions. Furthermore, reports that musicians show greater sensitivity than non-musicians, albeit only slightly, in neural measures of expectation violation (see § 8.2.2) are consistent with the hypothesis that musicians find unpredictable events more unexpected, and predictable events more expected, than non-musicians.

Two studies have been set up as direct tests of the expert prediction hypothesis. In their probe-tone experiment (covered in Chapter 3), Hansen and Pearce (2014) compared effects of musical training on predictive uncertainty and unexpectedness in a design that explicitly distinguished musical contexts in which IDyOM made high- and low-entropy predictions. The hypothesis was that listeners would show greater prospective predictive uncertainty in the high-entropy than the low-entropy contexts, and that this effect would be enhanced for musicians. Participants gave explicit ratings of their uncertainty about the continuation of incomplete melodic fragments from English hymns and Schubert *lieder* (the latter selected to show greater musical complexity than the former) before rating how unexpected they found nine different single-note continuations of the context. As well as the explicit ratings of predictive uncertainty, this design also permitted an inferred measure of uncertainty based on the entropy of an estimated probability distribution derived from the probe-tone ratings. Age-matched musicians and non-musicians were recruited, using scores on the musical training subscale of the Gold-MSI as a criterion. Musicians scored 33 or more, while non-musicians scored 20 or less, corresponding respectively to the 66th and 32nd percentiles of a large sample of British individuals (Müllensiefen et al., 2014).

For the inferred measure of uncertainty, the results demonstrated that musicians exhibited lower levels of predictive uncertainty overall. Furthermore, while both groups

[2]Note that this holds only for musical events covering a full range of expectedness as in the probe-tone paradigm, where actual musical events are replaced by a wide range of alternatives, including some with very high and others with very low probability. When the range of events is restricted to those that actually appear in music, thus avoiding very-low-probability events, musicians appear to show greater levels of expectedness than non-musicians (Pearce et al., 2010c).

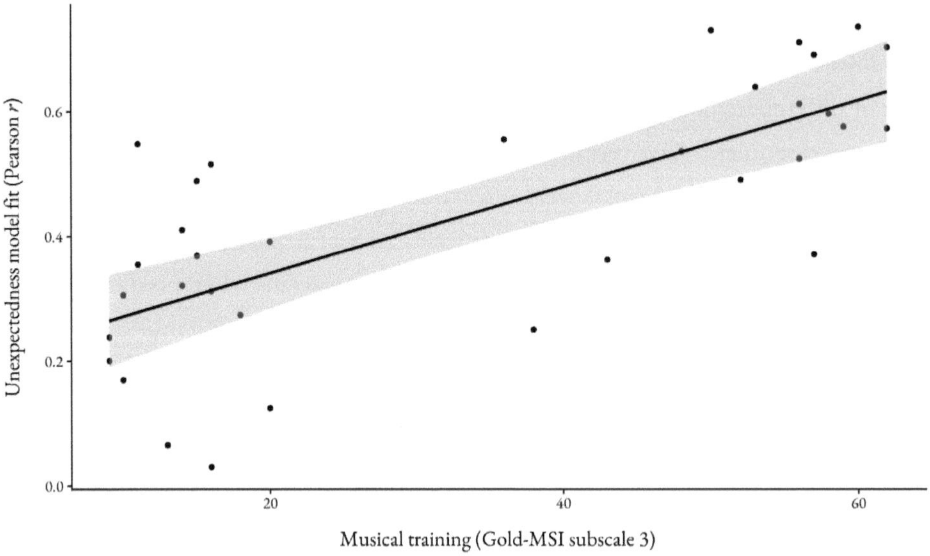

Figure 8.1: Correlation between musical training (Gold-MSI subscale 3; Müllensiefen et al., 2014) and the fit between unexpectedness ratings and IDyOM information content (Hansen & Pearce, 2014). The ordinate shows the Fisher Z-transformed correlation coefficient between IDyOM information content and unexpectedness ratings for each participant.

distinguished high- and low-entropy stimuli for the simpler hymns only the musicians did so for the more complex Schubert *lieder*. Musicians also showed lower levels of explicit uncertainty than non-musicians for the hymns. These results suggest that the musicians were able to make better use of the predictive cues in low-entropy contexts to generate sharper (i.e., more precise or focused) expectations for particular continuations. Accordingly, to a greater extent than non-musicians, musicians found the nine pitch continuations in the low-entropy condition more unexpected, on average, than in the high-entropy condition. This indicates that musicians show greater levels of prediction error resulting from their ability to generate predictions with greater certainty or precision. Finally, musicians' unexpectedness ratings correlated more strongly with IDyOM's information content than those of non-musicians. In fact, as shown in Figure 8.1, the correlation between unexpectedness and information content itself showed a significant positive correlation with the extent of musical training as assessed by the musical training subscale of the Gold-MSI.

These results are consistent with the expert prediction hypothesis and also suggest that IDyOM captures, at least to some extent, the predictive advantage shown by musicians. However, they do not specifically demonstrate that this advantage of musicianship is a result of musicians' greater schematic knowledge of the syntactic structure of a musical style. Therefore, Hansen et al. (2016) used the same experimental design to establish whether these effects could be related specifically to style-specific musical training. The experiment examined the uncertainty and unexpectedness ratings of 20 non-musicians, 20 classical musicians, and 22 jazz musicians at selected points in 24 solos by the jazz saxophonist Charlie Parker, comparing them to two long-term IDyOM models, the first trained on Charlie Parker solos and the second trained on the corpora shown in Table 3.2.

For recruitment, the inclusion criteria for musicians were that they regularly performed concerts or contributed to commercial recordings, receiving the majority of their income from performing or teaching within the relevant musical genre (jazz or classical), or were full-time performance degree students in classical music or jazz. The inclusion criteria for non-musicians were that they must never have had regular one-to-one music lessons, nor performed music in public after the age of 12. The two groups of musicians showed greater musical sophistication than the non-musicians but did not differ from each other, either overall or in terms of any of the five subscales of the Gold-MSI. The two groups of musicians both averaged above the 90th percentile for the musical training subscale, while non-musicians averaged below the 11th percentile (Müllensiefen et al., 2014). To distinguish jazz and classical musicians, a composite measure of jazz-specific musical experience was developed consisting of the total number of jazz concerts performed and attended as well as the number of jazz songs in their personal music collection and active performance repertoire. Unsurprisingly, jazz musicians scored much higher on this measure than classical musicians, but in spite of efforts to recruit classical musicians with as little jazz experience as possible, they also scored slightly higher on average than non-musicians. All three groups were matched for age.

The results demonstrated an advantage of style-specific expertise only for explicit uncertainty ratings, not for the measure of inferred uncertainty or the unexpectedness ratings. Specifically, only the jazz musicians found the high-entropy contexts to be more uncertain than the low-entropy contexts. Furthermore, the explicit uncertainty ratings of jazz musicians correlated significantly more strongly with the IDyOM model trained on Charlie Parker solos than those of classical musicians. Finally, as shown in Figure 8.2, explicit uncertainty correlated significantly with inferred uncertainty only for jazz musicians, not for classical musicians or non-musicians, suggesting an effect of style-specific expertise on the association between implicit and explicit predictive processing. These results are consistent with an explicit, conscious effect of style-specific musical knowledge on expectations. The results for unexpectedness and inferred uncertainty largely replicated the findings of Hansen

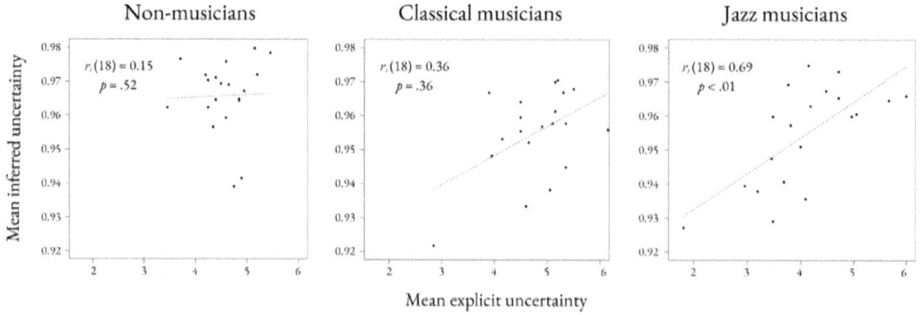

Figure 8.2: Correlations between mean explicit and inferred uncertainty for the three groups studied in Hansen et al. (2016). Created by Niels Chr. Hansen and reproduced under the terms of the Creative Commons attribution licence (CC-BY 4.0).

and Pearce (2014) in showing a higher correlation with the IDyOM model trained on Charlie Parker solos for musicians than non-musicians but no difference between classical and jazz musicians. Both classical and jazz musicians experienced lower inferred uncertainty and greater unexpectedness in low-entropy compared with high-entropy contexts whereas non-musicians' ratings did not differ. It remains possible that these effects, common to classical and jazz musicians, reflect an influence of implicit musical knowledge, at least in part.

Altogether, these results suggest that musical training augments the effects of passive exposure on musical expectation in accordance with the expert prediction hypothesis. Musical training leads to more detailed predictive models of the syntactic structure of musical styles that allow for low-entropy predictions, characterized by greater unexpectedness (prediction error) for notes that are stylistically unlikely to follow a given context. Non-musicians, on the other hand, generate flatter, higher-entropy predictions, suggesting less detailed cognitive models of musical syntax. Style-specific expertise seems to be distinguished from generalized musical expertise primarily in terms of enhanced formation of explicit predictions about the probability of different possible continuations of an incomplete melodic passage. It seems likely that this reflects explicit knowledge derived from formal training, consistent with the observation that experts across domains are able to reflect consciously on their domain of expertise (Feltovich et al., 2006). However, in addition to such effects of style-specific knowledge acquired explicitly through formal instruction and study, the results for the classical musicians are consistent with an effect of knowledge acquired implicitly through more intensive musical experience than non-musicians. Finally, while IDyOM does predict the expectations of non-musicians, it also seems to capture the better prediction performance exhibited by musicians.

8.3 Expertise-related effects on aesthetic experience of music

Given the relationships between expectation and aesthetic experience of music outlined in Chapter 5, any effects of musical expertise on expectation should confer a consequent effect of musicianship on emotional and aesthetic experience of music. The results reviewed in Chapter 5 suggest that unpredictable passages of music (with high information content) increase arousal and subjective perception of complexity while pleasure is associated with intermediate degrees of unpredictability. Meanwhile, the results reviewed in § 8.2 provide support for the expert prediction hypothesis, suggesting that musical expertise enables the generation of more certain predictions, which tend to be confirmed in actual music, thereby reducing subjective unpredictability.

Putting these results together, musicians would be expected to show preferences for music exhibiting greater objective complexity (i.e., higher information content) than non-musicians because their more accurate and precise predictive models render this music less unpredictable. Putting it another way, musicians should show peak pleasure (corresponding to the apex of the inverted-U shaped Wundt effect reviewed in Chapter 5) at higher levels of musical complexity than non-musicians, as shown in Figure 8.3. While it is difficult to pinpoint precisely the origin of this hypothesis, it has been a longstanding assumption in research on experimental aesthetics dating back at least to the 1950s and 1960s (Dember & Earl, 1957; Munsinger & Kessen, 1964; Vitz, 1966). The hypothesis was articulated most clearly and precisely by Orr and Ohlsson (2005) who also subjected it to detailed experimental scrutiny.

The experimental evidence pertaining to this hypothesis falls into four categories, differing in the operational definition of musical complexity. First, there is circumstantial evidence such as the finding by Rubin-Rabson (1940) of a positive correlation between degree of musical training and liking for music that was judged at the time to be "modern" (including music by Scriabin, Stravinsky, and Schoenberg). While this is indicative, it is difficult to be sure that the "modern" stimuli were any more complex (objectively or subjectively) than the other stimuli or, even if so, whether the increased liking by musicians was due to this increased complexity.

Second, some researchers have asked the same group of participants to report both pleasure and subjective complexity, looking for relationships between the two (e.g., North & Hargreaves, 1995; Orr & Ohlsson, 2001, 2005). This research tends to suggest little consistent effect of musical training on the apex of the inverted-U shaped relationship between complexity and pleasure. It seems likely that this reflects the use of a subjective measure of complexity. Somewhat paradoxically, this point was made by Orr and Ohlsson (2005) themselves, who argued that an effect of musicianship would be expected only for objective and not subjective measures of complexity. Their reasoning was that pleasure should show an inverted-U shaped relationship with an individual's subjective perception of complexity (the Wundt effect), regardless of musical training. However, they hypothesized that musical training would change the relationship between an individual's subjective perception of complexity and objective measures of complexity, such that intermediate, and maximally pleasurable, levels of subjective complexity correspond to greater levels of objective complexity for musicians than non-musicians. For an inverted-U shaped relationship between pleasure and perceived complexity for a given set of stimuli, increased musical training should shift the apex of perceived complexity of the stimuli rightwards towards greater complexity, as shown in Figure 8.3. The picture is complicated by the fact that the relationship between objective complexity and pleasure might also be affected by individual differences in, for example, openness to new experience (McCrae, 2007), need for complex cognition (Cacioppo & Petty, 1982; Webster & Kruglanski, 1994), or music reward (Mas-Herrero et al., 2013); and such individual differences might not be independent from musical training (David Huron, personal communication, 13 November 2023).

Third, there is research in which the experimenter has used their own subjective judgements, or those of other musicians, to define levels of complexity in their stimuli prior to collecting measures of pleasure from a separate cohort of participants (e.g., Burke & Gridley, 1990; Coggiola, 2004; Fung, 1996; Gordon & Gridley, 2013). In some ways, this is an improvement over taking simultaneous liking and complexity judgements from participants, since the complexity of the stimuli is assessed independently and consistently for all participants prior to the experimental session. A representative example is provided by Coggiola (2004) who examined aesthetic responses to four jazz recordings, ranked in terms of melodic complexity ratings of the improvised solos by five independent jazz experts. Participants were undergraduate and postgraduate music students divided into jazz musicians and non-jazz musicians depending on whether they had more or less than three years of instrumental jazz ensemble experience. The groups only differed for the most complex piece (Dr. Jekyll from the album Milestones by Miles Davis) for which jazz musicians reported stronger aesthetic experience than non-jazz musicians. While this approach allows the use of real musical examples, the operational definition of complexity leaves open the question of how

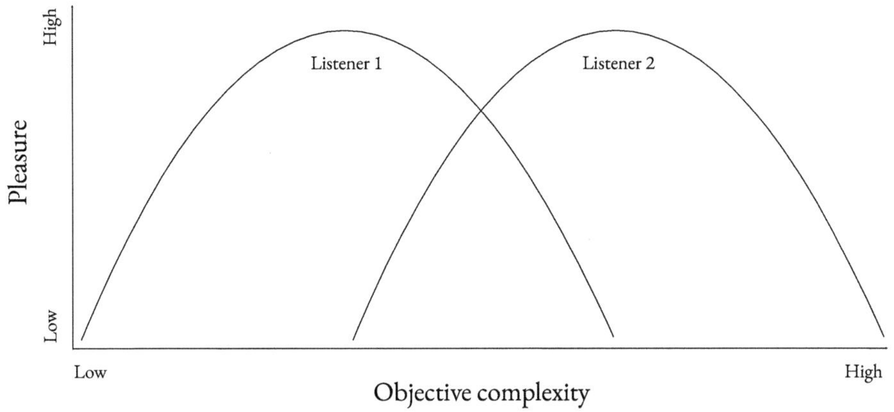

Figure 8.3: Schematic illustration of two individuals, each showing an inverted-U shaped relationship between objective complexity and pleasure (see Figure 5.1) but peaking at different levels of complexity. Listener 1 could be a non-musician with limited musical experience and Listener 2 a musician with greater musical experience. The musician shows peak pleasure shifted rightwards to greater degrees of objective complexity than the non-musician because their musical training reduces the subjectively perceived complexity of objectively more complex musical stimuli. In this scenario, both listeners would show equivalent inverted-U shaped relationships between subjective complexity and pleasure. Alternatively, the illustrated differences in peak pleasure could be related to individual differences in personality in which case corresponding differences in subjective complexity between the two listeners might also occur. Consider, for example, two individuals with equal amounts of musical training and the same mapping from objective to subjective complexity where Listener 1 has a personality type that is drawn to greater complexity (both objectively and subjectively defined) than Listener 2. Redrawn from Orr and Ohlsson (2005, Figure 1).

it is related to characteristics of the music and the stimuli are likely to vary in dimensions other than complexity, potentially confounding the effect of complexity on pleasure.

Finally, there is evidence from research that has used objective measures of complexity, derived from analysis of the musical stimuli. In the first such study, Paul Vitz (1966) set out to test both the Wundt effect (as it relates to stimulus variation, taken to be an indicator of complexity) and the extent to which it is influenced by expertise, based on the proposals of Dember and Earl (1957) amongst others. In particular, he hypothesized that:

> [Participants] with much training and interest in music should prefer a greater amount of variation than [participants] with little or no musical training and interest. The rationale is that with exposure to a stimulus, some of the variation is organized and stored by [the participant], and the perceived amount of variation in the stimulus declines. To keep the amount of perceived variation

> at the preferred level, [the participant] must seek out a new stimulus with a greater amount of objectively defined stimulus variation.
>
> (Vitz, 1966, p. 75)

To test this hypothesis, Vitz constructed tone sequences "sounding like simple music" (p. 75). They were composed of pure tones with 17 different frequencies (plus silence which was considered an 18th tone), four different levels of intensity and eight different durations. The sequences were constructed in six levels of increasing variation, each containing an expanded set of frequencies, intensities, and durations. Correspondingly, each level exhibited greater zeroth-order entropy. Rather charmingly from a present-day perspective (though it must have been painstaking at the time), the sequences were constructed using a random-number table and rendered by recording each pure tone frequency at each loudness on magnetic tape and then cutting and splicing sections of tape of different lengths together.

In his third experiment, Vitz (1966) obtained pleasantness ratings from 44 participants, divided into non-musicians (with no formal musical training and no more than a moderate interest in music) and musicians (with at least two years of formal training in music and a strong or very strong interest in music). The results demonstrated that musicians preferred greater stimulus complexity than non-musicians, with pleasure peaking at a greater level of stimulus variation. Musicians found sequences with low levels of complexity relatively less pleasurable than non-musicians and sequences with high levels of complexity relatively more pleasurable.

A difficulty with this study, as with many studies using objective measures of complexity, is that the artificial stimuli lack ecological validity, differing considerably from the music that most people listen to most of the time (see also Chapter 5). To address this concern, Simon and Wohlwill (1968) extended Vitz's experiments to actual musical passages. The stimuli consisted of four melodic themes, one composed especially for the study and the others taken from J. S. Bach's Well-tempered Clavier (Book 2, Prelude 11) and two pieces from Bartók's Mikrokosmos. These themes were considered to possess high internal variation (reflecting complexity). A musicologist created two additional versions of each theme which had moderate and low degrees of variation respectively. This was achieved by, for example, replacing changing time signatures with a single consistent metre (moderate variation) and removing rhythmic variation entirely (low variation), or removing non-diatonic (moderate variation) or non-triadic pitches (low variation). The participants were 12 musicians and 12 non-musicians selected from a group of 65 undergraduates on the basis of a questionnaire asking about extra-curricular formal musical training, participation in music groups, concert attendance, and musical preferences. The reported analysis shows that musicians preferred versions with high variation for three out of four stimuli (with no difference in preference for the fourth) while non-musicians preferred the versions with high variation for two stimuli, the versions with moderate or low variation for one stimulus with no difference in preference for the final stimulus. Because a direct comparison was not made between the musicians and non-musicians, these results provide at best weak evidence for an expertise effect.

Smith and Melara (1990) trod an intermediate path between Vitz's highly artificial auditory stimuli and the realistic (albeit recomposed) musical stimuli of Simon and Wohlwill. They developed harmonic sequences that varied across seven levels of syntactic prototypicality. The stimuli were nine-chord progressions starting with a tonic triad and ending on an authentic cadence which were constructed by replacing an increasing number of the intervening chords (from zero to six) in an original prototype that adhered to Western stylistic norms. The chords were chosen so as to follow common-practice conventions of harmony and voice leading such as avoiding parallel octaves and fifths, avoiding repeated bass notes, and appropriate resolution of leading tones. Musicians and non-musicians rated the stimuli for unusualness, complexity, interestingness, and pleasingness. The musicians were recruited in two groups: first, 16 undergraduate students of music; second, nine graduate music students and their professor. Overall the musicians had high average levels of training in music appreciation (3.7 years), music theory (4.5 years), and instrumental or vocal performance (17 years) while the non-musicians had low levels of training in these areas (0.5, 0.3, and 3.7 years respectively). Both musicians and non-musicians showed strong linear increases in unusualness and complexity with decreasing prototypicality, indicating once again that even non-musicians are sensitive to the manipulation of syntactic structure in music.

In spite of this, Smith and Melara found that the non-musicians and undergraduate musicians showed decreasing preference with decreasing prototypicality while the postgraduate musicians showed increasing preference with decreasing prototypicality. While these results do not bring evidence to bear on the Wundt effect directly, they do suggest that musical training can lead to preference for greater syntactic atypicality but also that this may be restricted to those with very advanced involvement in music.

A difficulty with Smith and Melara's experimental design is that it lacks a clear quantitative measure of syntactic prototypicality, relying instead on the rank order of the degree of manipulation, just like Simon and Wohlwill (1968). As argued in Chapter 5, this can be addressed by computational models like IDyOM, capable of providing measures of structural predictability that go far beyond the zeroth-order entropy measures used by Vitz (1966), which really only reflect the number of elements used to compose the stimulus sequences. In a study reported fully in Chapter 5, Ben Gold examined effects of musical training on the Wundt effect for 55 30 s excerpts from melodies in a wide range of styles including Baroque, Classical, Romantic, and 20th-century Western music, jazz, and traditional music (Gold et al., 2019). The results showed that participants who exhibited the Wundt effect tended to have higher levels of musical training, while their degree of musical sophistication (assessed by the Gold-MSI; Müllensiefen et al., 2014) correlated positively with the kurtosis of the Wundt effect, suggesting that musicians exhibit more focused preferences for intermediate degrees of predictability. However, the interpretation of these results is complicated by the observation that the kurtosis of the Wundt effect was correlated with its skew. Furthermore, although the range of musical sophistication was quite broad (from the second to the 91st percentile), the mean was quite low at approximately the 30th percentile (cf. Müllensiefen et al., 2014). Therefore, while this approach shows promise in using a meaningful computational measure of stylistic complexity to achieve a better balance of ecological validity and experimental control, it needs to be extended to samples with greater coverage of the full range of musical sophistication.

In summary, empirical results gathered across half a century of research are suggestive of musicians showing preferences for greater complexity, consistent with the hypothesized influence of musical expertise on the Wundt effect. However, the evidence remains far from conclusive. As I have tried to illustrate, this is due to several methodological difficulties in the experiments that have been conducted. First, many studies have failed to achieve a satisfactory balance between ecologically valid musical stimuli, important for achieving genuine aesthetic experiences, and the experimental control that comes from having a meaningful quantitative definition of complexity (see also Chapter 5). Second, many studies have failed to include both stimuli expressing a sufficiently broad range of musical complexity and a sample of participants covering a sufficiently broad range of musical expertise. Looking ahead, I believe the most promising strategy will be to take the approach of Gold et al. (2019), using sophisticated model-based measures of the complexity of realistic musical stimuli, and apply it to participants representing a broader range of music expertise and sophistication (assessed as a continuous dimension as in Hansen & Pearce, 2014) as well as stimuli representing a broader range of complexity, ideally across multiple musical cultures.

8.4 Expertise-related effects on perception and cognition

The evidence considered in §§ 8.2 and 8.3 suggests that musicians exhibit more accurate expectations than non-musicians, in accordance with the expert prediction hypothesis, and also, albeit less convincingly, preferences for greater complexity, consistent with a hypothesized shifting of the Wundt effect. A final question to consider is whether it is possible that these apparent effects of musical expertise on musical enculturation could arise from better non-musical abilities in musicians, potentially also resulting from musical training, rather than accumulated (implicit or explicit) knowledge of the syntactic structure of musical styles. Such abilities might include better auditory perception, allowing more accurate perceptual encoding of events, or better statistical learning ability, allowing more robust acquisition of internal representations of musical syntax.

There is a large literature on the effects of musical training on non-musical cognitive and perceptual abilities, most of which relies on correlational designs, comparing musicians with non-musicians (for detailed reviews, see Kraus & Chandrasekaran, 2010; Swaminathan & Schellenberg, 2019). An area of psychology relevant to the present enquiry is auditory perception and cognition: if musicians have more highly developed auditory abilities, they might achieve better performance than non-musicians on an expectation task, for example, simply because they are better able to discriminate and memorize the individual sounds and sound sequences involved (e.g., the contexts used for prediction) independently of any effect of musical training on enculturation through statistical learning. There is, in fact, evidence that musicians possess better auditory perceptual abilities than non-musicians. This includes, for example, verbal working, short-term and long-term memory (Franklin et al., 2008), and temporal information processing (Rammsayer & Altenmüller, 2006). These differences extend to speech processing where better abilities have been detected in musicians than non-musicians in, for example, perception of prosody (e.g., Thompson et al., 2004), phonological awareness (e.g., Chobert et al., 2011, 2014), and speech in noise perception (e.g., Coffey et al., 2017; Du & Zatorre, 2017).

As discussed in § 8.1, an advantage for musicians over non-musicians on a given measure provides correlational rather than causal evidence for an effect of musical training. Swaminathan and Schellenberg (2019) point out that while correlation does not necessarily entail causation, causation definitely does imply correlation. As they put it: "if music training causes improvements in cognitive abilities, one should rightly expect this effect to be evident in everyday life, such that individuals who take years of music lessons exhibit the documented positive effects" (Swaminathan & Schellenberg, 2019, p. 646). Therefore, negative evidence in the form of an absence of significant differences between musicians and non-musicians, although it might attract less attention, is worth paying attention to. Indeed some studies have failed to find differences in speech perception between musicians and non-musicians (e.g., Boebinger et al., 2017; Ruggles et al., 2014; Whiteford et al., 2024). Furthermore, in a systematic review of 13 longitudinal studies, Mckay (2021) found no evidence that musical training has a causal effect on speech perception ability in hearing-impaired listeners.

Another example of such evidence comes from a study conducted by Daniel Carey and Fred Dick, which included a series of experiments comparing the perceptual and cognitive abilities of musicians and non-musicians (Carey et al., 2015). The musicians were 24 violinists and 24 pianists, all but two of whom (one in each group) were currently taking or had completed a performance degree. The violinists had between 11 and 27.5 years of musical training, starting on average at the age of 5 and currently practised on their main instrument for between one and six hours per day, with an average of 10,928 total accumulated lifetime hours of practice. The pianists had between eight and 21 years of musical training, starting on average at the age of 5 and currently practised on their main instrument for between one and seven hours per day, with an average of 9,901 total accumulated lifetime hours of practice. A group of 24 non-musicians was also recruited who were matched to the musically trained groups for age, sex, and educational level. Seven of these non-musicians had no musical training whatsoever, while the remaining 17 had between three months and five years of musical training, starting on average at the age of 10, but had not played an instrument for between six and 19 years prior to the study.

These three groups of participants undertook a battery of perceptual and cognitive tasks. First, psychophysical discrimination of tones with different onset envelope rise time, amplitude modulation and frequency modulation showed that musicians had greater auditory sensitivity than non-musicians to manipulations of spectral and temporal envelope, with no difference emerging between violinists and pianists (somewhat surprisingly). However, a task in which participants judged which of two chords rendered in different tuning systems (including just and equal temperament) sounded most "in tune" showed that violinists strongly preferred just intonation while pianists showed preference for in tune (just or equal tempered) chords over chords with a mistuned note.

A set of less directly musical tasks was also included to assess more generalized differences between musicians and non-musicians. These included an auditory scene analysis task that involved identifying a target sound in an environmental scene (e.g., a "moo" in a background auditory scene taken from a farmyard); a sustained auditory attention task in which participants discriminated a target sound (a bird call) from eight non-targets (including a dog bark, a bicycle bell, and a camera shutter) in a sequence of 81 sounds; and, finally, a multimodal visual/auditory/spatial sequence reproduction task in which sequence length increased by one on each successful trial (starting from sequences of one up to a maximum of 21 tones). The results showed very little evidence for significant differences between musicians and non-musicians, with musicians showing marginally better sustained auditory attention but no advantage (or disadvantage) over non-musicians in environmental auditory scene analysis or multimodal sequence reproduction.

The results of this particular experiment suggest fairly well-circumscribed differences between musicians and non-musicians that are limited to fine-grained psychophysical (and in some cases instrument-specific) aspects of auditory processing. Results such as this are useful because they raise the possibility that differences between musicians and non-musicians observed in other studies may in fact have existed prior to onset of musical training—being one of a range of factors leading to continuation of musical training—rather than causal effects of musical training. Other evidence supports this perspective more directly. Swaminathan and Schellenberg (2020) found that individual differences on tests of speech perception and grammar in 91 children aged 6–9 were best predicted by IQ and musical aptitude (melody and rhythm discrimination and long-term memory for music) with duration of musical training having no effect after controlling for other factors. Furthermore, when 44 of those children returned five years later, individual differences in musical aptitude remained stable; while initial musical aptitude predicted subsequent duration of musical training, duration of musical training did not predict musical aptitude after five years once initial musical aptitude had been controlled for (Kragness et al., 2021). This is consistent with evidence from a study of adult monozygotic twins in which there was no correlation between intra-pair differences in amount of lifetime musical training (ranging as high as 20,228 hours) and performance on pitch, rhythm, and melody discrimination tasks as measures of musical aptitude (Mosing et al., 2014).

Overall, the evidence suggests that musicians may in some cases possess better auditory perceptual abilities than non-musicians but also that any such differences may exist prior to musical training rather than being a causal outcome of it. Regardless of the direction of causality, it remains possible that such differences could influence performance on tests of musical enculturation. Two complementary approaches to disentangling these effects seem promising. First, in comparisons of musicians and non-musicians on measures of enculturation (such as those considered in §§ 8.2 and 8.3), ensuring that such abilities are matched between the groups or at least measured and accounted for in the analysis would increase confidence that any differences observed result from enculturation rather than differences in perceptual ability. Second, in mapping out developmental trajectories of the effects of enculturation within a musical tradition (see Chapter 7) assessment of perceptual abilities alongside musical training (either by recording when it naturally occurs or by introducing it as an experimental manipulation) would allow robust comparison of the effects of enculturation through informal exposure and formal training.

Finally, it is also important to ask whether musicians and non-musicians differ in terms of their capability for statistical learning, especially given wide individual differences in modality-specific statistical learning ability (Frost et al., 2015). If musicians were better at auditory statistical learning, either as a result of musical training or as a pre-existing trait, this could account for any enhanced performance noted in §§ 8.2 and 8.3, including, for example, more robust schematic knowledge of their musical culture enabling more accurate and differentiated expectations. In Chapter 7, three artificial grammar learning studies were reviewed, providing evidence that listeners are capable of rapidly acquiring the statistical structure of auditory sequences generated by unfamiliar grammars. Two of these studies compared the statistical learning ability of musicians and non-musicians, in both cases finding no discernible differences in performance between the groups (Loui et al., 2010; Rohrmeier et al., 2011). The definitions of musicianship in these studies were fairly convincing: in Rohrmeier et al. (2011) the musicians had, on average, 10 years of music lessons and practised for seven hours per week compared with two years and zero hours for the non-musicians who had also not practised for 10 years on average; in Loui et al. (2010), the musicians had five to 14 years of music training whereas the non-musicians had no musical training outside normal school education. These findings have been replicated using an implicit measure of statistical learning in which errors during a recall task indicate points of low transition probability: two separate groups of Conservatory-level musicians (totalling 68 individuals) showed no difference in statistical learning compared with matched non-musicians without any training beyond ordinary school education (Carey et al., 2015; Krishnan et al., 2022).[3] On balance, it seems unlikely that musicians have any inherent advantage in auditory statistical learning over non-musicians that might influence the evidence considered in §§ 8.2 and 8.3.

[3] A counter-example is provided by the experiments of Oram and Cuddy (1995), discussed above, in which the probe-tone ratings of musicians showed greater influence of the frequency of the probe in the context than non-musicians for both diatonic and non-diatonic tone sets, although this relates to short-term dynamic statistical learning from the stimulus context rather than longer term schematic statistical learning.

8.5 Conclusion

Drawing firm conclusions on the psychological effects of musical expertise is complicated by a focus on correlational rather than causal designs (comparing musicians with non-musicians) and different operational definitions of musical expertise (mostly based on training as a proxy). Nonetheless, empirical results gathered over several decades are consistent with the expert prediction hypothesis, suggesting that musicians are able to use their greater knowledge of the syntactic structure of musical styles to generate more certain or specific expectations, which tend to be confirmed in real music but lead to greater prediction errors when real musical events are replaced by low-probability surrogates as in the probe-tone task. While there is concrete evidence that this additional knowledge is at least partly explicit, it remains possible that musicians may also possess greater implicit knowledge of musical structure than non-musicians. While IDyOM simulates the implicit expectations of non-musicians, it also appears to capture the more accurate expectations exhibited by musicians. If musicians possess more detailed and accurate probabilistic models of the syntactic structure of musical styles, we would expect them to show corresponding preference for greater levels of objective musical complexity, corresponding to a rightwards shifting of the apex of the Wundt curve. There is some evidence to suggest that this is the case but it remains incomplete due to a failure to find a satisfactory balance between ecologically valid musical stimuli, important for achieving genuine aesthetic experiences, and the experimental control that comes from having a meaningful quantitative definition of complexity, while also covering a sufficiently broad range of musical complexity and musical expertise. Finally, effects of musical expertise on enculturation are not likely to result from any differences in statistical learning ability between musicians and non-musicians but may be confounded by better auditory perception and cognition in musicians. Although the evidence here is not as clear cut as it might first appear, care needs to be taken to control for such potential differences between musicians and non-musicians.

9 Future Perspectives

Learning to Listen, Listening to Learn presents and evaluates a theory of music perception, expressed in computational terms based on the statistical learning and probabilistic prediction hypotheses set out in Chapter 1. These hypotheses state that listeners acquire internal models of musical structure during musical listening (including long-term exposure during enculturation and within individual listening episodes) and use these models to generate probabilistic predictions that inform their perception and appreciation of music. The empirical research considered in this book provides corroborating evidence for these hypotheses. In Chapter 3, it was shown that the theory can account for melodic pitch expectations while in Chapter 4 it was demonstrated that the theory generalizes to expectations for other prominent aspects of music—timing and harmony—as well as the psychological processes of metrical and tonal inference. Chapters 5 and 6 demonstrated that the same underlying mechanisms of statistical learning and probabilistic prediction can be extended to account for other psychological processes invoked while listening to music: emotional experience, complexity perception, aesthetic appreciation, memory, perceptual segmentation, and similarity perception.

Chapters 3–6 examined the predictions of the theory in the context of listeners and stimuli from Western musical cultures. However, a central prediction of the statistical learning hypothesis is that music perception should vary systematically as a function of musical experience. Chapter 7 examined research on differences in music perception between individuals with different musical experience, by systematically manipulating experience within an experimental session, comparing listeners from different musical cultures, or examining developmental changes in music perception with increasing exposure. The experimental evidence suggests not only that such differences exist but also that they can be predicted by computational models based on the statistical learning and probabilistic prediction hypotheses. Chapter 8 focused on differences in perception and experience of music as a function of musicianship, considering evidence that musicians possess more refined cognitive models of musical syntax, generating more specific and accurate expectations. The empirical evidence suggests that this may reflect both implicit knowledge acquired through greater domain experience and explicit declarative knowledge from formal training.

Throughout the book, rather than simply reporting empirical results that are consistent with the theory, the strategy has been to use a computational implementation of the theory (IDyOM) to simulate human perception. This brings two key benefits, whose significance is not to be underestimated (Johnson-Laird, 1983; Longuet-Higgins, 1981; Newell & Simon, 1976). First, it ensures that the theory is specified in sufficient detail to actually produce the behaviour it purports to explain. Second, it means that listeners' behaviour in empirical experiments can be quantitatively compared with the behaviour of the model. The computational approach facilitates the use of real-world stimuli (even if they are simplified) rather than stimuli that are experimentally manipulated to test particular hypotheses; in other words hypotheses about psychological mechanisms are implemented in the computational model and tested by comparing the model's responses with human responses to real-world stimuli. While ecological validity is often considered a lower priority for experimental design in many areas of psychology and neuroscience, the ability to use real-world stimuli is paramount in understanding perception of cultural phenomena such as music and language. Although the stimuli used in many of the simulations reported in this book were simplified in various ways (e.g., by extracting the melody or removing performance timing) they were generally not manipulated as they might have been in an experimental study, making it more likely that the results will generalize to real-world musical listening.

The remainder of this chapter discusses limitations and directions for further development.

9.1 Limits of statistical learning and probabilistic prediction

First, it is worth considering whether there are limits to the theory of statistical learning and probabilistic prediction as underlying mechanisms for music perception. There are, in fact, obvious aspects of music perception that go beyond these mechanisms. For example, it remains possible that perceptual expectations are influenced by other factors as well as probabilistic regularities acquired through statistical learning. These may reflect sensory memory mechanisms in which events are expected to remain proximal to the local context, maintaining pitch proximity in the case of melody (Schellenberg, 1997; Temperley, 2007) and spectral distance (Leman, 2000) in the case of harmony. There is empirical evidence for such effects on expectation that are not subsumed by statistical learning, though the models reflecting these effects are generally weaker predictors than IDyOM both for melody (Morgan et al., 2019; Pearce, 2005) and harmony (Cheung et al., 2023; Goldman et al., 2021; Harrison, 2019; Sears et al., 2019). Furthermore, as noted in Chapter 6, it seems likely that phrase-boundary perception and stream segregation rely on Gestalt principles (again, potentially related to constraints of sensory memory) in addition to any hypothesized effects of probabilistic prediction based on statistical learning. As a rather different example, while it was argued in Chapter 8 that musicians may possess stronger implicit models of musical structure, it is also undoubtedly the case that musicians possess explicit, declarative knowledge about music (the conscious expectations of Huron, 2006, pp. 235–237), acquired through deliberate study, that can be applied in a self-reflective way while listening attentively to music. While IDyOM appears to be capable of capturing some effects of explicit knowledge on expectation (see Chapter 8), it seems likely that not all such effects of explicit knowledge acquired by musicians operate via mechanisms of expectation.

Emotion induction provides another good example. Expectation is thought to be just one of several psychological mechanisms by which music can induce emotions in listeners Juslin (2019). The other mechanisms include: responses to extremes or fast changes in basic acoustic features such as loudness, dissonance, or timbre; emotions arising as a result of behavioural or physiological synchronisation with musical rhythms; conditioned associations between certain pieces of music and valence (e.g., a TV theme tune, Hepper, 1991); musical mimicry of linguistic or bodily expression of emotion (e.g., slow tempi and descending pitch contours might be associated with somatic or verbal expressions of sadness); and evoked imagery (e.g., a piece of program music might evoke imagery of a storm or a flock of birds). While emotions arising from acoustic changes or beat synchronisation could potentially be related to predictive processing (and therefore fall within scope of the present account), the remaining mechanisms have more to do with individual experiences of particular pieces of music or similarities between music and non-musical phenomena. The latter is also likely to play an important role in social dimensions of music perception and their relationship with individual identity.

A complete model of the aesthetic appreciation of music would have to include all these components, so the present work provides only a fragmentary picture. There is also work to be done to tie the different model components into a unified account of music perception, integrating melodic, rhythmic and harmonic expectations with tonality and metre induction, viewpoint selection, memory constraints, stream segregation, phrase segmentation and theme detection. In this respect, the present approach is less well developed than that of

Temperley (2009), for example, which combines metrical analysis, harmonic analysis and stream segregation as well as accepting expressively timed musical input. Putting the pieces of the current perspective together into an integrated model of music perception remains an important task for future research.

Finally, the statistical learning hypothesis proposes that internal cognitive representations of musical structure are learned. It is tempting to interpret this as meaning that individuals begin life with a complete *tabula rasa* as far as musical structure is concerned. However, no learning can take place without any kind of innate constraint (or inductive bias) and it is important to distinguish two different kinds of innate constraint on learning (Elman et al., 1996): *representational constraints* relate directly to the content that is learned while *architectural constraints* relate to the learning that takes place, which may indirectly constrain the content that is learned. Regarding representational constraints, the present approach assumes that listeners perceive individual events such as notes or chords while listening to music and are able to derive representations from those events in terms of pitch, timing and harmony. These include representations that relate an event to its predecessor (e.g., `interval` or `ioi`) and representations that relate an event to an inferred tonality or metre (e.g., `scale-degree` or `posinbar`). Importantly, these representational constraints do not cover sequential dependencies between events—i.e., anything meaning that the occurrence of one event at one point in a piece makes the occurrence of another event at another point more or less likely. The mechanisms of metrical and tonal inference allowing relative representations of pitch and timing are themselves subject to learning, as discussed in Chapters 4 and 7.

Stylistic regularities, reflecting probabilistic constraints on how musical events can be combined into a sequence, are posited to be learned by listeners through their experience of music. The architectural constraints on this process of learning are expressed in the capacity to: represent musical contexts of varying length; learn the frequency with which events appear in a given context; use these frequencies to generate conditional probabilities; and vary the context length to allow generalization to unseen contexts. The empirical results suggest that this combination of representational and architectural constraints provides a rather accurate simulation of many aspects of music perception. However, it is worth interrogating these hypothesized constraints in more detail. Without changing the simulation accuracy it is possible that listeners possess more powerful learning mechanisms (i.e., stronger architectural constraints) such that fewer innate representational constraints need to be assumed. Alternatively, it is possible that listeners possess stronger representational constraints such that less is learned and more is innately specified. This would align with traditional rule-based models of cognition but goes against the grain of the theoretical position presented here and also the weight of experimental data considered throughout the book (Chapter 7 especially). These questions are considered further in § 9.2 below.

9.2 Computational modelling

It is worth considering further the implications of the particular choice of modelling techniques used consistently throughout the book and whether any improvements could be made. IDyOM is based on Markov modelling enhanced with variable-order prediction, smoothing, long- and short-term models, and multiple viewpoints. This implies a psychological process in which predictions are made for the next event given the immediately preceding context. This conflicts with theoretical accounts (Lerdahl & Jackendoff, 1983; Schenker, 1935) as well as grammar-based computational models (Johnson-Laird, 1991; Rohrmeier, 2011; Steedman, 1996), supported by some experimental evidence (e.g., Koelsch et al., 2013), suggesting that listeners perceive long-range non-sequential dependencies in music. In other words, predictions for the next event might depend not (or not only) on the immediately preceding context but on events that occurred prior to that context. From the present perspective, the only difficulty with this approach is that computational models that explicitly represent long-range dependencies (e.g., context-free grammars) are hard to learn from data, requiring the specification of additional knowledge about musical structure whose psychological analogue is innate representational constraints.

It would go against the empirical evidence to rule out innate constraints on music perception—obvious candidates include pitch range, just-noticeable differences for musical attributes such as pitch and timing, memory limitations and so on. However, positing an innate constraint carries with it a certain amount of theoretical baggage. Is the constraint universal across cultures? Is it present in newborn infants? Is it affected by experience and, if so, how? What is the evolutionary explanation for its existence? From an experimental perspective, these are difficult questions to answer, especially when it can be shown that the constraint is not required for an accurate simulation of perception by models that learn through experience. There is methodological utility in taking a minimalist path to modelling, in which as little as possible is specified in advance and as much as possible is learned. This allows us to examine how well perception can be simulated with as few innate constraints as possible.[1]

The argument for traditional grammar-based models of music perception is further weakened by the development of neural network architectures whose generative capacity, given sufficient training data, suggests that they have internalized long-range dependencies even when trained on single-step-ahead prediction tasks. For example, the self-attention mechanism of the *transformer* architecture (Phuong & Hutter, 2022; Turner, 2024), when configured as an autoregressive decoder, learns a flexible combination of weights governing the influence of different events in the preceding context on the identity of the event being predicted.[2] The prediction task addressed by the autoregressive transformer is strictly equivalent to the prediction task performed by IDyOM. However, the transformer implements a powerful generalization of IDyOM's mechanism for generating predictions, which is based on an exact match between (some portion of) the current context and a context that has previously been observed. By contrast, the transformer is able to assign different weights to the different events making up the predictive context such that the weights are learned through training for increased prediction performance (i.e., with cross-entropy minimization as the objective function). Rather than requiring an exact match with the context or a suffix of the context (as IDyOM does), this allows for more generalized approximate matching of the predictive context (see also Bjare et al., 2022).

The idea of more generalized approximate matching can be related to the idea of threaded viewpoints, which allow IDyOM to create predictive contexts over non-contiguous events such as the beginnings and endings of groups (e.g., phrases) or metrical periods (e.g., bars). Also relevant here is the expectation network framework of Verosky (2019) which tracks the extent to which each scale degree has been preceded recently—but not necessarily immediately—by every other scale degree. This can be viewed as a weighted sum of *n*-grams (of any length) which start and end with the two scale degrees in question such that the first scale degree only appears once (also known as skip-grams). Verosky and Morgan (2021) tested this model on the empirical data of Morgan et al. (2019) in which musicians sang continuations of specially composed tonal melodies. The results showed that the expectation

[1]One could argue that the converse process—assuming everything is innate and nothing is learned—is also methodologically useful. However, learning is easy to demonstrate experimentally and examine in detail whereas it is extremely hard to conclusively demonstrate the existence of specific innate psychological constraints.

[2]A related approach, the continuous-code short-term model (CCSTM), replaces IDyOM's exact context matching for prediction with a scheme that matches the predictive context to a combination of previously experienced contexts weighted by their similarity to the predictive context (Bjare et al., 2022). While the CCSTM runs online like IDyOM's short-term model (STM), the similarity mapping is trained offline on a large corpus.

network accounted for significant unique variance in the continuations over and above that accounted for by IDyOM and the pitch model of Temperley (2007). This suggests the psychological relevance of non-sequential predictive contexts and motivates a perceptual examination of the more powerful data-driven context-matching ability of the transformer.

The transformer's multi-head attention mechanism is capable of learning multiple sets of weights for the predictive context, which are combined to generate the prediction. This gives it the potential for developing different representations of the context such that different continuations are predicted by each representation (or attention head, in the jargon). At an abstract level, this bears a strong resemblance to the use of multiple viewpoints for prediction. In some ways, it is more powerful in that the representations are learned and not dependent on hand-coding, though it remains to be seen whether there are constraints on the kinds of auditory representations that can be learned through multi-head self-attention. Like other neural network architectures, challenges for transformers include the difficulty of interpreting their behaviour and the fact that training involves orders of magnitude more data than human learners experience (Warstadt & Bowman, 2023), but there is no doubt that they have achieved extraordinary results in prediction and generation of both natural language (Devlin et al., 2018; OpenAI, 2024) and music (Bjare et al., 2022; Hawthorne et al., 2022), demonstrating learning of hierarchical syntactic structure and long-range dependencies.

Because the statistical learning and predictive processing hypotheses are rather general, there is no reason why the continuation of the research programme outlined in this book, should not substitute IDyOM with other predictive models, such as the transformer, that also instantiate these hypotheses. For present purposes, the pertinent question is whether such models would simulate perception of music more accurately than IDyOM regardless of prediction performance. In addressing this question, it will be important to ensure that these models are specified and trained in ways that are consistent with knowledge of human musical enculturation. Importantly, the quest for accurate simulations of perception can involve making models simpler and less powerful as well as more sophisticated and powerful. For example, in Chapter 6 it was shown that models with decaying frequency counts simulated human memory for auditory patterns more accurately than models without (Bianco et al., 2020; Harrison et al., 2020). Likewise, it is important to investigate whether or not computational models with better prediction performance than IDyOM also simulate music perception better when compared quantitatively with empirical data.

In this context, Kern et al. (2022) compared the performance of IDyOM (STM, LTM and BOTH configurations; see Table 2.1) with the Music Transformer (Huang et al., 2018) in predicting neural responses (EEG and MEG) to melodies taken from European art music.[3] The results showed that the greatest proportions of unique variance were explained by the IDyOM LTM and the Music Transformer with no significant difference between these models. This demonstrates quite clearly that better prediction performance does not necessarily translate into better simulation of psychological or neural mechanisms.

Finally, it is worth noting that to be consistent with the statistical learning and probabilistic prediction hypotheses, it is not necessary to generate probabilities in the formal sense, only to make predictions that can be interpreted probabilistically based on learning from experience. In other words, it is important that predictions for multiple continuations are generated and associated with relative weights based on experience (as opposed to deterministic predictions or multiple predictions with fixed weights). Although IDyOM computes full distributions over the alphabet of an attribute, I don't consider this a necessary component of the psychological simulation. Rather it is a useful means of weighting the contributions of short- and long-term models trained using different viewpoints and, from a methodological perspective, ensures that probabilities are comparable at all points in a simulation. It seems more likely that listeners' perceptual systems would make predictions for a handful of possible continuations at any given point in time and that this set of possibilities could vary depending on the context. Understanding how the set of predictive possibilities is constrained and how these constraints change over time is an interesting topic for future research.

9.3 Auditory representations

9.3.1 Low-level auditory processing

The computational modelling presented in this book assumes a representation of an auditory stimulus at the level of discrete events, possessing attributes such as pitch or onset time. Although there is good evidence that listeners perceive auditory events in these ways using similar representations, they are not handed these representations on a plate. Rather, the auditory system must extract such auditory objects from sound waves impinging on the ears.

[3]The stimulus set included melodies from pieces by Benjamin Britten, César Franck, C. P. E. Bach, J. S. Bach, Ludwig van Beethoven, Muzio Clementi, Ernesto Köhler, Georg Friedrich Handel, Joseph Haydn, W. A. Mozart, Modest Mussorgsky, and Pyotr Ilyich Tchaikovsky. Models were trained on the Monophonic Corpus of Complete Compositions (MCCC) (https://osf.io/dg7ms/) developed by Edward Hall consisting of 623 melodic renditions of compositions from the European tradition of art music from 1703 to 1934 as depicted in Figure 6.7 (see § 6.3.3 and Hall & Pearce, 2021).

The problem was captured vividly by Al Bregman (1990) as being analogous to trying to figure out what is happening on a lake with only two handkerchiefs fastened in channels at the waterside to go by. A complete computational simulation must start with the same input as the auditory system and build up to the kind of symbolic representations operated on by IDyOM. It is perfectly possible that the nature of these higher-level representations would be affected by an accurate simulation of the mechanisms of lower-level afferent auditory processing.

It is also important to consider the possibility of low-level auditory representations contributing directly to expectation independently of higher-level representations. In Chapter 4, it was observed that the periodicity pitch (PP) model (Leman, 2000) contributed significantly and independently alongside IDyOM (albeit to a lesser extent) in accounting for harmonic expectations (Cheung et al., 2023). The PP model simulates auditory processing of an acoustic signal from periphery to a local image, with a short time constant, representing the strength of different pitch periodicities in the auditory nerve which is correlated with a global image, with a longer time constant, reflecting the contents of echoic memory.

The PP model, like most models of low-level auditory processing, produces continuous-valued predictions. This contrasts with IDyOM's predictions for discrete, categorical attributes of auditory events, raising the question of how to simulate the combination of continuous and categorical attributes in generating auditory predictions. Peter Harrison (2019) has developed an approach that combines output from categorical (e.g., long-term and short-term IDyOM viewpoint models) and continuous predictors using an exponentiated weighted sum of the elements, where the weights are optimized by training on a corpus. Time series for continuous predictors are modelled with orthogonal polynomial functions of a specified degree n_p such that n_p terms are added to the set of predictors, allowing for non-linear effects of continuous predictors on the combined predictions. Harrison (2019) used quartic polynomials ($n_p = 4$) and considered various continuous attributes of chords including representations of dissonance (Harrison & Pearce, 2020; Hutchinson & Knopoff, 1978; Milne, 2013) and voice-leading distance (Tymoczko, 2006).

Skerritt-Davis and Elhilali (2021) proposed a rather different approach for predictive modelling of low-level auditory attributes. The Dynamic Regularity Extraction (D-REX) model acts on any continuous acoustic or perceptual feature sampled at discrete time steps. The model makes predictions from multiple context windows prior to the timestep to be predicted in an analogous way to the PPM smoothing mechanism implemented in IDyOM. Each context is represented by the sufficient statistics of a distribution fit to the context window where the choice of distribution (e.g., Gaussian, Poisson) is made by the user to match the input feature. These distributions are combined in a mixture weighted by a belief for each context based on how well it predicted past observations. The behaviour of the model was illustrated for a range of acoustic features including inter-aural level differences, pitch, spectral spread, spectral centroid, RMS energy, and transient onsets.

These are promising first steps but further research is required to investigate exactly how perceptual expectations for continuous low-level auditory attributes are generated and combined with expectations for the higher-level attributes emphasized in the present work.

9.3.2 Non-syntactic musical attributes

While the focus here has been on pitch (Chapter 3), rhythm, and harmony (Chapter 4), it is worth considering whether perceptual expectations for other attributes such as timbre, dynamics, and articulation can be simulated using the same modelling approach. Leonard Meyer (1989) made a distinction between musical attributes that carry syntactic relationships and those that vary parametrically, which I will refer to as syntactic and non-syntactic attributes.[4] Consider, for example, a melodic pattern in which the dominant (fifth scale degree) implies that the next pitch will be the tonic (first scale degree; see Chapter 4), or one in which an octave leap is followed by a descent to a pitch in between the extremes of the opening pair of pitches (see Chapter 3).[5] In these examples, for an appropriately enculturated listener, the context creates an expectation for the final pitch, even though it is not necessarily close in pitch to its predecessors. Note also that the implications remain intact when these patterns are transposed to different absolute pitch levels. Contrast these examples with dynamics (i.e., changes in intensity) for which it is usually difficult to identify patterns of dynamic level that predict the specific intensity of the next event. Rather, intensity often varies continuously, such that implications are created for small changes over time, with expectation-disrupting sudden changes in dynamics used to add emphasis and drama. Here, small and large are approximate categories allowing for considerable variability unlike pitch or timing, where a semitone or semiquaver difference could change entirely the tonal or metrical implications of the predictive context or continuation. Therefore, it may be that a different underlying model is needed to simulate perception of such non-syntactic parameters, with

[4] Meyer used the terms primary and secondary parameters, which might be interpreted as unfairly pejorative for the latter.

[5] I use these examples as intuitive ways of describing musical patterns for illustration but this should not be taken to suggest that these particular patterns, encoded as rule-based models, constitute models of cognition that could replace a more general psychological process of probabilistic prediction based on statistical learning.

the approaches discussed above (Harrison, 2019; Leman, 2000; Skerritt-Davis & Elhilali, 2021) providing attractive candidates for further investigation.

In Western musical cultures, pitch, rhythm, and harmony are often the primary syntactic attributes while dynamics, timbre, and articulation are often non-syntactic attributes. However, it is important to note that non-syntactic parameters can modulate the effects of syntactic parameters. For example, a performer can use expressive dynamics to emphasize metrical or phrase structure while orchestration can be used to distinguish syntactic patterns in different timbral voices. Furthermore, the status of a given musical attribute as syntactic or non-syntactic can vary between musical styles and even pieces. Although pitch is often employed as a syntactic parameter, it need not be and could vary parametrically, while dynamics or timbre, which are often non-syntactic parameters in Western musical traditions, can function as syntactic parameters in experimental electronic music (e.g., electroacoustic music). From a psychological perspective, listeners can learn statistical structure in timbral sequences (Tillmann & McAdams, 2004) and there is even some evidence that electroacoustic composers are capable of perceiving equivalence of intervals between instrumental timbres that are transposed to different parts of a perceptual timbre space (McAdams & Cunible, 1992). As another example, Olsen et al. (2016) examined the question of how phrases are perceived in sound-based music, finding significant effects of changes in intensity and timbre.

9.3.3 Selection and combination of auditory representations

A question remains over the way in which auditory attributes are identified and combined for prediction by perceptual systems. In much of the research covered here, IDyOM simulations have used a single viewpoint, motivated by psychological and musicological considerations (`interval⊗scale-degree` for pitch and `ioi-ratio⊗posinbar⊗barlength` for timing). This was a methodological decision to facilitate interpretability and consistency across individual studies.

As discussed in Chapter 1, David Huron (2006) argued that perceptual systems generate multiple representations of sensory events and selectively emphasize those that produce the most accurate predictions. In other words, listeners are equipped with a rather flexible representational system that can identify, for a given stimulus, information-bearing attributes—those that allow for better prediction performance—and attend to those in generating expectations. This putative attentional mechanism would maintain predictions for multiple attributes and dynamically switch focus between attributes depending on the presence of structure in the stimulus, so as to emphasize attributes that lead to the most accurate predictions. For example, in a strongly tonal piece, the perceptual system would focus on scale degree representations whereas in a non-tonal piece it might focus on chromatic pitch and in an electroacoustic piece on timbre and dynamics. As Huron notes, this mechanism would be expected to differ between listeners with different enculturation, since the most predictive representations might well differ between different musical cultures.

This hypothesis is consistent with IDyOM's multiple viewpoint framework and some of the research reviewed in this book has used a viewpoint selection mechanism to select viewpoints that maximize either prediction performance or fit to human behaviour (see Chapter 3). Indeed, results considered in § 3.5 provide some evidence for perceptual selection of representations for melodic expectation that increase the predictability of the music being listened to. In other cases, contributions of different viewpoints have been weighted, without selection, for optimized prediction performance (see Chapter 4).[6] These viewpoint selection or weighting schemes for optimized prediction performance can make the behaviour of the models less straightforwardly interpretable but they may better approximate a perceptual mechanism. Further implementation and testing of this hypothesis remain an interesting direction for future research since many of the details remain to be worked out and evaluated. How are new representations generated and under what constraints? To what extent is this dependent on experience? Over what timeframe is the attentional mechanism optimized or, relatedly, how much evidence is required to switch representational focus? Does the attentional mechanism filter out representations or simply downweight them? Is it plausible that multiple predictive processes are tracked for a collection of auditory attributes simultaneously in time? What processing constraints are imposed on this process?

9.4 Neural mechanisms

The hypothesized mechanisms of statistical learning and probabilistic prediction have been evaluated in this book primarily through computational simulation and behavioural experimentation. While this provides evidence for the corresponding psychological mechanisms, it is worth considering the neural processes associated with these mechanisms. There is, in fact, a long tradition of research examining predictive neural processing both in music (Rohrmeier & Koelsch, 2012) and language (Elman, 1990; Heilbron et al., 2022; Kuperberg & Jaeger, 2016; Schrimpf et al., 2021). Zatorre (2024) emphasizes the role of prediction throughout the auditory brain, including at subcortical levels of the auditory system (Parras

[6] Note that this hypothesis is also consistent with the multi-head self-attention mechanism of transformers, as discussed above.

et al., 2017), and within the ventral and dorsal auditory cortical pathways (Hickok & Poeppel, 2007; Rauschecker & Scott, 2009; Zatorre et al., 2007), as well as their interactions with the reward system (Berridge & Kringelbach, 2015).

For music, the evidence suggests that simple zeroth-order regularities generate surprise responses such as the mismatch negativity (see § 3.4) which can be localized to auditory cortex (Schönwiesner et al., 2007). Violations of harmonic expectation appear to activate superior temporal gyrus (STG) and inferior frontal gyrus (IFG) in both hemispheres with a tendency towards right lateralization (Tillmann et al., 2006). The left IFG contains Broca's area, which together with the posterior middle temporal gyrus (pMTG) is thought to play a key role in comprehension and production of syntax in speech (Matchin & Hickok, 2020). Furthermore, the volume of the right ventral arcuate fasciculus (AF), a white matter tract connecting IFG with STG (dorsal AF) and MTG (ventral AF), correlates with generalization performance in a pitch-based artificial grammar learning task (Loui et al., 2011). Accordingly, it has been suggested that the ventral auditory pathway from (right) STG to IFG is involved in the acquisition, representation, retention in memory, and deployment of musical syntax (Zatorre, 2024).

It has been proposed that temporal expectations are generated within the dorsal auditory pathway (Zatorre, 2024; Zatorre et al., 2007) projecting from auditory cortex via the posterior STG and inferior parietal lobule (IPL) to dorsal premotor cortex (dPMC). Evidence for the role of the motor system in auditory temporal perception comes from neuroimaging studies in which subcortical nuclei in the basal ganglia (especially the dorsal striatum) and cortical areas associated with motor planning (premotor cortex and Supplementary Motor Area, SMA) show greater activation for rhythms that strongly induce a beat compared with those that do not (Grahn & Brett, 2007; Teki et al., 2011). Patel and Iversen (2014) argued that this pathway supports an influence of motor simulation in dPMC on predictive processing of event timing. Cannon and Patel (2020) further proposed that the motor simulation consists of activation in SMA representing beat periods with the dorsal striatum chaining them together into a sequence to generate a repeated periodic motor plan. This mechanism could be used to represent multiple different hypothesized metrical structures underlying a temporal stimulus, whose simulated sensory feedback could be compared with sensory input to infer the actual metre (see Chapter 4). This has been modelled computationally as a generative process whose parameters are updated through Bayesian inference (Cannon, 2021; Kaplan et al., 2022).

These proposed neural mechanisms are compatible with the Bayesian brain hypothesis (Knill & Pouget, 2004) according to which the brain possesses an internal generative model of the sensory environment explaining how sensory input (s) is generated by a set of hidden causes (h)—consisting of both a likelihood, $p(s|h)$ and a prior p(h)—which is used to infer the hidden causes given the sensory input, $p(h|s)$. An influential instantiation of this hypothesis (Clark, 2013; Friston, 2005) proposes that perception of noisy input involves a hierarchical predictive process in which top-down generative predictions are compared with sensory input at a cascading series of hierarchical levels, representing a gradient of spatial or temporal scales. The comparison at each level yields a prediction error (sometimes weighted by the precision, or variance, of the sensory input), which is passed up the hierarchy, reflecting the portion of the sensory signal still to be accounted for. For the generative model, selection of a hidden cause at one level constrains those at the level below such that the posterior at one level can be used as the prior at the level below. This reflects the hierarchical organization of sensory cortices, with bottom-up and top-down signals coded in different layers of cortical columns (Friston, 2005; Keller & Mrsic-Flogel, 2018).

The psychological approach presented in this book can be viewed as an instance of the Bayesian brain hypothesis; IDyOM simulates a generative psychological process that estimates the probability of incoming events, given an internal model of the sensory domain. The inference models covered in Chapter 4 extend this generative process with hidden causes—metre and tonality—that are inferred for incoming musical sequences. However, it is worth noting some differences between the present account and the theory of predictive coding specifically (without de-emphasizing the many similarities, especially in terms of theoretical motivation). While predictive coding is generally applied to the problem of perceiving current sensory input given a noisy and ambiguous perceptual process (e.g., Rao & Ballard, 1999), the present approach focuses on prediction of the next event in an auditory sequence, given the preceding context (which has been assumed for simplicity to be noise free, except for the memory decay modelling covered in Chapter 6). Furthermore, predictive coding generally assumes Gaussian models of continuous sensory dimensions whereas the present approach assumes discrete probabilistic models of categorical sensory dimensions (but see § 9.3 for a discussion of how predictive models of continuous and discrete attributes can be combined). This is important because many of the variables considered (e.g., `scale-degree`, `posinbar`, or `root-sd`) have clearly non-Gaussian, multimodal distributions.

There are also differences between the generative models typically used in predictive coding and those introduced here. For example, the internal model in predictive coding is generally defined with respect to the recent past of a specific sensory environment, whereas an important component of the present account is an internal model based on a long-term process of statistical learning across the lifespan which is combined with learning over shorter and more recent timescales. In addition, a central aspect of predictive coding is the hierarchical structure of the generative model corresponding to a hierarchical perceptual process whereas the present account is mostly confined to a single perceptual level. Although viewpoints with differing degrees of abstraction are considered, they are used to predict at a single level of perceptual representation in which auditory events have discrete properties such as onset time and pitch. That said, the models presented here could be viewed as occupying one level in the predictive coding hierarchy, with information content and entropy reflecting prediction error and precision respectively, while the models of tonal and metrical inference discussed in Chapter 4 introduce a hierarchical Bayesian element to the generative model.

9.5 Beyond music: Language, dance and other cultural domains

By virtue of the fact that it draws on general-purpose cognitive mechanisms of statistical learning and probabilistic prediction combined with exposure to a cultural domain, the approach presented here naturally extends to perception of other cultural domains with time-varying communicative structure such as language, dance, and motion picture (with all of which, of course, music is often combined).

There is a great deal of theoretical and empirical research on language perception along these lines, including the role of statistical learning (Frost et al., 2019; Saffran & Kirkham, 2018) and probabilistic prediction (Elman, 1990; Hale, 2006; Kuperberg & Jaeger, 2016; Levy, 2008). For example, Smith and Levy (2013) found that word reading times decreased systematically with increasing conditional probability of the word estimated by a probabilistic language model with similar properties to IDyOM (a smoothed trigram model trained on the British National Corpus—a large corpus of written and spoken English). Schrimpf et al. (2021) examined a wide variety of predictive language models (including transformers) in terms of their ability to predict neural responses (fMRI, ECoG) during reading of printed text or listening to speech. The results showed that, across models, better performance on a prediction task (next-word prediction accuracy) but not other language tasks (e.g., grammaticality, sentence similarity) was correlated with the model's fit to the neural and behavioural responses (reading times). These findings suggest that language is processed using probabilistic predictive mechanisms based on statistical learning that are very similar to the mechanisms proposed here for music perception.

It remains to be seen whether the approach can generalise to other cultural domains in which visual sequencing plays a role. IDyOM has been extended to prediction of visual elements in a simple audiovisual domain involving tones and horizontally moving lines, for which information content influenced perception of complexity for both visual and auditory modalities (Clemente et al., 2025). Perception and appreciation of dance (Christensen & Jola, 2015) provides an interesting potential case study, showing distinctive differences across cultures. As a preliminary sketch, a dance performance could be coded as a sequence of positions or movements between positions. An individual watching the performance might generate expectations for the next movement given the preceding sequence of movements based both on their long-term cultural experience of dance and their perception of repeated patterns within the performance itself. For example, non-dancers expressed greater enjoyment for sequences of dance movements by an individual dancer when the timing was more variable but also more predictable (Orlandi et al., 2020). How this simple scenario extends to perception of choreography with multiple performers remains an open, but interesting, question. Just as polyphonic music is not represented entirely as a parallel sequence of independent voices, dancers could not be represented as individual time series without considering the interactions between the dancers and the global configuration. Furthermore, as with music, aesthetic experience is also influenced by many aspects of expressive performance that go beyond sequential event structure, which would need to be accommodated. There is also the question of the beholder's attention, which might be drawn to different parts of the visual scene in a way that changes dynamically during a performance. The multiple viewpoint mechanisms applied in the present work to music (see § 9.3) could be adapted to visuomotor sequences to simulate processing of multiple aspects of a dance performance. These aspects could vary between dance styles and cultures shaping the experience of enculturated observers, their internal predictive models, and expectations for continuation.

9.6 Creativity: Musical performance and composition

The focus in this book has been on simulating perception of music, including the role of predictive mechanisms in aesthetic experience. The generative counterpart to aesthetic experience is creativity, which can be viewed as the psychological process of creating aesthetic experiences for an audience. Does the present account have any implications for composition and performance of music? Well, IDyOM is a generative model in that it specifies probability distributions governing the properties of sequences of musical events. By sampling from a trained model, it is possible to generate music that reflects the Markov regularities of the training corpus. This usually produces poor output in the absence of other constraints that bias the output towards higher probability parts of the compositional space (e.g., representational constraints or probability thresholds to avoid low-probability events). This might provide a reasonable model of a listener humming or whistling an improvised tune. However, I think it provides a poor model of composition more broadly.

Psychological research on the process (as distinguished from the output) of musical composition is much less well established than research on music perception. In his book, *The musical mind*, John Sloboda (1985) considers four sources of evidence: first, the history of particular compositions as evidenced in manuscripts, sketches, and notebooks; second, composers' own writing about their compositional processes; third, live observation of composers in the act of composition; fourth, observation and description of improvisatory performance. All of these sources have severe limitations and none provides anything close to a complete picture. Sketches are not extant for many composers, inference from manuscripts is limited to serendipitous events such as Mozart's changing of ink during transcription of The Magic Flute, introspection provides a limited window onto mental processes. However, the picture that emerges is of a process in which ideas are generated, tested, sometimes rejected outright, often modified, and iteratively combined into a provisional structure for the entire piece in which details are gradually filled in, sometimes over time spans of years and with extensive use of the score or an instrument as an external workspace. This process depends on implicit schematic knowledge of a musical style (of the kind considered in detail in this book) that can be used to generate ideas but also an explicit repertoire of compositional devices (which falls outside the scope of the present approach).[7] Judgements are made about intermediate stages of the emerging composition which motivate subsequent decisions in the journey to the final form. These features bear significant overlap with expert human performance in other domains (see Feltovich et al., 2006).

It is clear from this picture that an accurate simulation of the psychological process of composition goes beyond the mechanisms of statistical learning and probabilistic prediction

[7]One might argue that the generative ability of transformers calls this statement into question but I would counter that regardless of the quality of the output, the generative process implemented in the transformer is very different from that employed by the majority of composers.

that I have argued play an important role in music perception. However, composers and performers are also listeners, with their own predictive mechanisms based on musical enculturation. If a composer intends to challenge the expectations of their audience, thereby introducing tension into the listening experience, they must have an internal model of what their audience's expectations will be. Similarly, a performer introducing expressive timing and dynamics needs to be able to predict the effect that these will have on processing of syntactic parameters by their audience. For improvisatory musical styles, where the performer is also a real-time composer, these two scenarios are combined. I would hypothesize that composers and performers by virtue of being experienced listeners, enculturated in a musical style, possess the same internal predictive models as their in-culture audience which can be used to evaluate the perceptual effects that their musical creations or interpretations will have. Therefore, the mechanisms presented and evaluated in this book may have some (albeit circumscribed) role to play in understanding composition, improvisation, and performance.

9.7 Conclusion

Chapter 1 considered cultural domains as dynamic systems in which the set of phenomena constituting the culture is passed on through learning and developed from generation to generation. The research programme outlined here represents an attempt to understand perception in the context of this dynamic process of cultural evolution for the domain of music. This has resulted in a formal and extensively evaluated account of how listeners acquire knowledge of a musical culture through experience so as to be able to perceive and appreciate it in ways that are appropriate to the culture. The statistical learning and probabilistic prediction hypotheses provide concrete hypothesized mechanisms for a broad range of psychological processes required for successful perception of the music of a culture. The empirical research presented—both experiments with human listeners and computational simulations—provides corroborating evidence for these hypotheses. A complementary computational understanding of the creative process of musical composition would open the door to an integrated simulation of the cultural evolution of music. Extending these ideas to other forms of art—literary, performing, and visual—would pave the way to a better understanding of the relationships between these core components of human cultural experience, which seem likely to have shared intertwined histories of cultural evolution.

References

Aarden, B. (2003). *Dynamic melodic expectancy* [Doctoral dissertation, Ohio State University].

Agres, K. R., Abdallah, S. A., & Pearce, M. T. (2018). Information-theoretic properties of auditory sequences dynamically influence expectation and memory. *Cognitive Science, 42*, 43–76.

Albrecht, J., & Shanahan, D. (2013). The use of large corpora to train a new type of key-finding algorithm: An improved treatment of the minor mode. *Music Perception, 31*, 59–67.

Allopenna, P. D., Magnuson, J. S., & Tanenhaus, M. K. (1998). Tracking the time course of spoken word recognition using eye movements: Evidence for continuous mapping models. *Journal of Memory and Language, 439*, 419–439.

Alluri, V., Toiviainen, P., Jääskeläinen, I. P., Glerean, E., Sams, M., & Brattico, E. (2012). Large-scale brain networks emerge from dynamic processing of musical timbre, key and rhythm. *NeuroImage, 59*, 3677–3689.

Altmann, E. M., & Gray, W. D. (2002). Forgetting to remember: The functional relationship of decay and interference. *Psychological Science, 13*, 27–33.

Altmann, E. M., & Schunn, C. D. (2012). Decay versus interference: A new look at an old interaction. *Psychological Science, 23*, 1435–1437.

Altmann, G. T. M., & Kamide, Y. (1999). Incremental interpretation at verbs: Restricting the domain of subsequent reference. *Cognition, 73*, 247–264.

Anglada-Tort, M., Harrison, P. M. C., Lee, H., & Jacoby, N. (2023). Large-scale iterated singing experiments reveal oral transmission mechanisms underlying music evolution. *Current Biology, 33*, 1472–1486.

Arai, M., & Keller, F. (2013). The use of verb-specific information for prediction in sentence processing. *Language and Cognitive Processes, 28*, 3–8.

Aslin, R. N., Saffran, J. R., & Newport, E. L. (1998). Computation of conditional probability statistics by 8-month-old infants. *Psychological Science*, *9*, 321–324.

Atkinson, R. C., & Shiffrin, R. M. (1971). The control of short-term memory. *Scientific American*, *225*, 82–91.

Aylett, M., & Turk, A. (2004). The smooth signal hypothesis: A functional explanation for relationships between redundancy, prosodic prominence, and duration in spontaneous speech. *Language and Speech*, *47*, 31–56.

Bach, J. S. (1875). *371 vierstimmige choralgesänge von Johann Sebastian Bach* (A. Dörffel, Ed.; 4th ed.). Breitkopf & Härtel.

Balkwill, L. L., & Thompson, W. F. (1999). A cross-cultural investigation of the perception of emotion in music: Psychophysical and cultural cues. *Music Perception*, *17*, 43–64.

Balkwill, L. L., Thompson, W. F., & Matsunaga, R. (2004). Recognition of emotion in Japanese, Western, and Hindustani music by Japanese listeners. *Japanese Psychological Research*, *46*, 337–349.

Bannister, S. (2020). A vigilance explanation of musical chills? Effects of loudness and brightness manipulations. *Music & Science*, *3*, 1–17.

Bannister, S., & Eerola, T. (2018). Suppressing the chills: Effects of musical manipulation on the chills response. *Frontiers in Psychology*, *9*, 2046.

Barascud, N., Pearce, M. T., Griffiths, T. D., Friston, K. J., & Chait, M. (2016). Brain responses in humans reveal ideal observer-like sensitivity to complex acoustic patterns. *Proceedings of the National Academy of Sciences*, *113*, E616–E625.

Barlow, H. (2001). Redundancy reduction revisited. *Network: Computation in Neural Systems*, *12*, 241–253.

Bartlett, J. C., Halpern, A. R., & Dowling, W. J. (1995). Recognition of familiar and unfamiliar melodies in normal aging and Alzheimer's disease. *Memory & Cognition*, *23*, 531–546.

Begleiter, R., El-Yaniv, R., & Yona, G. (2004). On prediction using variable order Markov models. *Journal of Artificial Intelligence Research*, *22*, 385–421.

Bell, T. C., Cleary, J. G., & Witten, I. H. (1990). *Text compression*. Prentice Hall.

Benetos, E., & Holzapfel, A. (2013). Automatic transcription of Turkish makam music. *Proceedings of the 14th International Society for Music Information Retrieval Conference*, 355–360.

Berlyne, D. E. (1971). *Aesthetics and psychobiology*. Appleton-Century-Crofts.

Berridge, K. C., & Kringelbach, M. L. (2015). Pleasure systems in the brain. *Neuron*, *86*, 646–664.

Besson, M., & Faïta, F. (1995). An event-related potential (ERP) study of musical expectancy: Comparison of musicians with nonmusicians. *Journal of Experimental Psychology: Human Perception and Performance*, *21*, 1278–1296.

Besson, M., & Macar, F. (1987). An event-related potential analysis of incongruity in music and other non-linguistic contexts. *Psychophysiology*, *24*, 14–25.

Bharucha, J. J. (1987). Music cognition and perceptual facilitation: A connectionist framework. *Music Perception*, *5*, 1–30.

Bharucha, J. J. (1994). Tonality and expectation. In R. Aiello (Ed.), *Musical perceptions* (pp. 213–239). Oxford University Press.

Bharucha, J. J., & Stoeckig, K. (1986). Reaction time and musical expectancy: Priming of chords. *Journal of Experimental Psychology: Human Perception and Performance*, *12*, 403–410.

Bianco, R., Hall, E. T. R., Pearce, M. T., & Chait, M. (2023). Implicit auditory memory in older listeners: From encoding to 6-month retention. *Current Research in Neurobiology*, *5*, 100115.

Bianco, R., Harrison, P. M. C., Hu, M., Bolger, C., Picken, S., Pearce, M. T., & Chait, M. (2020). Long-term implicit memory for sequential auditory patterns in humans. *Elife*, *9*, e56073.

Biederman, I., & Vessel, E. (2006). Perceptual pleasure and the brain. *American Scientist*, *94*, 247.

Bigand, E., Delbé, C., Poulin-Charronnat, B., Leman, M., & Tillmann, B. (2014). Empirical evidence for musical syntax processing? Computer simulations reveal the contribution of auditory short-term memory. *Frontiers in Systems Neuroscience*, *8*, 94.

Bigand, E., & Parncutt, R. (1999). Perceiving musical tension in long chord sequences. *Psychological Research*, *62*, 237–254.

Bigand, E., & Poulin-Charronnat, B. (2006). Are we "experienced listeners"? A review of the musical capacities that do not depend on formal musical training. *Cognition*, *100*, 100–130.

Bigand, E., Poulin-Charronnat, B., Tillmann, B., Madurell, F., & D'Adamo, D. A. (2003). Sensory versus cognitive components in harmonic priming. *Journal of Experimental Psychology: Human Perception and Performance*, *29*, 159–171.

Bigand, E., Tillmann, B., & Poulin, B. (2001). The effect of harmonic context on phoneme monitoring in vocal music. *Cognition*, *81*, 11–20.

Bjare, M. R., Lattner, S., & Widmer, G. (2022). Differentiable short-term models for efficient online learning and prediction in monophonic music. *Transactions of the International Society for Music Information Retrieval*, *5*, 190–207.

Blakemore, S.-J., Wolpert, C. A. D., & Frith, C. (2000). Why can't you tickle yourself? *NeuroReport*, *11*, 11–16.

Blom, T., Feuerriegel, D., Johnson, P., Bode, S., & Hogendoorn, H. (2020). Predictions drive neural representations of visual events ahead of incoming sensory information. *Proceedings of the National Academy of Sciences*, *117*, 7510–7515.

Blood, A. J., & Zatorre, R. J. (2001). Intensely pleasurable responses to music correlate with activity in brain regions implicated in reward and emotion. *Proceedings of the National Academy of Sciences*, *98*, 11818–11823.

Boebinger, D., Evans, S., Rosen, S., Lima, C. F., Manly, T., & Scott, S. K. (2017). Musicians and non-musicians are equally adept at perceiving masked speech. *Journal of the Acoustical Society of America*, *378*, 378–387.

Boltz, M. G. (1999). The processing of melodic and temporal information: Independent or unified dimensions? *Journal of New Music Research*, *28*, 67–79.

Bouwer, F. L., Honing, H., & Slagter, H. A. (2020). Beat-based and memory-based temporal expectations in rhythm: Similar perceptual effects, different underlying mechanisms. *Journal of Cognitive Neuroscience*, *32*, 1221–1241.

Brattico, E., Tervaniemi, M., Näätänen, R., & Peretz, I. (2006). Musical scale properties are automatically processed in the human auditory cortex. *Brain Research*, *1117*, 162–174.

Bregman, A. S. (1990). *Auditory scene analysis: The perceptual organization of sound*. MIT Press.

Bregman, A. S., & Campbell, J. (1971). Primary auditory stream segregation and perception of order in rapid sequences of tones. *Journal of Experimental Psychology*, *89*, 244–249.

Brent, M. R. (1999). An efficient, probabilistically sound algorithm for segmentation and word discovery. *Machine Learning*, *34*, 71–105.

Brooks Jr., F. P., Hopkins, A. L., Neumann, P. G., & Wright, W. V. (1957). An experiment in musical composition. *IRE Transactions on Electronic Computers*, *EC-6*, 175–182.

Brown, G. D. A., & Chater, N. (2007). A temporal ratio model of memory. *Psychological Review*, *114*, 539–576.

Budd, M. (1995). *Values of art: Painting, poetry, and music*. Penguin.

Bunton, S. (1996). *On-line stochastic processes in data compression* [Doctoral dissertation, University of Washington].

Burgoyne, J. A. (2011). *Stochastic processes & database-driven musicology* [Doctoral dissertation, McGill University].

Burke, M. J., & Gridley, M. C. (1990). Musical preferences as a function of stimulus complexity and listeners' sophistication. *Perceptual and Motor Skills*, *7*, 687–690.

Burunat, I., Toiviainen, P., Alluri, V., Bogert, B., Ristaniemi, T., Sams, M., & Brattico, E. (2016). The reliability of continuous brain responses during naturalistic listening to music. *NeuroImage*, *124*, 224–231.

Cacioppo, J. T., & Petty, R. E. (1982). The need for cognition. *Journal of Personality and Social Psychology*, *42*, 116–131.

Cambouropoulos, E. (2001). The local boundary detection model (LBDM) and its application in the study of expressive timing. *Proceedings of the International Computer Music Conference*, 17–22.

Cannon, J. J. (2021). Expectancy-based rhythmic entrainment as continuous Bayesian inference. *PLOS Computational Biology*, *17*, e1009025.

Cannon, J. J., & Patel, A. D. (2020). How beat perception co-opts motor neurophysiology. *Trends in Cognitive Sciences*, *25*, 137–150.

Cantor, G. N. (1968). Children's "like-dislike" ratings of familiarized and unfamiliarized visual stimuli. *Journal of Experimental Child Psychology*, *6*, 651–657.

Caplin, W. E. (2004). The classical cadence: Conceptions and misconceptions. *Journal of the American Musicological Society*, *57*, 51–118.

Carey, D., Rosen, S., Krishnan, S., Pearce, M. T., Shepherd, A., Aydelott, J., & Dick, F. (2015). Generality and specificity in the effects of musical expertise on perception and cognition. *Cognition*, *137*, 81–105.

Carlsen, J. C. (1981). Some factors which influence melodic expectancy. *Psychomusicology*, *1*, 12–29.

Carlsen, J. C., Divenyi, P. I., & Taylor, J. A. (1970). A preliminary study of perceptual expectancy in melodic configurations. *Bulletin of the Council for Research in Music Education*, *22*, 4–12.

Carroll, N. (2000). Art and the domain of the aesthetic. *British Journal of Aesthetics*, *40*, 191–208.

Castellano, M. A., Bharucha, J. J., & Krumhansl, C. L. (1984). Tonal hierarchies in the music of North India. *Journal of Experimental Psychology: General*, *113*, 394–412.

Chang, H. W., & Trehub, S. E. (1977). Auditory processing of relational information by young infants. *Journal of Experimental Child Psychology*, *24*, 324–331.

Chater, N., & Vitányi, P. M. B. (2001). The generalized universal law of generalization. *Journal of Mathematical Psychology*, *47*, 346–369.

Chen, S. F., & Goodman, J. (1999). An empirical study of smoothing techniques for language modeling. *Computer Speech and Language*, *13*, 359–394.

Cherla, S. (2016). *Neural probabilistic models for melody prediction, sequence labelling and classification* [Doctoral dissertation, City University of London].

Cherla, S., Tran, S. N., Weyde, T., & d'Avila Garcez, A. (2015). Hybrid long-and short-term models of folk melodies. *Proceedings of the 16th International Society for Music Information Retrieval Conference*, 584–590.

Cheung, V. K. M., Harrison, P. M. C., Koelsch, S., Pearce, M. T., Friederici, A. D., & Meyer, L. (2023). Cognitive and sensory expectations independently shape musical expectancy and pleasure. *Philosophical Transactions of the Royal Society B*, *379*, 20220420.

Cheung, V. K. M., Harrison, P. M. C., Meyer, L., Pearce, M. T., Haynes, J.-D., & Koelsch, S. (2019). Uncertainty and surprise jointly predict musical pleasure and amygdala, hippocampus, and auditory cortex activity. *Current Biology*, *29*, 4084–4092.e4.

Chmiel, A., & Schubert, E. (2017). Back to the inverted-U for music preference: A review of the literature. *Psychology of Music*, *45*, 886–909.

Chobert, J., François, C., Velay, J.-L., & Besson, M. (2014). Twelve months of active musical training in 8- to 10-year-old children enhances the preattentive processing of syllabic duration and voice onset time. *Cerebral Cortex*, *24*, 956–967.

Chobert, J., Marie, C., François, C., Schön, D., & Besson, M. (2011). Enhanced passive and active processing of syllables in musician children. *Journal of Cognitive Neuroscience*, *23*, 3874–3887.

Christensen, J. F., & Jola, C. (2015). Moving towards ecological validity in empirical aesthetics of dance. In J. P. Huston, M. Nadal, F. Mora, L. F. Agnati & C. J. Cela-Conde (Eds.), *Art, aesthetics, and the brain* (pp. 223–260). Oxford University Press.

Christiansen, M. H. (2019). Implicit statistical learning: A tale of two literatures. *Topics in Cognitive Science*, *11*, 468–481.

Clark, A. (2013). Whatever next? Predictive brains, situated agents, and the future of cognitive science. *Behavioral and Brain Sciences*, *36*, 181–204.

Clarke, E. F. (1999). Rhythm and timing in music. In D. Deutsch (Ed.), *The psychology of music* (2nd ed., pp. 473–500). Academic Press.

Cleary, J. G., & Teahan, W. J. (1997). Unbounded length contexts for PPM. *The Computer Journal*, *40*, 67–75.

Cleary, J. G., & Witten, I. H. (1984). Data compression using adaptive coding and partial string matching. *IEEE Transactions on Communications*, *32*, 396–402.

Clemente, A., Board, F., Pearce, M. T., & Orgs, G. (2025). Complexity in audiovisual aesthetics: Liking for auditory, visual and audiovisual dynamic stimuli. *Psychology of Aesthetics, Creativity, and the Arts*.

Clemente, A., Vila-Vidal, M., Pearce, M. T., Aguiló, G., Corradi, G., Nadal, M., & Clemente, A. (2020). A set of 200 musical stimuli varying in balance, contour, symmetry, and complexity: Behavioral and computational assessments. *Behavior Research Methods*, *52*, 1491–1509.

Coffey, E. B. J., Mogilever, N. B., & Zatorre, R. J. (2017). Speech-in-noise perception in musicians: A review. *Hearing Research*, *352*, 49–69.

Coggiola, J. C. (2004). The effect of conceptual advancement in jazz music selections and jazz experience on musicians' aesthetic response. *Journal of Research in Music Education*, *52*, 29–42.

Cohen, J. E. (1962). Information theory and music. *Behavioral Science*, *7*, 137–163.

Cohen, J. E. (2003). Human population: The next half century. *Science*, *302*, 1172–1176.

Collins, T., Tillmann, B., Barrett, F. S., Delbé, C., & Janata, P. (2014). A combined model of sensory and cognitive representations underlying tonal expectations in music: From audio signals to behavior. *Psychological Review*, *121*, 33–65.

Conklin, D. (1990). *Prediction and entropy of music* [Doctoral dissertation, University of Calgary].

Conklin, D., & Cleary, J. G. (1988). Modelling and generating music using multiple viewpoints. *Proceedings of the First Workshop on AI and Music*, 125–137.

Conklin, D., & Witten, I. H. (1995). Multiple viewpoint systems for music prediction. *Journal of New Music Research*, *24*, 51–73.

Conway, A. R. A., Kane, M. J., Bunting, M. F., Hambrick, D. Z., Wilhelm, O., & Engle, R. W. (2005). Working memory span tasks: A methodological review and user's guide. *Psychonomic Bulletin & Review*, *12*, 769–786.

Cook, N. (1987). The perception of large-scale tonal closure. *Music Perception*, *5*, 197–205.

Coons, E., & Kraehenbuehl, D. (1958). Information as a measure of structure in music. *Journal of Music Theory*, *2*, 127–161.

Cooper, G., & Meyer, L. B. (1960). *The rhythmic structure of music*. University of Chicago Press.

Corrigall, K. A., & Trainor, L. J. (2010). Musical enculturation in preschool children: Acquisition of key and harmonic knowledge. *Music Perception*, *28*, 195–200.

Cover, T. M., & King, R. C. (1978). A convergent gambling estimate of the entropy of English. *IEEE Transactions on Information Theory*, *24*, 413–421.

Cowan, N. (1984). On short and long auditory stores. *Psychological Bulletin*, *96*, 341–370.

Cowan, N. (1988). Evolving conceptions of memory storage, selective attention, and their mutual constraints within the human information-processing system. *Psychological Bulletin*, *104*, 163–191.

Cowan, N. (2000). The magical number 4 in short-term memory: A reconsideration of mental storage capacity. *Behavioural and Brain Sciences*, *24*, 87–185.

Cowan, N. (2008). What are the differences between long-term, short-term, and working memory? In W. S. Sossin, J.-C. Lacaille, V. F. Castelluci & S. Belleville (Eds.), *Progress in brain research* (pp. 323–338, Vol. 169). Elsevier.

Cowans, P. J. (2006). *Probabilistic document modelling* [Doctoral dissertation, University of Cambridge].

Creighton, H. (1966). *Songs and ballads from Nova Scotia*. Dover.

Crowder, R. G. (1993). Auditory memory. In S. McAdams & E. Bigand (Eds.), *Thinking in sound: The cognitive psychology of human audition* (pp. 111–145). Oxford University Press.

Crozier, J. B. (1974). Verbal and exploratory responses to sound sequences varying in uncertainty level. In D. E. Berlyne (Ed.), *Studies in the New Experimental Aesthetics: Steps Towards an Objective Psychology of Aesthetic Appreciation* (pp. 27–90). Hemisphere Publishing Co.

Cuddy, L. L., & Badertscher, B. (1987). Recovery of the tonal hierarchy: Some comparisons across age and levels of musical experience. *Perception & Psychophysics*, *41*, 609–620.

Cuddy, L. L., & Lunney, C. A. (1995). Expectancies generated by melodic intervals: Perceptual judgments of melodic continuity. *Perception & Psychophysics*, *57*, 451–462.

Darwin, C. J. (1997). Auditory grouping. *Trends in Cognitive Sciences*, *1*, 327–333.

de Fleurian, R., & Pearce, M. T. (2021). Chills in music: A systematic review. *Psychological Bulletin*, *147*, 890–920.

Deliège, I. (1987). Grouping conditions in listening to music: An approach to Lerdahl & Jackendoff's grouping preference rules. *Music perception*, *4*, 325–359.

Demany, L., Armand, F., & Demany, L. (1985). The perceptual reality of tone chroma in early infancy. *Journal of the Acoustical Society of America*, *57*, 57–66.

Dember, W. N., & Earl, R. W. (1957). Analysis of exploratory, manipulatory, and curiosity behaviors. *Psychological Review*, *64*, 91–96.

Demorest, S. M., & Morrison, S. J. (2016). Quantifying culture: The cultural distance hypothesis of melodic expectancy. In J. Y. Chiao, S.-C. Li, R. Seligman & R. Turner (Eds.), *The Oxford handbook of cultural neuroscience* (pp. 183–194). Oxford University Press.

Demorest, S. M., Morrison, S. J., Beken, M. N., & Jungbluth, D. (2008). Lost in translation: An enculturation effect in music memory performance. *Music Perception*, *25*, 213–233.

Demorest, S. M., Morrison, S. J., Nguyen, V. Q., & Bodnar, E. N. (2016). The influence of contextual cues on cultural bias in music memory. *Music Perception*, *33*, 590–600.

Dennett, D. (1991). *Consciousness explained*. Back Bay Books.

Deouell, L. Y. (2007). The frontal generator of the mismatch negativity revisited. *Journal of Psychophysiology*, *21*, 188–203.

Derex, M., & Mesoudi, A. (2020). Cumulative cultural evolution within evolving population structures. *Trends in Cognitive Sciences*, *24*, 654–667.

Desain, P., Honing, H., van Thienen, H., & Windsor, L. (1998). Computational modelling of music cognition: Problem or solution. *Music Perception*, *16*, 151–166.

Devlin, J., Chang, M.-W., Lee, K., & Toutanova, K. (2018). BERT: Pre-training of deep bidirectional transformers for language understanding. *arXiv preprint*, arXiv:1810.04805v2.

Dewey, J. (1894). The theory of emotion (I): Emotional attitudes. *The Psychological Review*, *1*, 553–569.

Dewey, J. (1895). The theory of emotion (II): The significance of emotions. *Psychological Review*, *2*, 13–32.

Dewitt, L. A., & Robert, G. (1986). Recognition of novel melodies after brief delays. *Music Perception*, *3*, 259–274.

Di Liberto, G. M., Pelofi, C., Bianco, R., Patel, P., Shamma, S. A., Mehta, A. D., Herrero, J. L., & de Cheveigné, A. (2020). Cortical encoding of melodic expectations in human temporal cortex. *Elife*, *9*, e51784.

Dibben, N. (1994). The cognitive reality of hierarchic structure in tonal and atonal music. *Music Perception*, *12*, 1–25.

Dickie, G. (1964). The myth of the aesthetic attitude. *American Philosophical Quarterly*, *1*, 56–65.

Dowling, W. J. (1973). The perception of interleaved melodies. *Cognitive Psychology*, *5*, 322–337.

Dowling, W. J. (1978). Scale and contour: Two components in a theory of memory for melodies. *Psychological Review, 85*, 341–354.

Dowling, W. J. (1991). Tonal strength and melody recognition after long and short delays. *Perception & Psychophysics, 50*, 305–313.

Dowling, W. J. (1994). Melodic contour in hearing and remembering melodies. In R. Aiello & J. Sloboda (Eds.), *Musical Perceptions* (pp. 173–190). Oxford University Press.

Dowling, W. J., & Bartlett, J. C. (1981). The importance of interval information in long-term memory for melodies. *Psychomusicology, 1*, 30–49.

Dowling, W. J., & Harwood, D. L. (1986). *Music cognition*. Academic Press.

Downie, J. S. (2008). The music information retrieval evaluation exchange (2005–2007). *Acoustical Science and Technology, 29*, 247–255.

Downie, J. S., Ehmann, A. F., Bay, M., & Jones, M. C. (2010). The music information retrieval evaluation exchange: Some observations and insights. In Z. W. Raś & A. A. Wieczorkowska (Eds.), *Advances in music information retrieval* (pp. 93–115). Springer.

Du, Y., & Zatorre, R. J. (2017). Musical training sharpens and bonds ears and tongue to hear speech better. *Proceedings of the National Academy of Sciences, 114*, 13579–13584.

Ebcioğlu, K. (1988). An expert system for harmonizing four-part chorales. *Computer Music Journal, 12*, 43–51.

Eerola, T. (2016). Expectancy-violation and information-theoretic models of melodic complexity. *Empirical Musicology Review, 11*, 2–17.

Egermann, H., Fernando, N., Chuen, L., & McAdams, S. (2015). Music induces universal emotion-related psychophysiological responses: Comparing Canadian listeners to Congolese Pygmies. *Frontiers in Psychology, 6*, 1341.

Egermann, H., Pearce, M. T., Wiggins, G., & McAdams, S. (2013). Probabilistic models of expectation violation predict psychophysiological emotional responses to live concert music. *Cognitive, Affective & Behavioral Neuroscience, 13*, 533–553.

Eichenbaum, H., Otto, T., & Cohen, N. J. (1994). Two functional components of the hippocampal memory system. *Behavioural and Brain Sciences, 17*, 449–518.

Elhilali, M., & Shamma, S. A. (2008). A cocktail party with a cortical twist: How cortical mechanisms contribute to sound segregation. *Journal of the Acoustical Society of America, 124*, 3751–3771.

Elman, J. L. (1990). Finding structure in time. *Cognitive Science, 14*, 179–211.

Elman, J. L., Bates, E. A., Johnson, M. H., Karmiloff-Smith, A., Parisi, D., & Plunkett, K. (1996). *Rethinking innateness: A connectionist perspective on development*. MIT Press.

Endress, A. D., & Mehler, J. (2009). The surprising power of statistical learning: When fragment knowledge leads to false memories of unheard words. *Journal of Memory and Language, 60*, 351–367.

Enquist, M., Ghirlanda, S., Jarrick, A., & Wachtmeister, C. (2008). Why does human culture increase exponentially? *Theoretical Population Biology, 74*, 46–55.

Farach, M. (1997). Optimal suffix tree construction with large alphabets. *Proceedings of the 38th Annual Symposium on Fundamentals of Computer Science*, 137–143.

Fehér, O., Ljubičić, I., Suzuki, K., Okanoya, K., & Tchernichovski, O. (2016). Statistical learning in songbirds: From self- tutoring to song culture. *Philosophical Transactions of the Royal Society B, 372*, 20160053.

Feltovich, P. J., Prietula, M. J., & Ericsson, K. A. (2006). Studies of expertise from psychological perspectives. In *The Cambridge handbook of expertise and expert performance* (pp. 1–37, Vol. Cambridge). Cambridge University Press.

Ferreri, L., Mas-Herrero, E., Zatorre, R. J., Ripollés, P., Gomez-Andres, A., & Alicart, H. (2019). Dopamine modulates the reward experiences elicited by music. *Proceedings of the National Academy of Sciences, 116*, 3793–3798.

Fiser, J., & Aslin, R. N. (2002). Statistical learning of higher-order temporal structure from visual shape sequences. *Journal of Experimental Psychology: Learning, Memory, and Cognition, 28*, 458–467.

Fitch, W. T. (2006). The biology and evolution of music: A comparative perspective. *Cognition, 100*, 173–215.

Franco, A., & Destrebecqz, A. (2012). Chunking or not chunking? How do we find words in artificial language learning? *Advances in Cognitive Psychology, 8*, 144–154.

Frank, S. L., Otten, L. J., Galli, G., & Vigliocco, G. (2015). The ERP response to the amount of information conveyed by words in sentences. *Brain and Language, 140*, 1–11.

Frankland, B. W., & Cohen, A. J. (2004). Parsing of melody: Quantification and testing of the local grouping rules of Lerdahl and Jackendoff's A Generative Theory of Tonal Music. *Music Perception, 21*, 499–543.

Franklin, M. S., Moore, K. S., Rattray, K., & Moher, J. (2008). The effects of musical training on verbal memory. *Psychology of Music, 36*, 353–365.

Friston, K. J. (2005). A theory of cortical responses. *Philosophical Transactions of the Royal Society B, 360*, 815–836.

Fritz, T. (2013). The dock-in model of music culture and cross-cultural perception. *Music Perception, 30*, 511–516.

Fritz, T., Jentschke, S., Gosselin, N., Sammler, D., Peretz, I., Turner, R., Friederici, A. D., & Koelsch, S. (2009). Universal recognition of three basic emotions in music. *Current Biology*, *19*, 573–576.

Frost, R., Armstrong, B. C., & Christiansen, M. H. (2019). Statistical learning research: A critical review and possible new directions. *Psychological Bulletin*, *145*, 1128–1153.

Frost, R., Armstrong, B. C., Siegelman, N., & Christiansen, M. H. (2015). Domain generality versus modality specificity: The paradox of statistical learning. *Trends in Cognitive Sciences*, *19*, 117–125.

Fung, C. V. (1996). Musicians' and nonmusicians' preferences for world musics: Relation to musical characteristics and familiarity. *Journal of Research in Music Education*, *44*, 60–83.

Gardner, H. (1987). *The mind's new science: A history of the cognitive revolution*. Basic Books.

Gervain, J., Nespor, M., & Mazuka, R. (2008). Bootstrapping word order in prelexical infants: A Japanese-Italian cross-linguistic study. *Cognitive Psychology*, *57*, 56–74.

Giegerich, R., & Kurtz, S. (1997). From Ukkonen to McCreight and Weiner: A unifying view of linear-time suffix tree construction. *Algorithmica*, *19*, 331–353.

Gingras, B., Pearce, M. T., Goodchild, M., Dean, R. T., Wiggins, G., & McAdams, S. (2016). Linking melodic expectation to expressive performance timing and perceived musical tension. *Journal of Experimental Psychology: Human Perception and Performance*, *42*, 594–609.

Giroux, I., & Rey, A. (2009). Lexical and sublexical units in speech perception. *Cognitive Science*, *33*, 260–272.

Gold, B. P., Pearce, M. T., Mas-Herrero, E., Dagher, A., & Zatorre, R. J. (2019). Predictability and uncertainty in the pleasure of music: A reward for learning? *Journal of Neuroscience*, *39*, 9397–9409.

Goldman, A., Harrison, P. M. C., Jackson, T., & Pearce, M. T. (2021). Reassessing syntax-related ERP components using popular music chord sequences: A model-based approach. *Music Perception*, *39*, 118–144.

Goldwater, S., Griffiths, T. L., & Johnson, M. (2005). Interpolating between types and tokens by estimating power-law generators. *Proceedings of the 18th International Conference on Neural Information Processing Systems*, 459–466.

Goodchild, M., Wild, J., & McAdams, S. (2019). Exploring emotional responses to orchestral gestures. *Musicae Scientiae*, *23*, 25–49.

Gordon, J., & Gridley, M. C. (2013). Musical preferences as a function of stimulus complexity of piano jazz. *Creativity Research Journal*, *25*, 143–146.

Gordon, P. C., & Holyoak, K. J. (1983). Implicit learning and generalization of the "mere exposure" effect. *Journal of Personality and Social Psychology*, *45*, 492–500.

Gottlieb, J., & Oudeyer, P.-Y. (2018). Towards a neuroscience of active sampling and curiosity. *Nature Reviews Neuroscience*, *19*, 758–770.

Grahn, J. A., & Brett, M. (2007). Rhythm and beat perception in motor areas of the brain. *Journal of Cognitive Neuroscience*, *19*, 893–906.

Gregory, R. L. (1980). Perceptions as hypotheses. *Philosophical Transactions of the Royal Society B*, *290*, 181–97.

Grewe, O., Nagel, F., Kopiez, R., & Altenmüller, E. (2007). Listening to music as a re-creative process: Physiological, psychological and psychoacoustical correlates of chills and strong emotions. *Music Perception*, *24*, 297–314.

Griffiths, T. D., & Warren, J. D. (2002). The planum temporale as a computational hub. *Trends in Neurosciences*, *25*, 348–353.

Guhn, M., Hamm, A., & Zentner, M. (2007). Physiological and musico-acoustic correlates of the chill response. *Music Perception*, *24*, 473–484.

Hale, J. (2006). Uncertainty about the rest of the sentence. *Cognitive Science*, *30*, 643–672.

Hall, E. T. R. (2023). *Modelling perception of large-scale thematic structure in music* [Doctoral dissertation, Queen Mary University of London].

Hall, E. T. R., & Pearce, M. T. (2021). A model of large-scale thematic structure. *Journal of New Music Research*, *50*, 220–241.

Hall, E. T. R., & Pearce, M. T. (2025). Perception of thematic structure in music over short and long timescales. *Music Perception*.

Halpern, A. R. (2020). Processing of musical pitch, time, and emotion in older adults. In L. L. Cuddy, S. Belleville & A. Moussard (Eds.), *Music and the aging brain* (pp. 43–67). Academic Press.

Halpern, A. R., & Bartlett, J. C. (2010). Memory for melodies. In M. Riess Jones, R. R. Fay & A. N. Popper (Eds.), *Music perception* (pp. 233–258, Vol. 36). Springer New York.

Halpern, A. R., Kwak, S., Bartlett, J. C., & Dowling, W. J. (1996). Effects of aging and musical experience on the representation of tonal hierarchies. *Psychology and Aging*, *11*, 235–46.

Halpern, A. R., & Zatorre, R. J. (1999). When that tune runs through your head: A PET investigation of auditory imagery for familiar melodies. *Cerebral Cortex*, *9*, 697–704.

Halpern, A. R., Zioga, I., Shankleman, M., Lindsen, J., Pearce, M. T., & Bhattacharya, J. (2017). That note sounds wrong! Age-related effects in processing of musical expectation. *Brain and Cognition*, *113*, 1–9.

Hannon, E. E., Soley, G., & Ullal, S. (2012). Familiarity overrides complexity in rhythm perception: A cross-cultural comparison of American and Turkish listeners. *Journal of Experimental Psychology: Human Perception and Performance*, *38*, 543–548.

Hannon, E. E., & Trainor, L. J. (2007). Music acquisition: Effects of enculturation and formal training on development. *Trends in Cognitive Sciences*, *11*, 466–472.

Hannon, E. E., & Trehub, S. E. (2005a). Metrical categories in infancy and adulthood. *Psychological Science*, *16*, 48–55.

Hannon, E. E., & Trehub, S. E. (2005b). Tuning in to musical rhythms: Infants learn more readily than adults. *Proceedings of the National Academy of Sciences*, *102*, 12639–12643.

Hansen, N. C. (2016). *Predictive coding of musical expertise* [Doctoral dissertation, Aarhus University].

Hansen, N. C., Kragness, H. E., Vuust, P., Trainor, L. J., & Pearce, M. T. (2021). Predictive uncertainty underlies auditory boundary perception. *Psychological Science*, *32*, 1416–1425.

Hansen, N. C., & Pearce, M. T. (2014). Predictive uncertainty in auditory sequence processing. *Frontiers in Psychology*, *5*, 1052.

Hansen, N. C., Vuust, P., & Pearce, M. T. (2016). "If you have to ask, you'll never know": Effects of specialised stylistic expertise on predictive processing of music. *PLOS ONE*, *11*, e0163584.

Hanslick, E. (1854). *On the musically beautiful* (G. Payzant, Ed.; Trans.). Hackett Publishing Company.

Hard, B. M., Recchia, G., & Tversky, B. (2011). The shape of action. *Journal of Experimental Psychology: General*, *140*, 586–604.

Hardt, O., Nader, K., & Nadel, L. (2013). Decay happens: The role of active forgetting in memory. *Trends in Cognitive Sciences*, *17*, 109–118.

Harrison, P. M. C. (2019). *Modelling the perception and composition of Western musical harmony* [Doctoral dissertation, Queen Mary University of London].

Harrison, P. M. C., Bianco, R., Chait, M., & Pearce, M. T. (2020). PPM-Decay: A computational model of auditory prediction with memory decay. *PLOS Computational Biology*, *16*, e1008304.

Harrison, P. M. C., & Pearce, M. T. (2018). Dissociating sensory and cognitive theories of harmony perception through computational modeling. In R. Parncutt & S. Sattmann (Eds.), *Proceedings of the 15th International Conference on Music Perception and Cognition* (pp. 194–199). Centre for Systematic Musicology, University of Graz.

Harrison, P. M. C., & Pearce, M. T. (2020). Simultaneous consonance in music perception and composition. *Psychological Review, 127*, 216–244.

Harte, C. A., Sandler, M. B., Abdallah, S. A., & Gomez, E. (2005). Symbolic representation of musical chords: A proposed syntax for text annotations. *Proceedings of the 6th International Conference on Music Information Retrieval*, 66–71.

Hawthorne, C., Jaegle, A., Cangea, C., Borgeaud, S., Nash, C., Malinowski, M., Dieleman, S., Vinyals, O., Botvinick, M., Simon, I., Sheahan, H., Zeghidour, N., Alayrac, J.-B., Carreira, J., & Engel, J. (2022). General-purpose, long-context autoregressive modeling with Perceiver AR. *Proceedings of the 39th International Conference on Machine Learning*, 8535–8558.

Heilbron, M., Armeni, K., Schoffelen, J.-M., Hagoort, P., & Lange, F. P. D. (2022). A hierarchy of linguistic predictions during natural language comprehension. *Proceedings of the National Academy of Sciences, 119*, e2201968119.

Hepper, P. G. (1991). An examination of fetal learning before and after birth. *The Irish Journal of Psychology, 12*, 95–107.

Heyduk, R. G. (1975). Rated preference for musical compositions as it relates to complexity and exposure frequency. *Perception & Psychophysics, 17*, 84–90.

Hickok, G., & Poeppel, D. (2007). The cortical organization of speech processing. *Nature Reviews Neuroscience, 8*, 393–402.

Hiller, L., & Bean, C. (1966). Information theory analyses of four sonata expositions. *Journal of Music Theory, 10*, 96–137.

Hiller, L., & Fuller, R. (1967). Structure and information in Webern's Symphonie, Op . 21. *Journal of Music Theory, 11*, 60–115.

Hiller, L., & Isaacson, L. (1959). *Experimental music*. McGraw-Hill.

Hinton, G. E. (2002). Training products of experts by minimizing contrastive divergence. *Neural Computation, 1800*, 1771–1800.

Hodges, D. A. (2017). *A concise survey of music philosophy*. Taylor & Francis.

Holzapfel, A. (2015). Relation between surface rhythm and rhythmic modes in Turkish makam music. *Journal of New Music Research, 44*, 25–38.

Honing, H. (2003). The final ritard: On music, motion, and kinematic models. *Computer Music Journal, 27*, 66–72.

Honing, H. (2013). Structure and interpretation of rhythm in music. In D. Deutsch (Ed.), *The psychology of music* (3rd ed., pp. 369–404). Academic Press.

Honing, H., Cate, C., Peretz, I., & Trehub, S. E. (2015). Without it no music: Cognition, biology and evolution of musicality. *Philosophical Transactions of the Royal Society B*, *370*, 20140088.

Honing, H., & Ploeger, A. (2012). Cognition and the evolution of music: Pitfalls and prospects. *Topics in Cognitive Science*, *4*, 513–524.

Hopcroft, J. E., & Ullman, J. D. (1979). *Introduction to automata theory, languages and computation*. Addison-Wesley.

Hosken, F., Bechtold, T. A., Hoesl, F., Kilchenmann, L., & Senn, O. (2021). Drum groove corpora. *Empirical Musicology Review*, *16*, 114–123.

Howard, P. G. (1993). *The design and analysis of efficient lossless data compression systems* [Doctoral dissertation, Brown University].

Huang, C.-Z. A., Vaswani, A., Uszkoreit, J., Shazeer, N., Simon, I., Hawthorne, C., Dai, A. M., Hoffman, M. D., Dinculescu, M., & Eck, D. (2018). Music transformer: Generating music with long-term structure. *arXiv preprint*, arXiv:1809.04281v3.

Huettel, S. A., Mack, P. B., & McCarthy, G. (2002). Perceiving patterns in random series: Dynamic processing of sequence in prefrontal cortex. *Nature Neuroscience*, *5*, 485–490.

Hunt, R. H., & Aslin, R. N. (2001). Statistical learning in a serial reaction time task: Access to seperable statistical cues by individual learners. *Journal of Experimental Psychology: General*, *130*, 658–680.

Huron, D. (2001). Tone and voice: A derivation of the rules of voice-leading from perceptual principles. *Music Perception*, *19*, 1–64.

Huron, D. (2006). *Sweet anticipation: Music and the psychology of expectation*. MIT Press.

Huron, D. (2016). *Voice leading: The science behind a musical art*. MIT Press.

Hutchinson, W., & Knopoff, L. (1978). The acoustic component of Western consonance. *Interface*, *7*, 1–29.

Iseminger, G. (2003). Aesthetic experience. In J. Levinson (Ed.), *The Oxford handbook of aesthetics* (pp. 99–116). Oxford University Press.

Iversen, J. R., Patel, A. D., & Ohgushi, K. (2008). Perception of rhythmic grouping depends on auditory experience. *Journal of the Acoustical Society of America*, *124*, 2263.

Jackendoff, R. (1991). Musical parsing and musical affect. *Music Perception*, *9*, 199–229.

Jacoby, N., Polak, R., Grahn, J. A., Cameron, D. J., Lee, K. M., Godoy, R. A., Undurraga, E. A., Huanca, T., Thalwitzer, T., Doumbia, N., Goldberg, D., Margulis, E. H., Wong, P. C. M., Jure, L., Rocamora, M., Fujii, S., Savage, P. E., Ajimi, J., Konno, R., … McDermott, J. H. (2024). Commonality and variation in mental representations of music revealed by a cross-cultural comparison of rhythm priors in 15 countries. *Nature Human Behaviour*, *8*, 846–877.

Janata, P., Birk, J. L., Horn, J. D. V., Leman, M., Tillmann, B., & Bharucha, J. J. (2002). The cortical topography of tonal structures underlying Western music. *Science*, *298*, 2167–2171.

Jentschke, S., & Koelsch, S. (2009). Musical training modulates the development of syntax processing in children. *NeuroImage*, *47*, 735–744.

Johnson-Laird, P. N. (1983). *Mental models*. Harvard University Press.

Johnson-Laird, P. N. (1991). Jazz improvisation: A theory at the computational level. In P. Howell, R. West & I. Cross (Eds.), *Representing musical structure* (pp. 291–325). Academic Press.

Jones, M. R., & Boltz, M. G. (1989). Dynamic attending and responses to time. *Psychological Review*, *96*, 459–491.

Jones, M. R., Boltz, M. G., & Kidd, G. (1982). Controlled attending as a function of melodic and temporal context. *Perception & Psychophysics*, *32*, 211–218.

Jones, M. R., Moynihan, H., MacKenzie, N., & Puente, J. (2002). Temporal aspects of stimulus-driven attending in dynamic arrays. *Psychological Science*, *4*, 313–319.

Joshi, A. K., & Weir, D. (1990). *The convergence of mildly context-sensitive grammar formalisms* (tech. rep.). University of Pennsylvania.

Juslin, P. N. (2013). From everyday emotions to aesthetic emotions: Towards a unified theory of musical emotions. *Physics of Life Reviews*, *10*, 235–266.

Juslin, P. N. (2019). *Musical emotions explained: Unlocking the secrets of musical affect*. Oxford University Press.

Juslin, P. N., Barradas, G., & Eerola, T. (2015). From sound to significance: Exploring the mechanisms underlying emotional reactions to music. *The American Journal of Psychology*, *128*, 281–304.

Juslin, P. N., Sakka, L. S., Barradas, G., & Lartillot, O. (2022). Emotions, mechanisms, and individual differences in music listening: A stratified random sampling approach. *Music Perception*, *40*, 55–86.

Juslin, P. N., & Västfjäll, D. (2008). Emotional responses to music: The need to consider underlying mechanisms. *Behavioral and Brain Sciences*, *31*, 559–621.

Justus, T. C., & Bharucha, J. J. (2001). Modularity in musical processing: The automaticity of harmonic priming. *Journal of Experimental Psychology: Human Perception and Performance*, *27*, 1000–1011.

Justus, T. C., & Hutsler, J. J. (2005). Fundamental issues in the evolutionary psychology of music: Assessing innateness and domain specificity. *Music Perception*, *23*, 1–27.

Kaplan, T. (2024). *Probabilistic models of rhythmic expectation and synchronisation* [Doctoral dissertation, Queen Mary University of London].

Kaplan, T., Cannon, J. J., Jamone, L., & Pearce, M. T. (2022). Modeling enculturated bias in entrainment to rhythmic patterns. *PLOS Computational Biology*, *18*, e1010579.

Kaplan, T., Jamone, L., & Pearce, M. T. (2023). Probabilistic modelling of microtiming perception. *Cognition*, *239*, 105532.

Karno, M., & Konečni, V. J. (1992). The effects of structural interventions in the first movement of Mozart's symphony in G minor K. 550 on aesthetic preference. *Music Perception*, *10*, 63–72.

Katz, S. M. (1987). Estimation of probabilities from sparse data for the language model component of a speech recognizer. *IEEE Transactions on Acoustics, Speech and Signal Processing*, *35*, 400–401.

Keller, G. B., & Mrsic-Flogel, T. D. (2018). Predictive processing: A canonical cortical computation. *Neuron*, *100*, 424–435.

Kemp, G. (1999). The aesthetic attitude. *British Journal of Aesthetics*, *39*, 392–399.

Kern, P., Heilbron, M., Lange, F. P. D., & Spaak, E. (2022). Cortical activity during naturalistic music listening reflects short- range predictions based on long- term experience. *Elife*, *11*, e80935.

Kessler, E. J., Hansen, C., & Shepard, R. N. (1984). Tonal schemata in the perception of music in Bali and in the West. *Music Perception*, *2*, 131–165.

Kidd, C., & Hayden, B. Y. (2015). The psychology and neuroscience of curiosity. *Neuron*, *88*, 449–460.

Kidd, C., Piantadosi, S. T., & Aslin, R. N. (2012). The Goldilocks effect: Human infants allocate attention to visual sequences that are neither too simple nor too complex. *PLOS ONE*, *7*, e36399.

Kim, S.-G., Kim, J. S., & Chung, C. K. (2011). The effect of conditional probability of chord progression on brain response: An MEG study. *PLOS ONE*, *6*, e17337.

Kirby, S., Griffiths, T. L., & Smith, K. (2014). Iterated learning and the evolution of language. *Current Opinion in Neurobiology*, *28*, 108–114.

Kirby, S., Tamariz, M., Cornish, H., & Smith, K. (2015). Compression and communication in the cultural evolution of linguistic structure. *Cognition*, *141*, 87–102.

Kivy, P. (2001). *New essays on musical understanding*. Oxford University Press.

Kivy, P. (2017). On the recent remarriage of music to philosophy. *Journal of Aesthetics and Art Criticism*, *75*, 429–438.

Klarlund, M., Brattico, E., Pearce, M. T., Wu, Y., Vuust, P., Overgaard, M., & Du, Y. (2023). Worlds apart? Testing the cultural distance hypothesis in music perception of Chinese and Western listeners. *Cognition*, *235*, 105405.

Kneser, R., & Ney, H. (1995). Improved backing-off for m-gram language modelling. *Proceedings of the IEEE International Conference on Acoustics, Speech and Signal Processing*, *1*, 181–184.

Knill, D. C., & Pouget, A. (2004). The Bayesian brain: The role of uncertainty in neural coding and computation. *Trends in Neurosciences*, *27*, 712–719.

Knopoff, L., & Hutchinson, W. (1981). Information theory for musical continua. *Journal of Music Theory*, *25*, 17–44.

Knopoff, L., & Hutchinson, W. (1983). Entropy as a measure of style: The influence of sample length. *Journal of Music Theory*, *27*, 75–97.

Knutson, B., Adams, C. M., Fong, G. W., & Hommer, D. (2001). Anticipation of increasing monetary reward selectively recruits nucleus accumbens. *Journal of Neuroscience*, *21*, RC159.

Knutson, B., Fong, G. W., Bennett, S. M., Adams, C. M., & Hommer, D. (2003). A region of mesial prefrontal cortex tracks monetarily rewarding outcomes: Characterization with rapid event-related fMRI. *NeuroImage*, *18*, 263–272.

Koelsch, S. (2009). Music-syntactic processing and auditory memory: Similarities and differences between ERAN and MMN. *Psychophysiology*, *46*, 179–190.

Koelsch, S., Fritz, T., & Schlaug, G. (2008a). Amygdala activity can be modulated by unexpected chord functions during music listening. *NeuroReport*, *19*, 1815–1819.

Koelsch, S., Fritz, T., von Cramon, D. Y., Müller, K., & Friederici, A. D. (2006). Investigating emotion with music: An fMRI study. *Human Brain Mapping*, *27*, 239–250.

Koelsch, S., Gunter, T. T. C., Friederici, A. D., & Schröger, E. (2000). Brain indices of music processing: "Nonmusicians" are musical. *Journal of Cognitive Neuroscience*, *12*, 520–541.

Koelsch, S., & Jentschke, S. (2008). Short-term effects of processing musical syntax: An ERP study. *Brain Research*, *12*, 1–8.

Koelsch, S., & Jentschke, S. (2010). Differences in electric brain responses to melodies and chords. *Journal of Cognitive Neuroscience*, *22*, 2251–62.

Koelsch, S., Jentschke, S., Sammler, D., & Mietchen, D. (2007). Untangling syntactic and sensory processing: An ERP study of music perception. *Psychophysiology*, *44*, 476–490.

Koelsch, S., Kilches, S., Steinbeis, N., & Schelinski, S. (2008b). Effects of unexpected chords and of performer's expression on brain responses and electrodermal activity. *PLOS ONE, 3*, e2631.

Koelsch, S., Maess, B., Grossmann, T., & Friederici, A. D. (2003). Electric brain responses reveal gender differences in music processing. *Neuroreport, 14*, 709–713.

Koelsch, S., Rohrmeier, M. A., Torrecuso, R., & Jentschke, S. (2013). Processing of hierarchical syntactic structure in music. *Proceedings of the National Academy of Sciences, 110*, 15443–8.

Koelsch, S., & Sammler, D. (2008). Cognitive components of regularity processing in the auditory domain. *PLOS ONE, 3*, e2650.

Koelsch, S., Schmidt, B. H., & Kansok, J. (2002). Effects of musical expertise on the early right anterior negativity: An event-related brain potential study. *Psychophysiology, 39*, 657–663.

Kolinski, M. (1973). A cross-cultural approach to metro-rhythmic patterns. *Ethnomusicology, 17*, 494–506.

Korzeniowski, F., Sears, D. R. W., Widmer, G., Chou, J., & Williams, R. (2018). A large-scale study of language models for chord prediction. *Proceeedings of the IEEE International Conference on Acoustics, Speech and Signal Processing (ICASSP)*, 91–95.

Kraehenbuehl, D., & Coons, E. (1959). Information as a measure of the experience of music. *Journal of Aesthetics and Art Criticism, 17*, 510–522.

Kragness, H. E., Swaminathan, S., Cirelli, L. K., & Schellenberg, E. G. (2021). Individual differences in musical ability are stable over time in childhood. *Developmental Science, 24*, e13081.

Kranenburg, P. V., Bruin, M. D., Grijp, L. P., & Wiering, F. (2014). *Meertens online reports: The Meertens tune collections* (tech. rep.). Meertens Instituut. Amsterdam.

Kraus, N., & Chandrasekaran, B. (2010). Music training for the development of auditory skills. *Nature Reviews Neuroscience, 11*, 599–605.

Krishnan, S., Carey, D., Dick, F., & Pearce, M. T. (2022). Effects of statistical learning in passive and active contexts on reproduction and recognition of auditory sequences. *Journal of Experimental Psychology: General, 151*, 555–577.

Krumhansl, C. L. (1990). *Cognitive foundations of musical pitch*. Oxford University Press.

Krumhansl, C. L. (1995a). Effects of musical context on similarity and expectancy. *Systematische Musikwissenschaft, 3*, 211–250.

Krumhansl, C. L. (1995b). Music psychology and music theory: Problems and prospects. *Music Theory Spectrum*, *17*, 53–90.

Krumhansl, C. L. (2000). Tonality induction: A statistical approach applied cross-culturally. *Music Perception*, *17*, 461–479.

Krumhansl, C. L. (2010). Plink: "Thin slices" of music. *Music Perception*, *27*, 337–354.

Krumhansl, C. L., & Cuddy, L. L. (2010). A theory of tonal hierarchies in music. In M. R. Jones, R. R. Fay & A. N. Popper (Eds.), *Music perception* (pp. 51–88). Springer.

Krumhansl, C. L., & Keil, F. C. (1982). Acquisition of the hierarchy of tonal functions in music. *Memory & Cognition*, *10*, 243–251.

Krumhansl, C. L., & Kessler, E. J. (1982). Tracing the dynamic changes in perceived tonal organization in a spatial representation of musical keys. *Psychological Review*, *89*, 334–368.

Krumhansl, C. L., Louhivuori, J., Toiviainen, P., Järvinen, T., & Eerola, T. (1999). Melodic expectation in Finnish spiritual hymns: Convergence of statistical, behavioural and computational approaches. *Music Perception*, *17*, 151–195.

Krumhansl, C. L., & Shepard, R. N. (1979). Quantification of the hierarchy of tonal functions within a diatonic context. *Journal of Experimental Psychology: Human Perception and Performance*, *5*, 579–594.

Krumhansl, C. L., Toivanen, P., Eerola, T., Toiviainen, P., Järvinen, T., & Louhivuori, J. (2000). Cross-cultural music cognition: Cognitive methodology applied to North Sami yoiks. *Cognition*, *76*, 13–58.

Kuhn, R., & De Mori, R. (1990). A cache-based natural language model for speech recognition. *IEEE Transactions on Pattern Analysis and Machine Intelligence*, *12*, 570–583.

Kumar, S., Joseph, S., Gander, P. E., Barascud, N., Halpern, A. R., & Griffiths, T. D. (2016). A brain system for auditory working memory. *Journal of Neuroscience*, *36*, 4492–4505.

Kuperberg, G. R., & Jaeger, T. F. (2016). What do we mean by prediction in language comprehension? *Language, Cognition and Neuroscience*, *31*, 32–59.

Lahdelma, I., & Athanasopoulos, G. (2021). Sweetness is in the ear of the beholder: Chord preference across United Kingdom and Pakistani listeners. *Annals of the New York Academy of Sciences*, *1502*, 72–84.

Lamont, A., & Cross, I. (1994). Children's cognitive representations of musical pitch. *Music Perception*, *12*, 27–55.

Landy, L. (1994). *Experimental music notebooks*. Harwood Academic Publishers.

Langhabel, J., Lieck, R., Toussaint, M., & Rohrmeier, M. A. (2017). Feature discovery for sequential prediction of monophonic music. *Proceedings of the 18th International Society for Music Information Retrieval Conference*, 649–656.

Large, E. W., Herrera, J. A., & Velasco, M. J. (2015). Neural networks for beat perception in musical rhythm. *Frontiers in Systems Neuroscience*, *9*, 159.

Large, E. W., & Kolen, J. F. (1994). Resonance and the perception of musical meter. *Connection Science*, *6*, 177–208.

Larsson, N. J. (1996). Extended application of suffix trees to data compression. In J. A. Storer & M. Cohn (Eds.), *Proceedings of the IEEE Data Compression Conference* (pp. 190–199). IEEE Computer Society Press.

Lattner, S., Grachten, M., & Widmer, G. (2018). A predictive model for music based on learned interval representations. *Proceedings of the 19th International Society for Music Information Retrieval Conference*, 26–33.

Lee, C. S. (1991). The perception of metrical structure: Experimental evidence and a model. In P. Howell, R. West & I. Cross (Eds.), *Representing musical structure* (pp. 59–127). Academic Press.

Lehmann, A. C., & Gruber, H. (2006). Music. In K. A. Ericsson, N. Charness, P. J. Feltovich & R. R. Hoffman (Eds.), *The Cambridge handbook of expertise and expert performance* (pp. 457–470). Cambridge University Press.

Leman, M. (2000). An auditory model of the role of short-term memory in probe-tone ratings. *Music Perception*, *17*, 481–509.

Lerdahl, F., & Jackendoff, R. (1983). *A generative theory of tonal music*. MIT Press.

Lerdahl, F., & Krumhansl, C. L. (2007). Modeling tonal tension. *Music Perception*, *24*, 329–366.

Levinson, J. (1997). *Music in the moment*. Cornell University Press.

Levinson, J. (2006). Concatenationism, architectonicism, and the appreciation of music. *Association Revue Internationale de Philisophie*, *4*, 505–514.

Levinson, J. (2009). The aesthetic appreciation of music. *British Journal of Aesthetics*, *49*, 415–425.

Levy, R. (2008). Expectation-based syntactic comprehension. *Cognition*, *16*, 1126–1177.

Lewandowsky, S., Oberauer, K., & Brown, G. D. A. (2009). No temporal decay in verbal short-term memory. *Trends in Cognitive Sciences*, *13*, 120–126.

Lewin, D. (1987). *Generalized musical intervals and transformations*. Yale University Press.

Lewontin, R. (1998). The evolution of cognition: Questions we will never answer. In D. Scarborough & S. Sternberg (Eds.), *An invitation to cognitive science, volume IV: Methods, models and conceptual issues* (2nd ed., pp. 107–132). MIT Press.

Li, M., Chen, X., Li, X., Ma, B., & Vitányi, P. M. B. (2004). The similarity metric. *IEEE Transactions on Information Theory*, *50*, 3250–3264.

Loewenstein, G. (1994). The psychology of curiosity: A review and reinterpretation. *Psychological Bulletin*, *116*, 75–98.

London, J. M. (2012). *Hearing in time* (2nd ed.). Oxford University Press.

London, J. M., Polak, R., & Jacoby, N. (2017). Rhythm histograms and musical meter: A corpus study of Malian percussion music. *Psychonomic Bulletin & Review*, *24*, 474–480.

Longuet-Higgins, H. C. (1979). The perception of music. *Proceedings of the Royal Society of London*, *205*, 307–322.

Longuet-Higgins, H. C. (1981). Artificial intelligence—a new theroretical psychology ? *Cognition*, *10*, 197–200.

Longuet-Higgins, H. C., & Lee, C. S. (1984). The rhythmic interpretation of monophonic music. *Music Perception*, *1*, 424–441.

Longuet-Higgins, H. C., & Steedman, M. J. (1971). On interpreting Bach. In B. Meltzer & D. Michie (Eds.), *Machine intelligence 6* (pp. 221–241). Edinburgh University Press.

Loui, P., Li, H. C., & Schlaug, G. (2011). White matter integrity in right hemisphere predicts pitch-related grammar learning. *NeuroImage*, *55*, 500–507.

Loui, P., Wessel, D. L., & Kam, C. L. H. (2010). Humans rapidly learn grammatical structure in a new musical scale. *Music Perception*, *27*, 377–388.

Lumaca, M., & Baggio, G. (2017). Cultural transmission and evolution of melodic structures in multi-generational signaling games. *Artificial Life*, *23*, 406–423.

Lynch, M. P., Eilers, R. E., Oller, D. K., & Urbano, R. C. (1990). Innateness, experience, and music perception. *Psychological Science*, *1*, 272–276.

Manzara, L. C., Witten, I. H., & James, M. (1992). On the entropy of music: An experiment with Bach chorale melodies. *Leonardo*, *2*, 81–88.

Marin, M. M. (2009). Effects of early musical training on musical and linguistic syntactic abilities. *Annals of the New York Academy of Sciences*, *1169*, 187–190.

Marmel, F., & Tillmann, B. (2009). Tonal priming beyond tonics. *Music Perception*, *26*, 211–222.

Marmel, F., Tillmann, B., & Delbé, C. (2010). Priming in melody perception: Tracking down the strength of cognitive expectations. *Journal of Experimental Psychology: Human Perception and Performance, 36*, 1016–1028.

Marmel, F., Tillmann, B., & Dowling, W. J. (2008). Tonal expectations influence pitch perception. *Perception & Psychophysics, 70*, 841–852.

Marr, D. (1982). *Vision*. W. H. Freeman.

Marvin, E. W., & Brinkman, A. (1999). The effect of modulation and formal manipulation on perception of tonic closure by expert listeners. *Music Perception, 16*, 389–407.

Masataka, N. (2006). Preference for consonance over dissonance by hearing newborns of deaf parents and of hearing parents. *Developmental Science, 9*, 46–50.

Mas-Herrero, E., Maini, L., Sescousse, G., & Zatorre, R. J. (2021). Common and distinct neural correlates of music and food-induced pleasure: A coordinate-based meta-analysis of neuroimaging studies. *Neuroscience and Biobehavioral Reviews, 123*, 61–71.

Mas-Herrero, E., Marco-Pallares, J., Lorenzo-Seva, U., Zatorre, R. J., & Rodriguez-Fornells, A. (2013). Individual differences in music reward experiences. *Music Perception, 31*, 118–138.

Mas-Herrero, E., Pla-Juncà, F., Ferreri, L., Cardona, G., Riba, J., Zatorre, R. J., Valle, M., Antonijoan, R. M., & Rodriguez-Fornells, A. (2023). The role of opioid transmission in music-induced pleasure. *Annals of the New York Academy of Sciences, 1520*, 105–114.

Matchin, W., & Hickok, G. (2020). The cortical organization of syntax. *Cerebral Cortex, 30*, 1481–1498.

Matsunaga, R., Hartono, P., Yokosawa, K., & Abe, J.-I. (2020). The development of sensitivity to tonality structure of music: Evidence from Japanese children raised in a simultaneous and unbalanced bi-musical environment. *Music Perception, 37*, 225–239.

Matsunaga, R., Yasuda, T., Johnson-Motoyama, M., Hartono, P., Yokosawa, K., & Abe, J.-I. (2018). A cross-cultural comparison of tonality perception in Japanese, Chinese, Vietnamese, Indonesian, and American listeners. *Psychomusicology, 28*, 178–188.

Matthews, T. E., Witek, M. A. G., Lund, T., Vuust, P., & Penhune, V. B. (2020). The sensation of groove engages motor and reward networks. *NeuroImage, 214*, 116768.

McAdams, S., & Cunible, J.-c. (1992). Perception of timbral analogies. *Philosophical Transactions of the Royal Society B, 336*, 383–389.

McAuley, J. D. (1995). *Perception of time as phase: Toward an adaptive oscillator model of rhythmic pattern processing* [Doctoral dissertation, Indiana University].

McAuley, J. D., Stevens, C. J., & Humphreys, M. S. (2004). Play it again: Did this melody occur more frequently or was it heard more recently? The role of stimulus familiarity in episodic recognition of music. *Acta Psychologica*, *116*, 93–108.

McCauley, S. M., & Christiansen, M. H. (2019). Language learning as language use: A cross-linguistic model of child language development. *Psychological Review*, *126*, 1–51.

McCrae, R. R. (2007). Aesthetic chills as a universal marker of openness to experience. *Motivation and Emotion*, *31*, 5–11.

McDermott, J. H., Schultz, A. F., Undurraga, E. A., & Godoy, R. A. (2016). Indifference to dissonance in native Amazonians reveals cultural variation in music perception. *Nature*, *25*, 21–25.

Mckay, C. M. (2021). No evidence that music training benefits speech perception in hearing-impaired listeners: A systematic review. *Trends in Hearing*, *25*, 1–16.

Mckone, E. (1995). Short-term implicit memory for words and nonwords. *Journal of Experimental Psychology: Learning, Memory, and Cognition*, *21*, 1108–1126.

Mesoudi, A. (2021). Cultural selection and biased transformation: Two dynamics of cultural evolution. *Philosophical Transactions of the Royal Society B*, *376*, 20200053.

Meyer, L. B. (1956). *Emotion and meaning in music*. University of Chicago Press.

Meyer, L. B. (1957). Meaning in music and information theory. *Journal of Aesthetics and Art Criticism*, *15*, 412–424.

Meyer, L. B. (1961). On rehearing music. *Journal of the American Musicological Society*, *14*, 257–267.

Meyer, L. B. (1973). *Explaining music: Essays and explorations*. University of California Press.

Meyer, L. B. (1989). *Style and music: History, theory and ideology*. University of Chicago Press.

Mill, R. W., Bőhm, T. M., Bendixen, A., Winkler, I., & Denham, S. L. (2013). Modelling the emergence and dynamics of perceptual organisation in auditory streaming. *PLOS Computational Biology*, *9*, e1002925.

Miller, G. A. (1956). The magical number seven, plus or minus two: Some limits on our capacity for processing information. *Psychological Review*, *63*, 81–97.

Miller, G. A. (1958). Free recall of redundant strings of letters. *Journal of Experimental Psychology*, *56*, 485–491.

Miller, G. A. (2003). The cognitive revolution: A historical perspective. *Trends in Cognitive Sciences*, *7*, 141–144.

Miller, G. A., & Buckhout, R. (1973). *Psychology: The science of mental life*. Harper & Rowe.

Miller, G. A., & Heise, G. A. (1950). The trill threshold. *The Journal of the Acoustical Society of America*, *22*, 637–638.

Milne, A. J. (2013). *A computational model of the cognition of tonality* [Doctoral dissertation, The Open University].

Milne, A. J., Laney, R., & Sharp, D. B. (2015). A spectral pitch class model of the probe tone data and scalic tonality. *Music Perception*, *32*, 364–393.

Milne, A. J., Sethares, W. A., Laney, R., & Sharp, D. B. (2011). Modelling the similarity of pitch collections with expectation tensors. *Journal of Mathematics and Music*, *5*, 1–20.

Milne, A. J., Smit, E. A., Sarvasy, H. S., & Dean, R. T. (2023). Evidence for a universal association of auditory roughness with musical stability. *PLOS ONE*, *18*, e0291642.

Miranda, R. A., & Ullman, M. T. (2007). Double dissociation between rules and memory in music: An event-related potential study. *NeuroImage*, *38*, 331–345.

Miton, H., Wolf, T., Vesper, C., Knoblich, G., & Sperber, D. (2020). Motor constraints influence cultural evolution of rhythm. *Proceedings of the Royal Society B*, *287*, 20202001.

Moffat, A. (1990). Implementing the PPM data compression scheme. *IEEE Transactions on Communications*, *38*, 1917–1921.

Moffat, A., Neal, R. M., & Witten, I. H. (1998). Arithmetic coding revisited. *ACM Transactions on Information Systems*, *16*, 256–294.

Morgan, E., Fogel, A., Nair, A., & Patel, A. D. (2019). Statistical learning and Gestalt-like principles predict melodic expectations. *Cognition*, *189*, 23–34.

Mori, K. (2022). Decoding peak emotional responses to music from computational acoustic and lyrical features. *Cognition*, *222*, 105010.

Morrison, S. J., Demorest, S. M., & Stambaugh, L. A. (2008). Enculturation effects in music cognition: The role of age and music complexity. *Journal of Research in Music Education*, *56*, 118–129.

Mosing, M. A., Madison, G., Pedersen, N. L., Kuja-Halkola, R., & Ullén, F. (2014). Practice does not make perfect: No causal effect of music practice on music ability. *Psychological Science*, *25*, 1795–1803.

Müllensiefen, D., Gingras, B., Musil, J., & Stewart, L. (2014). The musicality of non-musicians: An index for assessing musical sophistication in the general population. *PLOS ONE*, *9*, e89642.

Munsinger, H., & Kessen, W. (1964). Uncertainty, structure, and preference. *Psychological Monographs: General and Applied*, *78*, 1–24.

Näätänen, R., Paavilainen, P., Rinne, T., & Alho, K. (2007). The mismatch negativity (MMN) in basic research of central auditory processing: A review. *Clinical Neurophysiology, 118*, 2544–2590.

Narmour, E. (1977). *Beyond Schenkerism: The need for alternatives in music analysis*. University of Chicago Press.

Narmour, E. (1990). *The analysis and cognition of basic melodic structures: The implication-realization model*. University of Chicago Press.

Narmour, E. (1992). *The analysis and cognition of melodic complexity: The implication-realization model*. University of Chicago Press.

Neiworth, J. J., London, J. M., Flynn, M. J., Rupert, D. D., Alldritt, O., & Hyde, C. (2017). Artificial grammar learning in tamarins (Saguinus oedipus) in varying stimulus contexts. *Journal of Comparative Psychology, 131*, 128–138.

Newell, A., & Simon, H. A. (1976). Computer science as empirical enquiry: Symbols and search. *Communications of the ACM, 19*, 113–126.

North, A. C., & Hargreaves, D. J. (1995). Subjective complexity, familiarity and liking for popular music. *Psychomusicology, 14*, 77–93.

Olsen, K. N., Dean, R. T., & Leung, Y. (2016). What constitutes a phrase in sound-based music? A mixed-methods investigation of perception and acoustics. *PLOS ONE, 11*, e0167643.

Omigie, D., Pearce, M. T., & Stewart, L. (2012). Tracking of pitch probabilities in congenital amusia. *Neuropsychologia, 50*, 1483–1493.

Omigie, D., Pearce, M. T., Williamson, V. J., & Stewart, L. (2013). Electrophysiological correlates of melodic processing in congenital amusia. *Neuropsychologia, 51*, 1749–1762.

OpenAI. (2024). GPT-4 technical report. *arXiv preprint*, arXiv:2303.08774v6.

Oram, N., & Cuddy, L. L. (1995). Responsiveness of Western adults to pitch-distributional information in melodic sequences. *Psychological Research, 57*, 103–118.

Orlandi, A., Cross, E. S., & Orgs, G. (2020). Timing is everything: Dance aesthetics depend on the complexity of movement kinematics. *Cognition, 205*, 104446.

Orr, M. G., & Ohlsson, S. (2001). The relationship between musical complexity and liking in jazz and bluegrass. *Psychology of Music, 29*, 108–127.

Orr, M. G., & Ohlsson, S. (2005). Relationship between complexity and liking as a function of expertise. *Music Perception, 22*, 583–611.

Oudeyer, P.-Y., Gottlieb, J., & Lopes, M. (2016). Intrinsic motivation, curiosity, and learning: Theory and applications in educational technologies. In B. Studer & S. Knecht (Eds.), *Motivation: Theory, neurobiology and applications* (pp. 257–284). Elsevier.

Paller, K. A., McCarthy, G., & Wood, C. (1992). Event-related potentials elicited by deviant endings to melodies. *Psychophysiology*, *29*, 202–206.

Palmer, C., & Krumhansl, C. L. (1987a). Independent temporal and pitch structures in determination of musical phrases. *Journal of Experimental Psychology: Human Perception and Performance*, *13*, 116–126.

Palmer, C., & Krumhansl, C. L. (1987b). Pitch and temporal contributions to musical phrase perception: Effects of harmony, performance timing and familiarity. *Perception & Psychophysics*, *41*, 505–518.

Palmer, C., & Krumhansl, C. L. (1990). Mental representations for musical meter. *Journal of Experimental Psychology: Human Perception and Performance*, *16*, 728–741.

Park, J. M., Chung, C. K., Kim, J. S., Lee, K. M., Seol, J., & Yi, S. W. (2018). Musical expectations enhance auditory cortical processing in musicians: A magnetoencephalography study. *Neuroscience*, *369*, 325–335.

Parncutt, R. (1988). Revision of Terhardt's psychoacoustical model of the root (s) of a musical chord. *Music Perception*, *6*, 65–94.

Parncutt, R. (1994). A perceptual model of pulse salience and metrical accent in musical rhythms. *Music Perception*, *11*, 409–464.

Parras, G. G., Nieto-Diego, J., Carbajal, G. V., Escera, C., & Malmierca, M. S. (2017). Hierarchical organization of prediction error. *Nature Communications*, *8*, 2148.

Patel, A. D. (2008). *Music, language and the brain*. Oxford University Press.

Patel, A. D. (2022). Human musicality and gene-culture coevolution: Ten concepts to guide productive exploration. In E. H. Margulis, D. Loughridge & P. Loui (Eds.), *The science-music borderlands: Reckoning with the past, imagining the future*. MIT Press.

Patel, A. D., & Demorest, S. M. (2013). Comparative music cognition: Cross-species and cross-cultural studies. In D. Deutsch (Ed.), *The psychology of music* (3rd ed., pp. 647–681). Academic Press.

Patel, A. D., Gibson, E., Ratner, J., Besson, M., & Holycomb, P. J. (1998). Processing syntactic relations in language and music: An event-related potential study. *Journal of Cognitive Neuroscience*, *10*, 717–733.

Patel, A. D., & Iversen, J. R. (2014). The evolutionary neuroscience of musical beat perception: The action simulation for auditory prediction (ASAP) hypothesis. *Frontiers in Systems Neuroscience*, *8*, 1–14.

Pearce, M. T. (2005). *The construction and evaluation of statistical models of melodic structure in music perception and composition* [Doctoral dissertation, City University of London].

Pearce, M. T. (2023). Music perception. In I. S. Johnsrude (Ed.), *The Oxford research encyclopedia of psychology*. Oxford University Press.

Pearce, M. T., Conklin, D., & Wiggins, G. (2005). Methods for combining statistical models of music. In U. K. Wiil (Ed.), *Computer music modelling and retrieval* (pp. 295–312). Springer.

Pearce, M. T., & Müllensiefen, D. (2017). Compression-based modelling of musical similarity perception. *Journal of New Music Research*, *46*, 135–155.

Pearce, M. T., Müllensiefen, D., & Wiggins, G. (2010a). Melodic grouping in music information retrieval: New methods and applications. In Z. W. Raś & A. A. Wieczorkowska (Eds.), *Advances in music information retrieval* (pp. 364–388). Springer.

Pearce, M. T., Müllensiefen, D., & Wiggins, G. (2010b). The role of expectation and probabilistic learning in auditory boundary perception: A model comparison. *Perception*, *39*, 1367–1391.

Pearce, M. T., Ruiz, M. H., Kapasi, S., Wiggins, G., & Bhattacharya, J. (2010c). Unsupervised statistical learning underpins computational, behavioural, and neural manifestations of musical expectation. *NeuroImage*, *50*, 302–313.

Peretz, I. (1989). Clustering in music: An appraisal of task factors. *International Journal of Psychology*, *24*, 157–178.

Perruchet, P. (2019). What mechanisms underlie implicit statistical learning? Transitional probabilities versus chunks in language learning. *Topics in Cognitive Science*, *11*, 520–535.

Perruchet, P., & Poulin-Charronnat, B. (2012). Beyond transitional probability computations: Extracting word-like units when only statistical information is available. *Journal of Memory and Language*, *66*, 807–818.

Phuong, M., & Hutter, M. (2022). Formal algorithms for transformers. *arXiv preprint*, arXiv:2207.09238v1.

Pickering, M. J., & Gambi, C. (2018). Predicting while comprehending language: A theory and review. *Psychological Bulletin*, *144*, 1002–1044.

Pinkerton, R. C. (1956). Information theory and melody. *Scientific American*, *194*, 77–86.

Plantinga, J., & Trainor, L. J. (2005). Memory for melody: Infants use a relative pitch code. *Cognition*, *98*, 1–11.

Platt, J. R. (1961). Beauty: Pattern and change. In D. W. Fiske & S. R. Maddi (Eds.), *Functions of varied experience* (pp. 402–430). The Dorsey Press.

Poli, F., Meyer, M., Mars, R. B., & Hunnius, S. (2022). Contributions of expected learning progress and perceptual novelty to curiosity-driven exploration. *Cognition*, *225*, 105119.

Polich, J. (2007). Updating P300: An integrative theory of P3a and P3b. *Clinical Neurophysiology, 118*, 2128–2148.

Politimou, N., Douglass-Kirk, P., Pearce, M. T., Stewart, L., & Franco, F. (2021). Melodic expectations in 5- and 6-year-old children. *Journal of Experimental Child Psychology, 203*, 105020.

Poulin-Charronnat, B., Bigand, E., Madurell, F., & Peereman, R. (2005). Musical structure modulates semantic priming in vocal music. *Cognition, 94*, B67–78.

Povel, D.-J. J., & Essens, P. J. (1985). Perception of temporal patterns. *Perception, 2*, 411–440.

Pressing, J. L. (2002). Black Atlantic rhythm: Its computational and transcultural foundations. *Music Perception, 19*, 285–310.

Prince, J. B. (2011). The integration of stimulus dimensions in the perception of music. *Quarterly Journal of Experimental Psychology, 64*, 2125–2152.

Prince, J. B., & Loo, L.-M. (2017). Surface and structural effects of pitch and time on global melodic expectancies. *Psychological Research, 81*, 255–270.

Prince, J. B., Schmuckler, M. A., & Thompson, W. F. (2009a). The effect of task and pitch structure on pitch-time interactions in music. *Memory & Cognition, 37*, 368–381.

Prince, J. B., Thompson, W. F., & Schmuckler, M. A. (2009b). Pitch and time, tonality and meter: How do musical dimensions combine? *Journal of Experimental Psychology: Human Perception and Performance, 35*, 1598–1617.

Pylyshyn, Z. (1989). Computing in cognitive science. In M. I. Posner (Ed.), *Foundations of cognitive science* (pp. 51–91). MIT Press.

Quinn, I., & White, C. W. (2017). Corpus-derived key profiles are not transpositionally equivalent. *Music Perception, 34*, 531–540.

Quiroga-Martinez, D. R., Hansen, N. C., Højlund, A., Pearce, M. T., Brattico, E., & Vuust, P. (2019). Reduced prediction error responses in high-as compared to low-uncertainty musical contexts. *Cortex, 120*, 181–200.

Rameau, J.-P. (1722). *Traite de l'harmonie reduite a ses principes naturels*. J.B.C. Ballard.

Rammsayer, T., & Altenmüller, E. (2006). Temporal information processing in musicians and non-musicians. *Music Perception, 24*, 37–48.

Rao, R. P. N., & Ballard, D. H. (1999). Predictive coding in the visual cortex: A functional interpretation of some extra-classical receptive-field effects. *Nature Neuroscience, 2*, 79–87.

Rauschecker, J. P., & Scott, S. K. (2009). Maps and streams in the auditory cortex: Nonhuman primates illuminate human speech processing. *Nature Neuroscience, 12*, 718–725.

Ravignani, A., Delgado, T., & Kirby, S. (2017). Musical evolution in the lab exhibits rhythmic universals. *Nature Human Behaviour*, *1*, 0007.

Raz, C. (2018). An eighteenth-century theory of musical cognition? *Journal of Music Theory*, *62*, 205–248.

Reber, A. S. (1967). Implicit learning of artificial grammars. *Journal of Verbal Learning and Verbal Behavior*, *6*, 855–863.

Reber, R., Schwarz, N., & Winkielman, P. (2004). Processing fluency and aesthetic pleasure: Is beauty in the perceiver's processing experience? *Personality and Social Psychology Review*, *8*, 364–382.

Regnault, P., Bigand, E., & Besson, M. (2001). Different brain mechanisms mediate sensitivity to sensory consonance and harmonic context: Evidence from auditory event-related brain potentials. *Journal of Cognitive Neuroscience*, *13*, 241–255.

Repp, B. H. (1991). Some cognitive and perceptual aspects of speech and music. In J. Sundberg, L. Nord & R. Carlsen (Eds.), *Music, Language, Speech and Brain* (pp. 257–258). Palgrave.

Roads, C., & Wieneke, P. (1979). Grammars as representations for music. *Computer Music Journal*, *3*, 48–55.

Rohrmeier, M. A. (2011). Towards a generative syntax of tonal harmony. *Journal of Mathematics and Music*, *5*, 35–53.

Rohrmeier, M. A., & Koelsch, S. (2012). Predictive information processing in music cognition. A critical review. *International Journal of Psychophysiology*, *83*, 164–175.

Rohrmeier, M. A., Rebuschat, P., & Cross, I. (2011). Incidental and online learning of melodic structure. *Consciousness and Cognition*, *20*, 214–22.

Rubin-Rabson, G. (1940). The influence of age, intelligence, and training on reactions to classic and modern music. *The Journal of General Psychology*, *22*, 413–429.

Ruggles, D. R., Freyman, R. L., & Oxenham, A. J. (2014). Influence of musical training on understanding voiced and whispered speech in noise. *PLOS ONE*, *9*, e86980.

Russo, F. A., & Cuddy, L. L. (1999). A common origin for vocal accuracy and melodic expectancy: Vocal constraints. *Journal of the Acoustical Society of America*, *105*, 1217–1217.

Saffran, J. R. (2003). Absolute pitch in infancy and adulthood: The role of tonal structure. *Developmental Science*, *6*, 35–43.

Saffran, J. R., Aslin, R. N., & Newport, E. L. (1996). Statistical learning by 8-month-old infants. *Science*, *274*, 1926–1928.

Saffran, J. R., Johnson, E. K., Aslin, R. N., & Newport, E. L. (1999). Statistical learning of tone sequences by human infants and adults. *Cognition*, *70*, 27–52.

Saffran, J. R., & Kirkham, N. Z. (2018). Infant statistical learning. *Annual Review of Psychology*, *69*, 181–203.

Salamon, J., & Gomez, E. (2012). Melody extraction from polyphonic music signals using pitch contour characteristics. *IEEE Transactions on Acoustics, Speech and Signal Processing*, *20*, 1759–1770.

Salimpoor, V. N., Benovoy, M., Larcher, K., Dagher, A., & Zatorre, R. J. (2011). Anatomically distinct dopamine release during anticipation and experience of peak emotion to music. *Nature Neuroscience*, *14*, 257–262.

Salimpoor, V. N., Zald, D. H., Zatorre, R. J., Dagher, A., & Mcintosh, A. R. (2015). Predictions and the brain: How musical sounds become rewarding. *Trends in Cognitive Sciences*, *19*, 86–91.

Sandberg, E. H. (1999). Cognitive constraints on the development of hierarchical spatial organization skills. *Cognitive Development*, *14*, 597–619.

Santolin, C., & Saffran, J. R. (2018). Constraints on statistical learning across species. *Trends in Cognitive Sciences*, *22*, 52–63.

Sapp, C. S. (2011). *Computational methods for the analysis of musical structure* [Doctoral dissertation, Stanford University].

Sauvé, S. (2018). *Prediction in polyphony: Modelling musical auditory scene analysis* [Doctoral dissertation, Queen Mary University of London].

Sauvé, S., & Pearce, M. T. (2019). Information-theoretic modeling of perceived musical complexity. *Music Perception*, *37*, 165–178.

Sauvé, S., Sayad, A., Dean, R. T., & Pearce, M. T. (2018). Effects of pitch and timing expectancy on musical emotion. *Psychomusicology*, *28*, 17–39.

Schaffrath, H. (1995). The Essen folksong collection. In D. Huron (Ed.), *Database containing 6,255 folksong transcriptions in the Kern format and a 34-page research guide [computer database]*. CCARH.

Schellenberg, E. G. (1996). Expectancy in melody: Tests of the implication-realization model. *Cognition*, *58*, 75–125.

Schellenberg, E. G. (1997). Simplifying the implication-realization model of melodic expectancy. *Music Perception*, *14*, 295–318.

Schellenberg, E. G., Adachi, M., Purdy, K. T., & McKinnon, M. C. (2002). Expectancy in melody: Tests of children and adults. *Journal of Experimental Psychology: General*, *131*, 511–537.

Schellenberg, E. G., Bigand, E., Poulin-Charronnat, B., Garnier, C., & Stevens, C. J. (2005). Children's implicit knowledge of harmony in Western music. *Developmental Science*, *8*, 551–566.

Schellenberg, E. G., Iverson, P., & McKinnon, M. C. (1999). Name that tune: Identifying popular recordings from brief excerpts. *Psychonomic Bulletin & Review*, *6*, 641–646.

Schellenberg, E. G., & Trehub, S. E. (2003). Good pitch memory is widespread. *Psychological Science*, *14*, 262–266.

Schenker, H. (1935). *Free composition* (O. Jonas, Ed.; Trans.). Longman.

Schmuckler, M. A. (1989). Expectation in music: Investigation of melodic and harmonic processes. *Music Perception*, *7*, 109–150.

Schmuckler, M. A. (1990). The performance of global expectations. *Psychomusicology*, *9*, 122–147.

Schönwicsncr, M., Novitski, N., Pakarinen, S., Carlson, S., Tervaniemi, M., & Näätänen, R. (2007). Heschl's gyrus, posterior superior temporal gyrus, and mid-ventrolateral prefrontal cortex have different roles in the detection of acoustic changes. *Journal of Neurophysiology*, *97*, 2075–2082.

Schrimpf, M., Asher, I., Tuckute, G., Kauf, C., & Hosseini, E. A. (2021). The neural architecture of language: Integrative modeling converges on predictive processing. *Proceedings of the National Academy of Sciences*, *118*, e2105646118.

Schubert, E., & Pearce, M. T. (2016). A new look at musical expectancy: The veridical versus the general in the mental organisation of music. In R. Kronland-Martinet, M. Aramaki & S. Ystad (Eds.), *Music, mind and embodiment* (pp. 358–370, Vol. 9617). Springer.

Schultz, W. (2016). Dopamine reward prediction-error signalling: A two-component response. *Nature Reviews Neuroscience*, *17*, 183–195.

Schultz, W., Dayan, P., & Montague, R. (1997). A neural substrate of prediction and reward. *Science*, *275*, 1593–1599.

Sears, D. R. W., Caplin, W. E., & McAdams, S. (2014). Perceiving the classical cadence. *Music Perception*, *31*, 397–417.

Sears, D. R. W., Korzeniowski, F., & Widmer, G. (2018). Evaluating language models of tonal harmony. *Proceedings of the 19th International Society for Music Information Retrieval Conference*, 211–217.

Sears, D. R. W., Pearce, M. T., Spitzer, J., Caplin, W. E., & McAdams, S. (2019). Expectations for tonal cadences: Sensory and cognitive priming effects. *Quarterly Journal of Experimental Psychology*, *72*, 1422–1438.

Seashore, C. E. (1938). *Psychology of music*. McGraw-Hill.

Senn, O., Hoesl, F., Jerjen, R., Bechtold, T. A., Kilchenmann, L., Rose, D., & Alessandri, E. (2023). A stimulus set of 40 popular music drum patterns with perceived complexity measures. *Music & Science*, *6*, 1–17.

Senn, O., Kilchenmann, L., Bechtold, T. A., & Hoesl, F. (2018). Groove in drum patterns as a function of both rhythmic properties and listeners' attitudes. *PLOS ONE*, *13*, e0199604.

Serafine, M. L., Glassman, N., & Overbeeke, C. (1989). The cognitive reality of hierarchic structure in music. *Music Perception*, *6*, 397–430.

Sescousse, G., Caldú, X., Segura, B., & Dreher, J.-C. (2013). Processing of primary and secondary rewards: A quantitative meta-analysis and review of human functional neuroimaging studies. *Neuroscience and Biobehavioral Reviews*, *37*, 681–696.

Shanahan, D., & Shanahan, E. (2014). The Densmore collection of Native American songs: A new corpus for studies of effects of geography, language, and social function on folksong. *Proceedings of the 13th International Conference for Music Perception and Cognition*, 206–209.

Shannon, C. E. (1948). A mathematical theory of communication. *Bell System Technical Journal*, *27*, 379–656.

Shannon, C. E. (1951). Prediction and entropy of printed English. *Bell System Technical Journal*, *30*, 50–64.

Shepard, R. N. (1982). Structural representations of musical pitch. In D. Deutsch (Ed.), *Psychology of music* (pp. 343–390). Academic Press.

Shepard, R. N. (1986). Toward a universal law of generalization for psychological science. *Science*, *237*, 1317–1323.

Shusterman, R. (1997). The end of aesthetic experience. *Journal of Aesthetics and Art Criticism*, *55*, 29–41.

Siegelman, N., Bogaerts, L., Armstrong, B. C., & Frost, R. (2019). What exactly is learned in visual statistical learning? Insights from Bayesian modeling. *Cognition*, *192*, 104002.

Siegelman, N., Bogaerts, L., Christiansen, M. H., & Frost, R. (2017). Towards a theory of individual differences in statistical learning. *Philosophical Transactions of the Royal Society B*, *372*, 20160059.

Siegelman, N., Bogaerts, L., Elazar, A., Arciuli, J., & Frost, R. (2018). Linguistic entrenchment: Prior knowledge impacts statistical learning performance. *Cognition, 177*, 198–213.

Simon, C. R., & Wohlwill, J. F. (1968). An experimental study of the role of expectation and variation in music. *Journal of Research in Music Education, 16*, 227–238.

Skerritt-Davis, B., & Elhilali, M. (2021). Computational framework for investigating predictive processing in auditory perception. *Journal of Neuroscience Methods, 360*, 109177.

Sloboda, J. A. (1985). *The musical mind: The cognitive psychology of music*. Oxford Science Press.

Sloboda, J. A. (1991). Musical structure and emotional response: Some empirical findings. *Psychology of Music, 19*, 110–120.

Sloboda, J. A., Davidson, J. W., Howe, M. J. A., & Moore, D. G. (1996). The role of practice in the development of performing musicians. *British Journal of Psychology, 87*, 287–309.

Smit, E. A., & Rathcke, T. V. (2024). The role of native language and beat perception ability in the perception of speech rhythm. *Psychonomic Bulletin & Review*.

Smith, J. D., & Melara, R. J. (1990). Aesthetic preference and syntactic prototypicality in music: 'Tis the gift to be simple. *Cognition, 34*, 279–298.

Smith, N. J., & Levy, R. (2013). The effect of word predictability on reading time is logarithmic. *Cognition, 128*, 302–319.

Snyder, B. (2016). Memory for music. In S. Hallam, I. Cross & M. Thaut (Eds.), *The Oxford handbook of music psychology* (2nd ed., pp. 167–180). Oxford University Press.

Solberg, R. T., & Dibben, N. (2019). Peak experiences with electronic dance music: Subjective experiences, physiological responses, and musical characteristics of the break routine. *Music Perception, 36*, 371–389.

Sowa, J. A. (1956). *A machine to compose music*. Oliver Garfeld Co.

Speer, J. R., & Meeks, U. P. (1985). School children's perception of pitch in music. *Psychomusicology, 5*, 49–56.

Stahl, A. E., & Feigenson, L. (2015). Observing the unexpected enhances infants' learning and exploration. *Science, 348*, 91–94.

Steedman, M. J. (1996). The blues and the abstract truth: Music and mental models. In A. Garnham & J. Oakhill (Eds.), *Mental models in cognitive science* (pp. 305–318). Erlbaum.

Stein, T., & Peelen, M. V. (2015). Content-specific expectations enhance stimulus detectability by increasing perceptual sensitivity. *Journal of Experimental Psychology: General, 144*, 1089–1104.

Steinbeis, N., Koelsch, S., & Sloboda, J. A. (2006). The role of harmonic expectancy violations in musical emotions: Evidence from subjective, physiological and neural responses. *Journal of Cognitive Neuroscience, 18*, 1380–1393.

Steinruecken, C., Ghahramani, Z., & MacKay, D. J. C. (2015). Improving PPM with dynamic parameter updates. *Proceedings of the Data Compression Conference*, 193–202.

Stobart, H., & Cross, I. (2000). The Andean anacrusis? Rhythmic structure and perception in Easter songs of Northern Potosí, Bolivia. *British Journal of Ethnomusicology, 9*, 63–94.

Summerfield, C., & Egner, T. (2016). Feature-based attention and expectation. *Trends in Cognitive Sciences, 20*, 401–404.

Swaminathan, S., & Schellenberg, E. G. (2019). Music training and cognitive abilities: Associations, causes, and consequences. In M. H. Thaut & D. A. Hodges (Eds.), *The Oxford handbook of music and the brain* (pp. 645–670). Oxford University Press.

Swaminathan, S., & Schellenberg, E. G. (2020). Musical ability, music training, and language ability in childhood. *Journal of Experimental Psychology: Learning, Memory, and Cognition, 46*, 2340–2348.

Tan, S.-L., Spackman, M. P., & Peaslee, C. L. (2006). The effects of repeated exposure on liking and judgments of musical unity of intact and patchwork compositions. *Music Perception, 23*, 407–421.

Tax, D. M. J., van Breukelen, M., Duin, R. P. W., & Kittler, J. (2000). Combining multiple classifiers by averaging or by multiplying? *Pattern Recognition, 33*, 1475–1485.

Teh, Y. W. (2006). A hierarchical Bayesian language model based on Pitman-Yor processes. *Proceedings of the 21st International Conference on Computational Linguistics and the 44th annual meeting of the Association for Computational Linguistics*, 985–992.

Teki, S., Grube, M., Kumar, S., & Griffiths, T. D. (2011). Distinct neural substrates of duration-based and beat-based auditory timing. *Journal of Neuroscience, 31*, 3805–3812.

Temperley, D. (1999). What's key for key? The Krumhansl-Schmuckler key-finding algorithm reconsidered. *Music Perception, 17*, 65–100.

Temperley, D. (2001). *The cognition of basic musical structures*. MIT Press.

Temperley, D. (2007). *Music and probability*. MIT Press.

Temperley, D. (2009). A unified probabilistic model for polyphonic music analysis. *Journal of New Music Research, 38*, 3–18.

Temperley, D. (2019). Uniform information density in music. *Music Theory Online, 25*.

Ten, A., Kaushik, P., Oudeyer, P.-Y., & Gottlieb, J. (2021). Humans monitor learning progress in curiosity-driven exploration. *Nature Communications, 12*, 5972.

Thiessen, E. D. (2017). What's statistical about learning? Insights from modelling statistical learning as a set of memory processes. *Philosophical Transactions of the Royal Society B*, *372*, 20160056.

Thompson, W. F., Cuddy, L. L., & Plaus, C. (1997). Expectancies generated by melodic intervals: Evaluation of principles of melodic implication in a melody-completion task. *Perception & Psychophysics*, *59*, 1069–1076.

Thompson, W. F., Schellenberg, E. G., & Husain, G. (2004). Decoding speech prosody: Do music lessons help? *Emotion*, *4*, 46–64.

Thompson, W. F., & Stainton, M. (1998). Expectancy in Bohemian folk song melodies: Evaluation of implicative principles for implicative and closural intervals. *Music Perception*, *15*, 231–252.

Tierney, A. T., Russo, F. A., & Patel, A. D. (2011). The motor origins of human and avian song structure. *Proceedings of the National Academy of Sciences*, *108*, 15510–15515.

Tillmann, B., & Bigand, E. (1996). Does formal musical structure affect perception of musical expressiveness? *Psychology of Music*, *24*, 3–17.

Tillmann, B., & Bigand, E. (2010). Musical structure processing after repeated listening: Schematic expectations resist veridical expectations. *Musicae Scientiae*, *SI*, 33–47.

Tillmann, B., Koelsch, S., Escoffier, N., Bigand, E., Lalitte, P., Friederici, A. D., & von Cramon, D. Y. (2006). Cognitive priming in sung and instrumental music: Activation of inferior frontal cortex. *NeuroImage*, *31*, 1771–1782.

Tillmann, B., & McAdams, S. (2004). Implicit learning of musical timbre sequences: Statistical regularities confronted with acoustical (dis)similarities. *Journal of Experimental Psychology: Learning, Memory, and Cognition*, *30*, 1131–42.

Tillmann, B., & Poulin-Charronnat, B. (2010). Auditory expectations for newly acquired structures. *Quarterly Journal of Experimental Psychology*, *63*, 1646–1664.

Tinbergen, N. (1963). On aims and methods of ethology. *Zeitschrift für Tierpsychologie*, *20*, 410–433.

Trainor, L. J., & Trehub, S. E. (1992). A comparison of infants' and adults' sensitivity to Western musical structure. *Journal of Experimental Psychology: Human Perception and Performance*, *18*, 394–402.

Trainor, L. J., & Trehub, S. E. (1994). Key membership and implied harmony in Western tonal music: Developmental perspectives. *Perception & Psychophysics*, *56*, 125–132.

Trehub, S. E., Cohen, A. J., Thorpe, L. A., & Morrongiello, B. A. (1986). Development of the perception of musical relations: Semitone and diatonic structure. *Journal of Experimental Psychology: Human Perception and Performance, 12*, 295–301.

Tschantz, A., Seth, A. K., & Buckley, C. L. (2020). Learning action-oriented models through active inference. *PLOS Computational Biology, 16*, e1007805.

Turk-Browne, N. B., Junge, J. A., & Scholl, B. J. (2005). The automaticity of visual statistical learning. *Journal of Experimental Psychology: General, 134*, 552–564.

Turner, R. E. (2024). An introduction to transformers. *arXiv preprint*, arXiv:2304.10557v5.

Tversky, A., & Gati, I. (1982). Similarity, separability and the triangle inequality. *Psychological Review, 89*, 123–154.

Tymoczko, D. (2006). The geometry of musical chords. *Science, 313*, 72–74.

Typke, R., Wiering, F., & Veltkamp, R. C. (2005). Evaluating the earth mover's distance for measuring symbolic melodic similarity. *Paper presented at the Annual Music Information Retrieval Evaluation exchange (MIREX) as part of the 6th International Conference on Music Information Retrieval (ISMIR).*

Ukkonen, E. (1995). On-line construction of suffix trees. *Algorithmica, 14*, 249–260.

van Noorden, L. (1975). *Temporal coherence in the perception of tone sequences* [Doctoral dissertation, Eindhoven University of Technology].

van der Weij, B. (2020). *Experienced listeners: Modeling the influence of long-term musical exposure on rhythm perception* [Doctoral dissertation, University of Amsterdam].

van der Weij, B., Pearce, M. T., & Honing, H. (2017). A probabilistic model of meter perception: Simulating enculturation. *Frontiers in Psychology, 8*, 824.

van der Weij, B., Pearce, M. T., & Honing, H. (2022). Computational modelling of rhythm perception and the role of enculturation. In D. Shanahan, J. A. Burgoyne & I. Quinn (Eds.), *The Oxford handbook of music and corpus studies*. Oxford University Press.

Verhoef, T., Kirby, S., & Boer, B. D. (2014). Emergence of combinatorial structure and economy through iterated learning with continuous acoustic signals. *Journal of Phonetics, 43*, 57–68.

Verleger, R. (1990). P3-evoking wrong notes: Unexpected, awaited or arousing? *International Journal of Neuroscience, 55*, 171–179.

Verosky, N. J. (2019). Corpus-based learning of tonal expectations with expectation networks. *Journal of New Music Research, 48*, 145–158.

Verosky, N. J., & Morgan, E. (2021). Pitches that wire together fire together: Scale degree associations across time predict melodic expectations. *Cognitive Science, 45*, e13037.

Vieillard, S., Peretz, I., Gosselin, N., Khalfa, S., Gagnon, L., & Bouchard, B. (2008). Happy, sad, scary and peaceful musical excerpts for research on emotions. *Cognition and Emotion*, *22*, 720–752.

Vitz, P. C. (1966). Affect as a function of stimulus variation. *Journal of Experimental Psychology*, *71*, 74–79.

Vitz, P. C. (1972). Preference for tones as a function of frequency (hertz) and intensity (decibels). *Perception & Psychophysics*, *11*, 84–88.

von Helmholtz, H. (1863). *On the sensations of tone as a physiological basis for the theory of music* (E. Ellis, Ed.; Trans.). Dover.

von Hippel, P. T., & Huron, D. (2000). Why do skips precede reversals? The effects of tessitura on melodic structure. *Music Perception*, *18*, 59–85.

Vuust, P., Ostergaard, L., Pallesen, K. J., Bailey, C., & Roepstorff, A. (2009). Predictive coding of music—Brain responses to rhythmic incongruity. *Cortex*, *45*, 80–92.

Warstadt, A., & Bowman, S. R. (2023). What artificial neural networks can tell us about human language acquisition. In S. Lappin & J.-P. Bernardy (Eds.), *Algebraic Structures in Natural Language* (pp. 17–59). Taylor & Francis.

Webster, D. M., & Kruglanski, A. W. (1994). Individual differences in need for cognitive closure. *Journal of Personality and Social Psychology*, *67*, 1049–1062.

Whiteford, K. L., Baltzell, L. S., Chiu, M., Cooper, J. K., Faucher, S., Goh, P. Y., Hagedorn, A., Irsik, V. C., Irvine, A., Lim, S.-J., Mesik, J. L., Mesquita, B., Oakes, B., Rajappa, N., Roverud, E., Schrlau, A. E., Hedger, S. C. V., Bharadwaj, H. M., Johnsrude, I. S., … Oxenham, A. J. (2024). Musical training does not enhance neural sound encoding at early stages of the auditory system: A large-scale multisite investigation. *bioRxiv preprint*, 2024.09.02.610856.

Wickelgren, W. A. (1973). The long and the short of memory. *Psychological Bulletin*, *80*, 425–438.

Widdess, R. (2015). North India. In M. Church (Ed.), *The other classical musics: 15 great traditions* (pp. 139–161). Boydell Press.

Willems, R. M., Frank, S. L., Nijhof, A. D., Hagoort, P., & Van Den Bosch, A. (2016). Prediction during natural language comprehension. *Cerebral Cortex*, *26*, 2506–2516.

Winkler, I., Háden, G. P., Ladinig, O., Sziller, I., & Honing, H. (2009). Newborn infants detect the beat in music. *Proceedings of the National Academy of Sciences*, *106*, 2468–2471.

Witek, M. A. G., Clarke, E. F., Wallentin, M., Kringelbach, M. L., & Vuust, P. (2014). Syncopation, body-movement and pleasure in groove music. *PLOS ONE*, *9*, e94446.

Witten, I. H., & Bell, T. C. (1991). The zero-frequency problem: Estimating the probabilities of novel events in adaptive text compression. *IEEE Transactions on Information Theory*, *37*, 1085–1094.

Wittgenstein, L. (1966). *Lectures and conversations on aesthetics, psychology, and religious belief* (C. Barrett, Ed.). Blackwell.

Wong, P. C. M., Roy, A., & Margulis, E. H. (2009). Bimusicalism: The implicit dual enculturation of cognitive and affective systems. *Music Perception*, *27*, 81–88.

Wundt, W. M. (1874). *Grundzüge der physiologischen psychologie*. Wilhelm Engelmann.

Yoshida, K. A., Iversen, J. R., Patel, A. D., Mazuka, R., Nito, H., Gervain, J., & Werker, J. F. (2010). The development of perceptual grouping biases in infancy: A Japanese-English cross-linguistic study. *Cognition*, *115*, 356–361.

Youngblood, J. E. (1958). Style as information. *Journal of Music Theory*, *2*, 24–35.

Zajonc, R. B. (1968). Attitudinal effects of mere exposure. *Journal of Personality and Social Psychology*, *9*, 1–27.

Zatorre, R. J. (2024). *From perception to pleasure: The neuroscience of music and why we love it*. Oxford University Press.

Zatorre, R. J., Chen, J. L., & Penhune, V. B. (2007). When the brain plays music: Auditory-motor interactions in music perception and production. *Nature Reviews Neuroscience*, *8*, 547–558.

Zioga, I., Harrison, P. M. C., Pearce, M. T., Bhattacharya, J., Di Bernardi Luft, C., & Luft, B. (2020). From learning to creativity: Identifying the behavioural and neural correlates of learning to predict human judgements of musical creativity. *NeuroImage*, *206*, 116311.

Author Index

Subject Index